Graduate Texts in Mathematics 230

Graduate Texts in Mathematics

Series Editors:

Sheldon Axler
San Francisco State University, San Francisco, CA, USA

Kenneth Ribet
University of California, Berkeley, CA, USA

Advisory Board:

Colin Adams, *Williams College, Williamstown, MA, USA*
Alejandro Adem, *University of British Columbia, Vancouver, BC, Canada*
Ruth Charney, *Brandeis University, Waltham, MA, USA*
Irene M. Gamba, *The University of Texas at Austin, Austin, TX, USA*
Roger E. Howe, *Yale University, New Haven, CT, USA*
David Jerison, *Massachusetts Institute of Technology, Cambridge, MA, USA*
Jeffrey C. Lagarias, *University of Michigan, Ann Arbor, MI, USA*
Jill Pipher, *Brown University, Providence, RI, USA*
Fadil Santosa, *University of Minnesota, Minneapolis, MN, USA*
Amie Wilkinson, *University of Chicago, Chicago, IL, USA*

Graduate Texts in Mathematics bridge the gap between passive study and creative understanding, offering graduate-level introductions to advanced topics in mathematics. The volumes are carefully written as teaching aids and highlight characteristic features of the theory. Although these books are frequently used as textbooks in graduate courses, they are also suitable for individual study.

For further volumes:
http://www.springer.com/series/136

Daniel W. Stroock

An Introduction to Markov Processes

Second Edition

Springer

Daniel W. Stroock
Department of Mathematics
Massachusetts Institute of Technology
Cambridge, MA, USA

ISSN 0072-5285 ISSN 2197-5612 (electronic)
Graduate Texts in Mathematics
ISBN 978-3-662-51782-6 ISBN 978-3-642-40523-5 (eBook)
DOI 10.1007/978-3-642-40523-5
Springer Heidelberg New York Dordrecht London

Mathematics Subject Classification (2010): 60-01, 60J10, 60J27, 60J28, 37L40

Printed on acid-free paper

Springer is part of Springer Science+Business Media (www.springer.com)

This book is dedicated to my longtime colleague:

Richard A. Holley

Preface

To some extent, it would be accurate to summarize the contents of this book as an intolerably protracted description of what happens when either one raises a transition probability matrix $\mathbf{P}$ (i.e., all entries $(\mathbf{P})_{ij}$ are non-negative and each row of $\mathbf{P}$ sums to 1) to higher and higher powers or one exponentiates $\mathbf{R}(\mathbf{P}-\mathbf{I})$, where $\mathbf{R}$ is a diagonal matrix with non-negative entries. Indeed, when it comes right down to it, that is all that is done in this book. However, I, and others of my ilk, would take offense at such a dismissive characterization of the theory of Markov chains and processes with values in a countable state space, and a primary goal of mine in writing this book was to convince its readers that our offense would be warranted.

The reason why I, and others of my persuasion, refuse to consider the theory here as no more than a subset of matrix theory is that to do so is to ignore the pervasive role that probability plays throughout. Namely, probability theory provides a model which both motivates and provides a context for what we are doing with these matrices. To wit, even the term "transition probability matrix" lends meaning to an otherwise rather peculiar set of hypotheses to make about a matrix. Specifically, it suggests that we think of the matrix entry $(\mathbf{P})_{ij}$ as giving the probability that, in one step, a system in state i will make a transition to state j. Moreover, if we adopt this interpretation for $(\mathbf{P})_{ij}$, then we must interpret the entry $(\mathbf{P}^n)_{ij}$ of $\mathbf{P}^n$ as the probability of the same transition in n steps. Thus, as $n \to \infty$, $\mathbf{P}^n$ is encoding the long time behavior of a randomly evolving system for which $\mathbf{P}$ encodes the one-step behavior, and, as we will see, this interpretation will guide us to an understanding of $\lim_{n\to\infty}(\mathbf{P}^n)_{ij}$. In addition, and perhaps even more important, is the role that probability plays in bridging the chasm between mathematics and the rest of the world. Indeed, it is the probabilistic metaphor which allows one to formulate mathematical models of various phenomena observed in both the natural and social sciences. Without the language of probability, it is hard to imagine how one would go about connecting such phenomena to $\mathbf{P}^n$.

In spite of the propaganda at the end of the preceding paragraph, this book is written from a mathematician's perspective. Thus, for the most part, the probabilistic metaphor will be used to elucidate mathematical concepts rather than to provide mathematical explanations for non-mathematical phenomena. There are two reasons for my having chosen this perspective. First, and foremost, is my own background. Although I have occasionally tried to help people who are engaged in various sorts of applications, I have not accumulated a large store of examples which are easily translated into terms which are appropriate for a book at this level. In fact, my experience has taught me that people engaged in applications are more than competent to handle the routine problems that they encounter, and that they come to someone like me only as a last resort. As a consequence, the questions which they ask me tend to be quite difficult and the answers to those few which I can solve usually involve material which is well beyond the scope of the present book. The second reason for my writing this book in the way that I have is that I think the material itself is of sufficient interest to stand on its own. In spite of what funding agencies would have us believe, mathematics *qua* mathematics is a worthy intellectual endeavor, and I think there is a place for a modern introduction to stochastic processes which is unabashed about making mathematics its top priority.

I came to this opinion after several semesters during which I taught the introduction to stochastic processes course offered by the M.I.T. department of mathematics. The clientele for that course has been an interesting mix of undergraduate and graduate students, less than half of whom concentrate in mathematics. Nonetheless, most of the students who stay with the course have considerable talent and appreciation for mathematics, even though they lack the formal mathematical training which is requisite for a modern course in stochastic processes, at least as such courses are now taught in mathematics departments to their own graduate students. As a result, I found no ready-made choice of text for the course. On the one hand, the most obvious choice is the classic text *A First Course in Stochastic Processes*, either the original one by S. Karlin or the updated version [4] by S. Karlin and H. Taylor. Their book gives a no nonsense introduction to stochastic processes, especially Markov processes, on a countable state space, and its consistently honest, if not always easily assimilated, presentation of proofs is complemented by a daunting number of examples and exercises. On the other hand, when I began, I feared that adopting Karlin and Taylor for my course would be a mistake of the same sort as adopting Feller's book for an undergraduate introduction to probability, and this fear prevailed the first two times I taught the course. However, after using, and finding wanting, two derivatives of Karlin's classic, I took the plunge and assigned Karlin and Taylor's book. The result was very much the one which I predicted: I was far more enthusiastic about the text than were my students.

In an attempt to make Karlin and Taylor's book more palatable for the students, I started supplementing their text with notes in which I tried to couch the proofs in terms which I hoped they would find more accessible, and my efforts were rewarded with a quite positive response. In fact, as my notes became more and more extensive and began to diminish the importance of the book, I decided to convert them into what is now this book, although I realize that my decision to do so may have been stupid. For one thing, the market is already close to glutted with books that purport to cover this material. Moreover, some of these books are quite popular, although my experience with them leads me to believe that their popularity is not always correlated with the quality of the mathematics they contain. Having made that pejorative comment, I will not make public which are the books that led me to this conclusion. Instead, I will only mention the books on this topic, besides Karlin and Taylor's, which I very much like. J. Norris's book [5] is an excellent introduction to Markov processes which, at the same time, provides its readers with a good place to exercise their measure-theoretic skills. Of course, Norris's book is only appropriate for students who have measure-theoretic skills to exercise. On the other hand, for students who possess those skills, his book is a place where they can see measure theory put to work in an attractive way. In addition, Norris has included many interesting examples and exercises which illustrate how the subject can be applied. For more advanced students, an excellent treatment of Markov chains on a general state space can be found in the book [6] by D. Revuz.

The present book includes most of the mathematical material contained in [5], but the proofs here demand much less measure theory than his do. In fact, although I have systematically employed measure theoretic terminology (Lebesgue's dominated convergence theorem, the monotone convergence theorem, etc.), which is explained in Chap. 7, I have done so only to familiarize my readers with the jargon that they will encounter if they delve more deeply into the subject. In fact, because the state spaces in this book are countable, the applications which I have made of Lebesgue's theory are, with one notable exception, entirely trivial. The one exception is that I need to know the existence of countably infinite families of mutually independent random variables. In Sect. 7.2 I discuss how one goes about proving their existence, but, as distinguished from the first edition of this text, I do not go into details and instead refer to the treatment in [8]. Be that as it may, the reader who is ready to accept that such families exist has little need to consult Chap. 7 except for terminology and the derivation of a few essentially obvious facts about series.

The organization of this book should be more or less self-evident from the table of contents. In Chap. 1, I give a bare hands treatment of the basic facts, with particular emphasis on recurrence and transience, about nearest neighbor random walks on the square, d-dimensional lattice $\mathbb{Z}^d$. Chapter 2 introduces the study of ergodic properties, and this becomes the central theme which ties together Chaps. 2 through 6. In Chap. 2, the stochastic processes under consideration are Markov chains (i.e., the time parameter is discrete), and the driving force behind the development there is an idea which was introduced by Doeblin. Restricted as the applicability of Doeblin's idea may be, it has the enormous advantage over the material in Chaps. 4 and 5 that it provides an estimate on the rate at which the chain is converging to its equilibrium distribution. Chapter 3 begins with the classification of states in terms of recurrence and transience and then introduces some computational techniques for computing stationary probabilities. As an application, the final section of Chap. 3 gives a proof that Wilson's algorithm works and that it can be used to derive Kirchhoff's matrix tree theorem. The contents of this section are based on ideas that I learned from S. Sternberg, who learned them from M. Kozdron, who in turn learned them from G. Lawler. It was Kozdron who had the idea of using Wilson's algorithm to derive Kirchhoff's theorem.

In Chap. 4, I study the ergodic properties of Markov chains that do not necessarily satisfy Doeblin's condition. The main result here is the one summarized in equation

(4.1.15). Even though it is completely elementary, the derivation of (4.1.15), is, without doubt, the most demanding piece of analysis in the entire book. So far as I know, every proof of (4.1.15) requires work at some stage. In supposedly "simpler" proofs, the work is hidden elsewhere (either measure theory, as in [5] and [6], or in operator theory, as in [2]). The treatment given here, which is a re-working of the one in [4] based on Feller's renewal theorem, demands nothing more of the reader than a thorough understanding of arguments involving limits superior, limits inferior, and their role in proving that limits exist. In Chap. 5, Markov chains are replaced by continuous-time Markov processes (still on a countable state space). I do this first in the case when the rates are bounded and therefore problems of possible explosion do not arise. Afterwards, I allow for unbounded rates and develop criteria, besides boundedness, which guarantee non-explosion. The remainder of the chapter is devoted to transferring the results obtained for Markov chains in Chaps. 2 and 4 to the continuous-time setting.

Aside from Chap. 7, which is more like an appendix than an integral part of the book, the book ends with Chap. 6. The goal in Chap. 6 is to obtain quantitative results, reminiscent of, if not as strong as, those in Chap. 2, when Doeblin's theory either fails entirely or yields rather poor estimates. The new ingredient in Chap. 6 is the assumption that the chain or process is reversible (i.e., the transition probability is self-adjoint in the L^2-space of its stationary distribution), and the engine which makes everything go is the associated Dirichlet form. In the final section, the power of the Dirichlet form methodology is tested in an analysis of the Metropolis (a.k.a. as simulated annealing) algorithm. Finally, as I said before, Chap. 7 is an appendix in which the ideas and terminology of Lebesgue's theory of measure and integration are reviewed. Sect. 7.2.1.

I have finally reached the traditional place reserved for thanking those individuals who, either directly or indirectly, contributed to this book. The principal direct contributors are the many students who suffered with various and spontaneously changing versions of this book. I am particularly grateful to Adela Popescu whose careful reading of the first edition brought to light many minor and a few major errors that have been removed and, undoubtedly, replaced by new ones. In addition, I am grateful to Sternberg and Kozdron for introducing me to the ideas in Sect. 3.3.

Thanking, or even identifying, the indirect contributors is trickier. Indeed, they include all the individuals, both dead and alive, from whom I received my education, and I am not about to bore you with even a partial list of who they were or are. Nonetheless, there is one person who, over a period of more than ten years, patiently taught me to appreciate the sort of material treated here. Namely, Richard A. Holley, to whom I have dedicated this book, is a *true probabilist*. To wit, for Dick, intuitive understanding usually precedes his mathematically rigorous comprehension of a probabilistic phenomenon. This statement should lead no one to to doubt Dick's powers as a rigorous mathematician. On the contrary, his intuitive grasp of probability theory not only enhances his own formidable mathematical powers, it has saved me and others from blindly pursuing flawed lines of reasoning. As all who have worked with him know, reconsider what you are saying if ever, during some diatribe into which you have launched, Dick quietly says "I don't follow that." In addition to his mathematical prowess, every one of Dick's many students will attest to his wonderful generosity. I was not his student, but I was his colleague, and I can assure you that his generosity is not limited to his students.

Cambridge, MA, USA
August 2013

Daniel W. Stroock

Contents

List of Notations

$\mathbb{Z}$ & $\mathbb{Z}^+$	set of all integers and the subset of positive integers
$\mathbb{N}$	set of non-negative integers
$\lfloor t \rfloor$	integer part $\max\{n \in \mathbb{Z} : n \leq t\}$ of $t \in \mathbb{R}$
$\#S$	number of elements in the set S
$A\complement$	complement of the set A
$\mathbf{1}_A$	indicator function of the set A: $\mathbf{1}_A(x) = 1$ if $x \in A$ and $\mathbf{1}_A(x) = 0$ if $x \notin A$
$F \restriction S$	restriction of the function F to the set S
$a \wedge b$ & $a \vee b$	minimum and the maximum of $a, b \in \mathbb{R}$
a^+ & a^-	positive part $a \vee 0$ & negative part $(-a) \vee 0$ of $a \in \mathbb{R}$
$i \to j$ & $i \leftrightarrow j$	state j is accessible from i & communicates with i. See Sect. 3.1
$\delta_{i,j}$	Kronecker delta: $\delta_{i,j}$ is 1 or 0 depending on whether i is equal or unequal to j
$\boldsymbol{\delta}_i$	point measure at $i \in \mathbb{S}$: $(\boldsymbol{\delta}_i)_j = \delta_{i,j}$ for $j \in \mathbb{S}$
$\mathbb{E}[X, A]$	expected value of X on the event A. See Sect. 6.2
$\mathbb{E}[X \mid A]$ & $\mathbb{E}[X \mid \Sigma]$	conditional expectatation value of X given the event A & the σ-algebra Σ. See Sect. 6.4
$\langle \varphi \rangle_\pi$	alternative notations for $\sum_{i \in \mathbb{S}} \varphi(i)(\boldsymbol{\pi})_i$. See (5.1.4)
$\langle \varphi, \psi \rangle_\pi$ & $\|\varphi\|_{2,\pi}$	inner product and norm in $L^2(\boldsymbol{\pi})$. See Sect. 5.1.2
$\mathrm{Stat}(\mathbf{P})$	set of stationary distribution for the transition probability matrix $\mathbf{P}$. See Sect. 4.1.3
$\|\boldsymbol{\mu}\|_{\mathrm{v}}$	variation norm of the row vector $\boldsymbol{\mu}$. See (2.1.5)
$\|f\|_{\mathrm{u}}$	uniform norm of the function f. See (2.1.10)
$\|\mathbf{M}\|_{\mathrm{u,v}}$	uniform-variation norm of the matrix $\mathbf{M}$. See (4.1.1)
$\mathrm{Var}_{\boldsymbol{\mu}}(f)$	variance of f relative to the probability vector $\boldsymbol{\mu}$

Chapter 1
Random Walks, a Good Place to Begin

The purpose of this chapter is to discuss some examples of Markov processes that can be understood even before the term "Markov process" is. Indeed, anyone who has been introduced to probability theory will recognize that these processes all derive from consideration of elementary "coin tossing."

1.1 Nearest Neighbor Random Walks on $\mathbb{Z}$

Let p be a fixed number from the open interval $(0, 1)$, and suppose that[1] $\{B_n : n \in \mathbb{Z}^+\}$ is a sequence of $\{-1, 1\}$-valued, identically distributed *Bernoulli random variables*[2] which are 1 with probability p. That is, for any $n \in \mathbb{Z}^+$ and any $E \equiv (\epsilon_1, \dots, \epsilon_n) \in \{-1, 1\}^n$,

$$\mathbb{P}(B_1 = \epsilon_1, \dots, B_n = \epsilon_n) = p^{N(E)} q^{n-N(E)} \quad \text{where } q \equiv 1 - p \quad \text{and}$$
$$N(E) \equiv \#\{m : \epsilon_m = 1\} = \frac{n + S_n(E)}{2} \quad \text{when } S_n(E) \equiv \sum_1^n \epsilon_m. \tag{1.1.1}$$

Next, set

$$X_0 = 0 \quad \text{and} \quad X_n = \sum_{m=1}^n B_m \quad \text{for } n \in \mathbb{Z}^+. \tag{1.1.2}$$

The above family of random variables $\{X_n : n \in \mathbb{N}\}$ is often called a *nearest neighbor random walk* on $\mathbb{Z}$. Nearest neighbor random walks are examples of

---

[1] $\mathbb{Z}$ is used to denote the set of all integers, of which $\mathbb{N}$ and $\mathbb{Z}^+$ are, respectively, the non-negative and positive members.

[2] For historical reasons, mutually independent random variables which take only two values are often said to be Bernoulli random variables. A mathematically rigorous proof that infinitely many exist can be found in Sect. 2.2 of [8].

D.W. Stroock, *An Introduction to Markov Processes*, Graduate Texts in Mathematics 230, DOI 10.1007/978-3-642-40523-5_1, © Springer-Verlag Berlin Heidelberg 2014

Markov processes, but the description that I have just given is the one which would be given in elementary probability theory, as opposed to a course, like this one, devoted to stochastic processes. When studying stochastic processes the description should emphasize the dynamic nature of the family. Thus, a stochastic process oriented description might replace (1.1.2) by

$$\mathbb{P}(X_0 = 0) = 1 \quad \text{and}$$
$$\mathbb{P}(X_n - X_{n-1} = \epsilon \mid X_0, \dots, X_{n-1}) = \begin{cases} p & \text{if } \epsilon = 1 \\ q & \text{if } \epsilon = -1, \end{cases} \tag{1.1.3}$$

where $\mathbb{P}(X_n - X_{n-1} = \epsilon \mid X_0, \dots, X_{n-1})$ denotes the *conditional probability* (cf. Sect. 7.4.1) that $X_n - X_{n-1} = \epsilon$ given $\sigma(\{X_0, \dots, X_{n-1}\})$. Certainly (1.1.3) is more dynamic a description than the one in (1.1.2). Specifically, it says that the process starts from 0 at time $n = 0$ and proceeds so that, at each time $n \in \mathbb{Z}^+$, it moves one step forward with probability p or one step backward with probability q, independent of where it has been before time n.

1.1.1 Distribution at Time n

In this subsection I will present two approaches to computing $\mathbb{P}(X_n = m)$. The first computation is based on the description given in (1.1.2). From (1.1.2) it is clear that $\mathbb{P}(|X_n| \le n) = 1$. In addition, it is obvious that

$$n \text{ odd} \implies \mathbb{P}(X_n \text{ is odd}) = 1 \quad \text{and} \quad n \text{ even} \implies \mathbb{P}(X_n \text{ is even}) = 1.$$

Finally, given $m \in \{-n, \dots, n\}$ with the same parity as n and a string $E = (\epsilon_1, \dots, \epsilon_n) \in \{-1, 1\}^n$ with (cf. (1.1.1)) $S_n(E) = m$, $N(E) = \frac{n+m}{2}$ and so

$$\mathbb{P}(B_1 = \epsilon_1, \dots, B_n = \epsilon_n) = p^{\frac{n+m}{2}} q^{\frac{n-m}{2}}.$$

Hence, because, when $\binom{\ell}{k} \equiv \frac{\ell!}{k!(\ell-k)!}$ is the *binomial coefficient* "ℓ choose k," there are $\binom{n}{\frac{m+n}{2}}$ such strings E, we see that

$$\mathbb{P}(X_n = m) = \binom{n}{\frac{m+n}{2}} p^{\frac{n+m}{2}} q^{\frac{n-m}{2}} \tag{1.1.4}$$
if $m \in \mathbb{Z}$, $|m| \le n$, and m has the same parity as n and is 0 otherwise.

Our second computation of the same probability will be based on the more dynamic description given in (1.1.3). To do this, we introduce the notation $(P^n)_m \equiv \mathbb{P}(X_n = m)$. Obviously, $(P^0)_m = \delta_{0,m}$, where $\delta_{k,\ell}$ is the *Kronecker symbol* which is 1 when $k = \ell$ and 0 otherwise. Further, from (1.1.3), we see that $\mathbb{P}(X_n = m)$ equals

$$\begin{aligned} &\mathbb{P}(X_{n-1} = m - 1 \;\&\; X_n = m) + \mathbb{P}(X_{n-1} = m + 1 \;\&\; X_n = m) \\ &= p\mathbb{P}(X_{n-1} = m - 1) + q\mathbb{P}(X_{n-1} = m + 1). \end{aligned}$$

That is,

$$(P^0)_m = \delta_{0,m} \quad \text{and} \quad (P^n)_m = p(P^{n-1})_{m-1} + q(P^{n-1})_{m+1}. \tag{1.1.5}$$

Obviously, (1.1.5) provides a complete, albeit implicit, prescription for computing the numbers $(P^n)_m$, and one can easily check that the numbers given by (1.1.4) satisfy this prescription. Alternatively, one can use (1.1.5) plus induction on n to see that $(P^n)_m = 0$ unless $m = 2\ell - n$ for some $0 \le \ell \le n$ and that $(C^n)_0 = (C^n)_n = 1$ and $(C^n)_\ell = (C^{n-1})_{\ell-1} + (C^{n-1})_\ell$ for $1 \le \ell < n$ where $(C^n)_\ell \equiv p^{-\ell}q^{n-\ell}(P^n)_{2\ell-n}$. In other words, the family $\{(C^n)_\ell : n \in \mathbb{N} \,\&\, 0 \le \ell \le n\}$ are given by Pascal's triangle and are therefore the binomial coefficients.

1.1.2 Passage Times via the Reflection Principle

More challenging than the computation in Sect. 1.1.1 is that of finding the distribution of the first passage time to a point $a \in \mathbb{Z}$. That is, given $a \in \mathbb{Z} \setminus \{0\}$, set[3]

$$\zeta^{\{a\}} = \inf\{n \ge 1 : X_n = a\} \quad (\equiv \infty \text{ when } X_n \ne a \text{ for any } n \ge 1). \tag{1.1.6}$$

Then $\zeta^{\{a\}}$ is the *first passage time* to a, and our goal here is to find its distribution. Equivalently, we want an expression for $\mathbb{P}(\zeta^{\{a\}} = n)$, and clearly, by the considerations in Sect. 1.1.1, we need only worry about n's which satisfy $n \ge |a|$ and have the same parity as a.

Again I will present two approaches to this problem, here based on (1.1.2) and in Sect. 1.1.5 on (1.1.3). To carry out the one based on (1.1.2), assume that $a \in \mathbb{Z}^+$, suppose that $n \in \mathbb{Z}^+$ has the same parity as a, and observe first that

$$\mathbb{P}(\zeta^{\{a\}} = n) = \mathbb{P}(X_n = a \,\&\, \zeta^{\{a\}} > n-1) = p\mathbb{P}(\zeta^{\{a\}} > n-1 \,\&\, X_{n-1} = a-1).$$

Hence, it suffices for us to compute $\mathbb{P}(\zeta_a > n-1 \,\&\, X_{n-1} = a-1)$. For this purpose, note that for any $E \in \{-1,1\}^{n-1}$ with $S_{n-1}(E) = a-1$, the event $\{(B_1, \dots, B_{n-1}) = E\}$ has probability $p^{\frac{n+a}{2}-1}q^{\frac{n-a}{2}}$. Thus,

$$\mathbb{P}(\zeta^{\{a\}} = n) = \mathcal{N}(n,a)p^{\frac{n+a}{2}}q^{\frac{n-a}{2}} \tag{$*$}$$

where $\mathcal{N}(n,a)$ is the number of $E \in \{-1,1\}^{n-1}$ with the properties that $S_m(E) \le a-1$ for $0 \le m \le n-1$ and $S_{n-1}(E) = a-1$. That is, everything comes down to the computation of $\mathcal{N}(n,a)$. Alternatively, since $\mathcal{N}(n,a) = \binom{n-1}{\frac{n+a}{2}-1} - \mathcal{N}'(n,a)$, where $\mathcal{N}'(n,a)$ is the number of $E \in \{-1,1\}^{n-1}$ such that $S_{n-1}(E) = a-1$ and $S_m(E) \ge a$ for some $1 \le m \le n-1$, we need only compute $\mathcal{N}'(n,a)$. For this purpose we will use a beautiful argument known as the *reflection*

[3] As the following indicates, I take the infimum over the empty set to be $+\infty$.

principle. Namely, consider the set $P(n,a)$ of paths $(S_0,\dots,S_{n-1})\in\mathbb{Z}^n$ with the properties that $S_0=0$, $S_m-S_{m-1}\in\{-1,1\}$ for $1\le m\le n-1$, and $S_m\ge a$ for some $1\le m\le n-1$. Clearly, $\mathcal{N}'(n,a)$ is the number of paths in the set $L(n,a)$ consisting of those $(S_0,\dots,S_{n-1})\in P(n,a)$ for which $S_{n-1}=a-1$, and, as an application of the reflection principle, we will show that the set $L(n,a)$ has the same number of elements as the set $U(n,a)$ whose elements are those paths $(S_0,\dots,S_{n-1})\in P(n,a)$ for which $S_{n-1}=a+1$. Since $(S_0,\dots,S_{n-1})\in U(n,a)$ if and only if $S_0=0$, $S_m-S_{m-1}\in\{-1,1\}$ for all $1\le m\le n-1$, and $S_{n-1}=a+1$, the condition that $S_m\ge a$ is redundant. Thus, we already know how to count them: there are $\binom{n-1}{\frac{n+a}{2}}$ of them. Hence, all that remains is to make the aforementioned application of the reflection principle. To this end, for a given $\mathbf{S}=(S_0,\dots,S_{n-1})\in P(n,a)$, let $\ell(\mathbf{S})$ be the smallest $0\le k\le n-1$ for which $S_k\ge a$, and define the *reflection* $\mathfrak{R}(\mathbf{S})=(\hat{S}_0,\dots,\hat{S}_{n-1})$ of $\mathbf{S}$ so that $\hat{S}_m=S_m$ if $0\le m\le\ell(\mathbf{S})$ and $\hat{S}_k=2a-S_k$ if $\ell(\mathbf{S})<m\le n-1$. Clearly, $\mathfrak{R}$ maps $L(n,a)$ into $U(n,a)$ and $U(n,a)$ into $L(n,a)$. In addition, $\mathfrak{R}$ is its own inverse: its composition with itself is the identity map. Therefore, as a map from $L(n,a)$ to $U(n,a)$, $\mathfrak{R}$ it must be both one-to-one and onto, and so $L(n,a)$ and $U(n,a)$ have the same numbers of elements.

We have now shown that $\mathcal{N}'(n,a)=\binom{n-1}{\frac{n+a}{2}}$ and therefore that

$$\mathcal{N}(n,a)=\binom{n-1}{\frac{n+a}{2}-1}-\binom{n-1}{\frac{n+a}{2}}.$$

Finally, after plugging this into (∗), we arrive at

$$\mathbb{P}\big(\zeta^{\{a\}}=n\big)=\left[\binom{n-1}{\frac{n+a}{2}-1}-\binom{n-1}{\frac{n+a}{2}}\right]p^{\frac{n+a}{2}}q^{\frac{n-a}{2}},$$

which simplifies to the remarkably simple expression

$$\mathbb{P}\big(\zeta^{\{a\}}=n\big)=\frac{a}{n}\binom{n}{\frac{n+a}{2}}p^{\frac{n+a}{2}}q^{\frac{n-a}{2}}=\frac{a}{n}\mathbb{P}(X_n=a).$$

The computation when $a<0$ can be carried out either by repeating the argument just given or, after reversing the roles of p and q, applying the preceding result to $-a$. However one arrives at it, the general result is that

$$a\neq 0 \implies \mathbb{P}\big(\zeta^{\{a\}}=n\big)=\frac{|a|}{n}\binom{n}{\frac{n+a}{2}}p^{\frac{n+a}{2}}q^{\frac{n-a}{2}}=\frac{|a|}{n}\mathbb{P}(X_n=a) \qquad (1.1.7)$$

for $n\ge|a|$ with the same parity as a and is 0 otherwise.

1.1.3 Some Related Computations

Although the formula in (1.1.7) is elegant, it is not particularly transparent. For example, it is not at all evident how one can use it to determine whether

$\mathbb{P}(\zeta^{\{a\}} < \infty) = 1$. To carry out this computation, let $a > 0$ be given, and represent $\zeta^{\{a\}}$ as $f_a(B_1, \ldots, B_n, \ldots)$, where f_a is the function which maps $\{-1, 1\}^{\mathbb{Z}^+}$ into $\mathbb{Z}^+ \cup \{\infty\}$ so that, for each $n \in \mathbb{N}$,

$$f_a(\epsilon_1, \ldots, \epsilon_n, \ldots) > n \quad \iff \quad \sum_{\ell=1}^{m} \epsilon_\ell < a \quad \text{for } 1 \le m \le n.$$

Because the event $\{\zeta^{\{a\}} = m\}$ depends only on $(B_1, \ldots, B_m)$ and

$$\zeta^{\{a\}} = m \quad \Longrightarrow \quad \zeta^{\{a+1\}} = m + \zeta^{\{1\}} \circ \Sigma^m$$
$$\text{where } \zeta_1 \circ \Sigma^m \equiv f_1(B_{m+1}, \ldots, B_{m+n}, \ldots), \quad (1.1.8)$$

$\{\zeta^{\{a\}} = m \ \& \ \zeta^{\{a+1\}} < \infty\} = \{\zeta^{\{a\}} = m\} \cap \{\zeta^{\{1\}} \circ \Sigma^m < \infty\}$, and $\{\zeta^{\{a\}} = m\}$ is independent of $\{\zeta^{\{1\}} \circ \Sigma^m < \infty\}$. In particular, this leads to

$$\begin{aligned}\mathbb{P}(\zeta^{\{a+1\}} < \infty) &= \sum_{m=1}^{\infty} \mathbb{P}(\zeta^{\{a\}} = m \ \& \ \zeta_{a+1} < \infty) \\ &= \sum_{m=1}^{\infty} \mathbb{P}(\zeta^{\{a\}} = m)\mathbb{P}(\zeta^{\{1\}} \circ \Sigma^m < \infty) \\ &= \mathbb{P}(\zeta^{\{1\}} < \infty) \sum_{m=1}^{\infty} \mathbb{P}(\zeta^{\{a\}} = m) = \mathbb{P}(\zeta^{\{1\}} < \infty)\mathbb{P}(\zeta^{\{a\}} < \infty),\end{aligned}$$

since $(B_{m+1}, \ldots, B_{m+n}, \ldots)$ and $(B_1, \ldots, B_n, \ldots)$ have the same distribution and therefore so do $\zeta^{\{1\}} \circ \Sigma^m$ and $\zeta^{\{1\}}$. The same reasoning applies equally well when $a < 0$, only now with -1 playing the role of 1. In other words, we have proved that

$$\mathbb{P}(\zeta^{\{a\}} < \infty) = \mathbb{P}(\zeta^{\{\mathrm{sgn}(a)\}} < \infty)^{|a|} \quad \text{for } a \in \mathbb{Z} \setminus \{0\}, \quad (1.1.9)$$

where $\mathrm{sgn}(a)$, the *signum* of a, is 1 or -1 according to whether $a > 0$ or $a < 0$. In particular, this shows that $\mathbb{P}(\zeta^{\{1\}} < \infty) = 1 \implies \mathbb{P}(\zeta^{\{a\}} < \infty) = 1$ and $\mathbb{P}(\zeta^{\{-1\}} < \infty) = 1 \implies \mathbb{P}(\zeta^{\{-a\}} < \infty) = 1$ for all $a \in \mathbb{Z}^+$.

In view of the preceding, we need only look at $\mathbb{P}(\zeta^{\{1\}} < \infty)$. Moreover, by the monotone convergence theorem, Theorem 7.1.9,

$$\mathbb{P}(\zeta^{\{1\}} < \infty) = \lim_{s \nearrow 1} \mathbb{E}\big[s^{\zeta^{\{1\}}}\big] = \lim_{s \nearrow 1} \sum_{n=1}^{\infty} s^{2n-1} \mathbb{P}(\zeta^{\{1\}} = 2n-1).$$

Applying (1.1.7) with $a = 1$, we know that

$$\mathbb{P}(\zeta^{\{1\}} = 2n-1) = \frac{1}{2n-1}\binom{2n-1}{n} p^n q^{n-1}.$$

Next, note that

$$\frac{1}{2n-1}\binom{2n-1}{n} = \frac{(2(n-1))!}{n!(n-1)!} = \frac{2^{n-1}}{n!}\prod_{m=1}^{n-1}(2m-1)$$
$$= \frac{4^{n-1}}{n!}\prod_{m=1}^{n-1}(m-\tfrac{1}{2}) = (-1)^{n-1}\frac{4^n}{2}\binom{\frac{1}{2}}{n},$$

where,[4] for any $\alpha \in \mathbb{R}$,

$$\binom{\alpha}{n} \equiv \begin{cases} 1 & \text{if } n = 0 \\ \frac{1}{n!}\prod_{m=0}^{n-1}(\alpha - m) & \text{if } n \in \mathbb{Z}^+ \end{cases}$$

is the *generalized binomial coefficient* which gives the coefficient of x^n in the Taylor's expansion of $(1+x)^\alpha$ around $x = 0$. Hence, since $4pq \le 1$,

$$\sum_{n=1}^{\infty} s^{2n-1}\mathbb{P}(\zeta^{\{1\}} = 2n-1) = -\frac{1}{2qs}\sum_{n=1}^{\infty}\binom{\frac{1}{2}}{n}(-4pqs^2)^n = \frac{1-\sqrt{1-4pqs^2}}{2qs},$$

and so

$$\mathbb{E}\big[s^{\zeta^{\{1\}}}\big] = \frac{1-\sqrt{1-4pqs^2}}{2qs} \quad \text{for } |s| < 1. \tag{1.1.10}$$

Obviously, by symmetry, one can reverse the roles of p and q to obtain

$$\mathbb{E}\big[s^{\zeta^{\{-1\}}}\big] = \frac{1-\sqrt{1-4pqs^2}}{2ps} \quad \text{for } |s| < 1. \tag{1.1.11}$$

By letting $s \nearrow 1$ in (1.1.10) and noting that $1 - 4pq = (p+q)^2 - 4pq = (p-q)^2$, we see that[5]

$$\lim_{s\nearrow 1}\mathbb{E}\big[s^{\zeta^{\{1\}}}\big] = \frac{1-|p-q|}{2q} = \frac{p \wedge q}{q},$$

and so

$$\mathbb{P}(\zeta^{\{1\}} < \infty) = \begin{cases} 1 & \text{if } p \ge q \\ \frac{p}{q} & \text{if } p < q. \end{cases}$$

---

[4]In the preceding, I have adopted the convention that $\prod_{j=k}^{\ell} a_j = 1$ if $\ell < k$.

[5]I use $a \wedge b$ to denote the minimum $\min\{a, b\}$ of $a, b \in \mathbb{R}$.

Of course, $\mathbb{P}(\zeta^{\{-1\}} < \infty)$ is given by the same formula, only with the roles of p and q reversed. Thus, by (1.1.12),

$$\mathbb{P}(\zeta^{\{a\}} < \infty) = \begin{cases} 1 & \text{if } a \in \mathbb{Z}^+ \text{ \& } p \geq q \text{ or } -a \in \mathbb{Z}^+ \text{ \& } p \leq q \\ (\frac{p}{q})^a & \text{if } a \in \mathbb{Z}^+ \text{ \& } p < q \text{ or } -a \in \mathbb{Z}^+ \text{ \& } p > q. \end{cases} \tag{1.1.12}$$

1.1.4 Time of First Return

Having gone to so much trouble to arrive at (1.1.12), it is only reasonable to draw from it a famous conclusion about the *recurrence* properties of nearest neighbor random walks on $\mathbb{Z}$. Namely, let

$$\rho_0 \equiv \inf\{n \geq 1 : X_n = 0\} \quad (\equiv \infty \text{ if } X_n \neq 0 \text{ for all } n \geq 1)$$

be the *time of first return* to 0. Then, by precisely the same sort of reasoning which allowed us to arrive at (1.1.9), we see that $\mathbb{P}(X_1 = 1 \text{ \& } \rho_0 < \infty) = p\mathbb{P}(\zeta_{-1} < \infty)$ and $\mathbb{P}(X_1 = -1 \text{ \& } \rho_0 < \infty) = q\mathbb{P}(\zeta^{\{1\}} < \infty)$, and so, by (1.1.12),

$$\mathbb{P}(\rho_0 < \infty) = 2(p \wedge q). \tag{1.1.13}$$

In other words, *the random walk* $\{X_n : n \geq 0\}$ *will return to* 0 *with probability* 1 *if and only if it is symmetric* in the sense that $p = \frac{1}{2}$.

By sharpening the preceding a little, one sees that $\mathbb{P}(X_1 = 1 \text{ \& } \rho_0 = 2n) = p\mathbb{P}(\zeta^{\{1\}} = 2n - 1)$ and $\mathbb{P}(X_1 = -1 \text{ \& } \rho_0 = 2n) = q\mathbb{P}(\zeta^{\{1\}} = 2n - 1)$, and so, by (1.1.10) and (1.1.11),

$$\mathbb{E}\big[s^{\rho_0}\big] = 1 - \sqrt{1 - 4pqs^2} \quad \text{for } |s| < 1. \tag{1.1.14}$$

Hence,

$$\mathbb{E}\big[\rho_0 s^{\rho_0}\big] = s\frac{d}{ds}\mathbb{E}\big[s^{\rho_0}\big] = \frac{4pqs^2}{\sqrt{1 - 4pqs^2}} \quad \text{for } |s| < 1,$$

and therefore, since[6] $\mathbb{E}[\rho_0 s^{\rho_0}] \nearrow \mathbb{E}[\rho_0, \rho_0 < \infty]$ as $s \nearrow 1$,

$$\mathbb{E}[\rho_0, \rho_0 < \infty] = \frac{4pq}{|p - q|},$$

which, in conjunction with (1.1.13), means that[7]

$$\mathbb{E}[\rho_0 \mid \rho_0 < \infty] = \frac{2(p \vee q)}{|p - q|} = 1 + \frac{1}{|p - q|}. \tag{1.1.15}$$

[6] When X is a random variable and A is an event, we will often use $\mathbb{E}[X, A]$ to denote $\mathbb{E}[X\mathbf{1}_A]$.

[7] $a \vee b$ is used to denote the maximum $\max\{a, b\}$ of $a, b \in \mathbb{R}$.

The conclusions drawn in the preceding provide significant insight into the behavior of nearest neighbor random walks on $\mathbb{Z}$. In the first place, they say that when the random walk is symmetric, it returns to 0 with probability 1 but the expected amount of time it takes to do so is infinite. Secondly, when the random walk is not symmetric, it will, with positive probability, fail to return. On the other hand, in the non-symmetric case, the behavior of the trajectories is interesting. Namely, (1.1.13) in combination with (1.1.15) say that either they fail to return at all or they return relatively quickly.

1.1.5 Passage Times via Functional Equations

I close this discussion of passage times for nearest neighbor random walks with a less computational derivation of (1.1.10). For this purpose, set $u_a(s) = \mathbb{E}[s^{\zeta^{\{a\}}}]$ for $a \in \mathbb{Z} \setminus \{0\}$ and $s \in (-1, 1)$. Given $a \in \mathbb{Z}^+$, we use the ideas in Sect. 1.1.3, especially (1.1.8), to arrive at

$$\begin{aligned} u_{a+1}(s) &= \sum_{m=1}^{\infty} s^m \mathbb{E}\big[s^{\zeta^{\{1\}} \circ \Sigma^m}, \zeta^{\{a\}} = m\big] = \sum_{m=1}^{\infty} s^m \mathbb{P}\big(\zeta^{\{a\}} = m\big)\mathbb{E}\big[s^{\zeta^{\{1\}} \circ \Sigma^m}\big] \\ &= \sum_{m=1}^{\infty} s^m \mathbb{P}(\zeta_a = m) u_1(s) = u_a(s) u_1(s). \end{aligned}$$

Similarly, if $-a \in \mathbb{Z}^+$, then $u_{a-1}(s) = u_a(s)u_{-1}(s)$. Hence

$$u_a(s) = u_{\operatorname{sgn}(a)}(s)^{|a|} \quad \text{for } a \in \mathbb{Z} \setminus \{0\} \text{ and } |s| < 1. \tag{1.1.16}$$

Continuing with the same line of reasoning, we also have

$$\begin{aligned} u_1(s) &= \mathbb{E}\big[s^{\zeta^{\{1\}}}, X_1 = 1\big] + \mathbb{E}\big[s^{\zeta^{\{1\}}}, X_1 = -1\big] \\ &= ps + qs\mathbb{E}\big[s^{\zeta_2 \circ \Sigma^1}\big] = ps + qsu_2(s) = ps + qsu_1(s)^2. \end{aligned}$$

Hence, by the quadratic formula,

$$u_1(s) = \frac{1 \pm \sqrt{1 - 4pqs^2}}{2qs}.$$

Because $\mathbb{P}(\zeta^{\{1\}}$ is odd$) = 1$, $u_1(-s) = -u_1(s)$. At the same time,

$$s \in (0, 1) \quad \Longrightarrow \quad \frac{1 + \sqrt{1 - 4pqs^2}}{2qs} > \frac{1 + \sqrt{1 - 4pq}}{2q} = \frac{p \vee q}{q} \geq 1.$$

Hence, since $s \in (0, 1) \implies u_1(s) < 1$, we can eliminate the "+" solution and thereby arrive at a second derivation of (1.1.10). In fact, after combining this with

(1.1.16), we have shown that

$$\mathbb{E}[s^{\zeta_a}] = \begin{cases} \left(\frac{1-\sqrt{1-4pqs^2}}{2qs}\right)^a & \text{if } a \in \mathbb{Z}^+ \\ \left(\frac{1-\sqrt{1-4pqs^2}}{2ps}\right)^{-a} & \text{if } -a \in \mathbb{Z}^+ \end{cases} \quad \text{for } |s| < 1. \tag{1.1.17}$$

In theory one can recover (1.1.7) by differentiating (1.1.17) n times with respect to s at 0, but the computation is tedious.

1.2 Recurrence Properties of Random Walks

In Sect. 1.1.4, we studied the time ρ_0 of first return of a nearest neighbor random walk to 0. As we will see in Chaps. 2 and 4, times of first return are critical (cf. Sect. 2.3.2) for an understanding of the long time behavior of random walks and related processes. Indeed, when the random walk returns to 0, it starts all over again. Thus, if it returns with probability 1, then the entire history of the walk will consist of epochs, each epoch being a sojourn which begins and ends at 0. Because it marks the time at which one epoch ends and a second, identically distributed, one begins, a time of first return is often called a *renewal time*, and the walk is said to be *recurrent* if $\mathbb{P}(\rho_0 < \infty) = 1$. Walks which are not recurrent are said to be *transient*.

In this section, we will discuss the recurrence properties of nearest neighbor random walks. Of course, we already know (cf. (1.1.13)) that a nearest neighbor random on $\mathbb{Z}$ is recurrent if and only if it is symmetric. Thus, our interest here will be in higher dimensional analogs. In particular, in the hope that it will be convincing evidence that recurrence is subtle, we will show that the recurrence of the nearest neighbor, symmetric random walk on $\mathbb{Z}$ persists when $\mathbb{Z}$ is replaced by $\mathbb{Z}^2$ but disappears in $\mathbb{Z}^3$.

1.2.1 Random Walks on $\mathbb{Z}^d$

To describe the analog on $\mathbb{Z}^d$ of a nearest neighbor random walk on $\mathbb{Z}$, we begin by thinking of $\mathbf{N}_1 \equiv \{-1, 1\}$ as the set of *nearest neighbors* in $\mathbb{Z}$ of 0. It should then be clear why the set $\mathbf{N}_d$ of nearest neighbors of the origin in $\mathbb{Z}^d$ consists of the $2d$ points in $\mathbb{Z}^d$ for which $(d-1)$ coordinates are 0 and the remaining coordinate is in $\mathbf{N}_1$. Next, we replace the $\mathbf{N}_1$-valued Bernoulli random variables in Sect. 1.1 by their d-dimensional analogs, that is: independent, identically distributed $\mathbf{N}_d$-valued random variables $\mathbf{B}_1, \dots, \mathbf{B}_n, \dots$.[8] Finally, a nearest neighbor random walk on $\mathbb{Z}^d$

[8]The existence of the $\mathbf{B}_n$'s can be seen as a consequence of Theorem 7.3.2. Namely, let $\{U_n : n \in \mathbb{Z}^+\}$ be a family of mutually independent random variables which are uniformly distributed on $[0, 1)$. Next, let $(\mathbf{k}_1, \dots, \mathbf{k}_{2d})$ be an ordering of the elements of $\mathbf{N}_d$, set $\beta_0 = 0$ and $\beta_m = \sum_{\ell=1}^{m} \mathbb{P}(\mathbf{B}_1 = \mathbf{k}_\ell)$ for $1 \le m \le 2d$, define $F : [0, 1) \longrightarrow \mathbf{N}_d$ so that $F \restriction [\beta_{m-1}, \beta_m) = \mathbf{k}_m$, and set $\mathbf{B}_n = F(U_n)$.

is a family $\{\mathbf{X}_n : n \geq 0\}$ of the form

$$\mathbf{X}_0 = \mathbf{0} \quad \text{and} \quad \mathbf{X}_n = \sum_{m=1}^{n} \mathbf{B}_m \quad \text{for } n \geq 1.$$

The equivalent, stochastic process oriented description of $\{\mathbf{X}_n : n \geq 0\}$ is

$$\begin{aligned} &\mathbb{P}(\mathbf{X}_0 = \mathbf{0}) = 1 \quad \text{and,} \quad \text{for } n \geq 1 \text{ and } \boldsymbol{\epsilon} \in \mathbf{N}_d, \\ &\mathbb{P}(\mathbf{X}_n - \mathbf{X}_{n-1} = \boldsymbol{\epsilon} \mid \mathbf{X}_0, \ldots, \mathbf{X}_{n-1}) = p_{\boldsymbol{\epsilon}}, \end{aligned} \tag{1.2.1}$$

where $p_{\boldsymbol{\epsilon}} \equiv \mathbb{P}(\mathbf{B}_1 = \boldsymbol{\epsilon})$. When $\mathbf{B}_1$ is uniformly distributed on $\mathbf{N}_d$, the random walk is said to be *symmetric*.

In keeping with the notation and terminology introduced above, we define the time $\rho_{\mathbf{0}}$ of first return to the origin to be n if $n \geq 1$, $\mathbf{X}_n = \mathbf{0}$, and $\mathbf{X}_m \neq \mathbf{0}$ for $1 \leq m < n$, and we take $\rho_{\mathbf{0}} = \infty$ if no such $n \geq 1$ exists. Also, we will say that the walk is *recurrent* or *transient* according to whether $\mathbb{P}(\rho_{\mathbf{0}} < \infty)$ is 1 or strictly less than 1.

1.2.2 An Elementary Recurrence Criterion

Given $n \geq 1$, let $\rho_{\mathbf{0}}^{(n)}$ be the time of the nth return to $\mathbf{0}$. That is, $\rho_{\mathbf{0}}^{(1)} = \rho_{\mathbf{0}}$ and, for $n \geq 2$,

$$\rho_{\mathbf{0}}^{(n-1)} < \infty \quad \Longrightarrow \quad \rho_{\mathbf{0}}^{(n)} = \inf\{m > \rho^{(n-1)} : \mathbf{X}_m = \mathbf{0}\}$$

and $\rho_{\mathbf{0}}^{(n-1)} = \infty \Longrightarrow \rho_{\mathbf{0}}^{(n)} = \infty$. Equivalently, if $g : (\mathbf{N}_d)^{\mathbb{Z}^+} \longrightarrow \mathbb{Z}^+ \cup \{\infty\}$ is determined so that

$$g(\boldsymbol{\epsilon}_1, \ldots \boldsymbol{\epsilon}_\ell, \ldots) > n \quad \text{if and only if} \quad \sum_{\ell=1}^{m} \boldsymbol{\epsilon}_\ell \neq \mathbf{0} \quad \text{for } 1 \leq m \leq n,$$

then $\rho_{\mathbf{0}} = g(\mathbf{B}_1, \ldots, \mathbf{B}_\ell, \ldots)$, and $\rho_{\mathbf{0}}^{(n)} = m \Longrightarrow \rho_{\mathbf{0}}^{(n+1)} = m + \rho_{\mathbf{0}} \circ \Sigma^m$ where $\rho_{\mathbf{0}} \circ \Sigma^m$ is equal to $g(\mathbf{B}_{m+1}, \ldots, \mathbf{B}_{m+\ell}, \ldots)$. In particular, this leads to

$$\begin{aligned} \mathbb{P}\big(\rho_{\mathbf{0}}^{(n+1)} < \infty\big) &= \sum_{m=1}^{\infty} \mathbb{P}\big(\rho_{\mathbf{0}}^{(n)} = m \;\&\; \rho_{\mathbf{0}} \circ \Sigma^m < \infty\big) \\ &= \mathbb{P}\big(\rho_{\mathbf{0}}^{(n)} < \infty\big)\mathbb{P}(\rho_{\mathbf{0}} < \infty), \end{aligned}$$

since $\{\rho_{\mathbf{0}}^{(n)} = m\}$ depends only on $(\mathbf{B}_1, \ldots, \mathbf{B}_m)$, and is therefore independent of $\rho_{\mathbf{0}} \circ \Sigma^m$, and the distribution of $\rho_{\mathbf{0}} \circ \Sigma^m$ is the same as that of $\rho_{\mathbf{0}}$. Thus, we have proved that

$$\mathbb{P}\big(\rho_{\mathbf{0}}^{(n)} < \infty\big) = \mathbb{P}(\rho_{\mathbf{0}} < \infty)^n \quad \text{for } n \geq 1. \tag{1.2.2}$$

One dividend of (1.2.2) is that it supports the epochal picture described above about the structure of recurrent walks. Namely, it says that *if the walk returns once to* $\mathbf{0}$ *with probability* 1, *then, with probability* 1, *it will do so infinitely often*. This observation has many applications. For example, it shows that if the mean value of ith coordinate of $\mathbf{B}_1$ is different from 0, then $\{\mathbf{X}_n : n \geq 0\}$ must be transient. To see this, use Y_n to denote the ith coordinate of $\mathbf{B}_n$, and observe that $\{Y_n - Y_{n-1} : n \geq 1\}$ is a sequence of mutually independent, identically distributed $\{-1,0,1\}$-valued random variables with mean value $\mu \neq 0$. By the strong law of large numbers (cf. Exercise 1.3.4 below), this means that $\frac{Y_n}{n} \longrightarrow \mu \neq 0$ with probability 1, which is possible only if $|\mathbf{X}_n| \geq |Y_n| \longrightarrow \infty$ with probability 1, and clearly this eliminates the possibility that, even with positive probability, $\mathbf{X}_n = \mathbf{0}$ infinitely often.

A second dividend of (1.2.2) is the following. Define

$$T_{\mathbf{0}} = \sum_{n=0}^{\infty} \mathbf{1}_{\{\mathbf{0}\}}(\mathbf{X}_n)$$

to be the total time that $\{\mathbf{X}_n : n \geq 0\}$ spends at the origin. Since $\mathbf{X}_0 = \mathbf{0}$, $T_{\mathbf{0}} \geq 1$. Moreover, for $n \geq 1$, $T_{\mathbf{0}} > n \iff \rho_{\mathbf{0}}^{(n)} < \infty$. Hence, by (1.2.2),

$$\mathbb{E}[T_{\mathbf{0}}] = \sum_{n=0}^{\infty} \mathbb{P}(T_{\mathbf{0}} > n) = 1 + \sum_{n=1}^{\infty} \mathbb{P}\big(\rho_{\mathbf{0}}^{(n)} < \infty\big) = 1 + \sum_{n=1}^{\infty} \mathbb{P}(\rho_{\mathbf{0}} < \infty)^n,$$

and so

$$\mathbb{E}[T_{\mathbf{0}}] = \frac{1}{1 - \mathbb{P}(\rho_{\mathbf{0}} < \infty)} = \frac{1}{\mathbb{P}(\rho_{\mathbf{0}} = \infty)}. \tag{1.2.3}$$

Before applying (1.2.3) to the problem of recurrence, it is interesting to note that $T_{\mathbf{0}}$ is a random variable for which the following peculiar dichotomy holds:

$$\begin{aligned} \mathbb{P}(T_{\mathbf{0}} < \infty) > 0 \quad &\Longrightarrow \quad \mathbb{E}[T_{\mathbf{0}}] < \infty \\ \mathbb{E}[T_{\mathbf{0}}] = \infty \quad &\Longrightarrow \quad \mathbb{P}(T_{\mathbf{0}} = \infty) = 1. \end{aligned} \tag{1.2.4}$$

Indeed, if $\mathbb{P}(T_{\mathbf{0}} < \infty) > 0$, then, with positive probability, $\mathbf{X}_n$ cannot be $\mathbf{0}$ infinitely often, and so, by (1.2.2), $\mathbb{P}(\rho_{\mathbf{0}} < \infty) < 1$, which, by (1.2.3), means that $\mathbb{E}[T_{\mathbf{0}}] < \infty$. On the other hand, if $\mathbb{E}[T_{\mathbf{0}}] = \infty$, then (1.2.3) implies that $\mathbb{P}(\rho_{\mathbf{0}} < \infty) = 1$ and therefore, by (1.2.2), that $\mathbb{P}(T_{\mathbf{0}} > n) = \mathbb{P}(\rho_{\mathbf{0}}^{(n)} < \infty) = 1$ for all $n \geq 1$. Hence (cf. (7.1.3)), $\mathbb{P}(T_{\mathbf{0}} = \infty) = 1$.

1.2.3 Recurrence of Symmetric Random Walk in $\mathbb{Z}^2$

The most frequent way that (1.2.3) gets applied to determine recurrence is in conjunction with the formula

$$\mathbb{E}[T_{\mathbf{0}}] = \sum_{n=0}^{\infty} \mathbb{P}(\mathbf{X}_n = \mathbf{0}). \tag{1.2.5}$$

Although the proof of (1.2.5) is essentially trivial (cf. Theorem 7.1.15):

$$\mathbb{E}[T_{\mathbf{0}}] = \mathbb{E}\left[\sum_{n=0}^{\infty} \mathbf{1}_{\{\mathbf{0}\}}(\mathbf{X}_n)\right] = \sum_{n=0}^{\infty} \mathbb{E}\big[\mathbf{1}_{\{\mathbf{0}\}}(\mathbf{X}_n)\big] = \sum_{n=0}^{\infty} \mathbb{P}(\mathbf{X}_n = \mathbf{0}),$$

in conjunction with (1.2.3) it becomes powerful. Namely, it says that

$$\{\mathbf{X}_n : n \geq 0\} \text{ is recurrent} \quad \text{if and only if} \quad \sum_{n=0}^{\infty} \mathbb{P}(\mathbf{X}_n = \mathbf{0}) = \infty, \tag{1.2.6}$$

and, since $\mathbb{P}(\mathbf{X}_n = \mathbf{0})$ is more amenable to estimation than quantities which involve knowing the trajectory at more than one time, this is valuable information.

In order to apply (1.2.6) to symmetric random walks, it is important to know that *when the walk is symmetric, then* $\mathbf{0}$ *is the most likely place for the walk to be at any even time*. To verify this, note that if $\mathbf{k} \in \mathbb{Z}^d$, then

$$\begin{aligned}
\mathbb{P}(\mathbf{X}_{2n} = \mathbf{k}) &= \sum_{\boldsymbol{\ell} \in \mathbb{Z}^d} \mathbb{P}(\mathbf{X}_n = \ell \;\&\; \mathbf{X}_{2n} - \mathbf{X}_n = \mathbf{k} - \boldsymbol{\ell}) \\
&= \sum_{\boldsymbol{\ell} \in \mathbb{Z}^d} \mathbb{P}(\mathbf{X}_n = \ell)\mathbb{P}(\mathbf{X}_{2n} - \mathbf{X}_n = \mathbf{k} - \boldsymbol{\ell}) \\
&= \sum_{\boldsymbol{\ell} \in \mathbb{Z}^d} \mathbb{P}(\mathbf{X}_n = \boldsymbol{\ell})\mathbb{P}(\mathbf{X}_n = \mathbf{k} - \boldsymbol{\ell}) \\
&\leq \left(\sum_{\boldsymbol{\ell} \in \mathbb{Z}^d} \mathbb{P}(\mathbf{X}_n = \boldsymbol{\ell})^2\right)^{\frac{1}{2}} \left(\sum_{\boldsymbol{\ell} \in \mathbb{Z}^d} \mathbb{P}(\mathbf{X}_n = \mathbf{k} - \boldsymbol{\ell})^2\right)^{\frac{1}{2}} = \sum_{\boldsymbol{\ell} \in \mathbb{Z}^d} \mathbb{P}(\mathbf{X}_n = \boldsymbol{\ell})^2,
\end{aligned}$$

where, in the passage to the last line, we have applied Schwarz's inequality (cf. Exercise 1.3.1 below). Up to this point we have not used symmetry. However, if the walk is symmetric, then $\mathbb{P}(\mathbf{X}_n = \ell) = \mathbb{P}(\mathbf{X}_n = -\ell)$, and so the last line of the preceding can be continued as

$$\begin{aligned}
&\sum_{\boldsymbol{\ell} \in \mathbb{Z}^d} \mathbb{P}(\mathbf{X}_n = \boldsymbol{\ell})\mathbb{P}(\mathbf{X}_n = -\ell) \\
&\qquad = \sum_{\boldsymbol{\ell} \in \mathbb{Z}^d} \mathbb{P}(\mathbf{X}_n = \boldsymbol{\ell})\mathbb{P}(\mathbf{X}_{2n} - \mathbf{X}_n = -\boldsymbol{\ell}) = \mathbb{P}(\mathbf{X}_{2n} = \mathbf{0}).
\end{aligned}$$

Thus,

$$\{\mathbf{X}_n : n \geq 0\} \text{ symmetric} \quad \Longrightarrow \quad \mathbb{P}(\mathbf{X}_{2n} = \mathbf{0}) = \max_{\mathbf{k}\in\mathbb{Z}^d} \mathbb{P}(\mathbf{X}_{2n} = \mathbf{k}). \tag{1.2.7}$$

To develop a feeling for how these considerations get applied, we begin by using them to give a second derivation of the recurrence of the nearest neighbor, symmetric random walk on $\mathbb{Z}$. For this purpose, note that, because $\mathbb{P}(|X_n| \leq n) = 1$, (1.2.7) implies that

$$1 = \sum_{\ell=-2n}^{2n} \mathbb{P}(X_{2n} = \ell) \leq (4n+1)\mathbb{P}(X_{2n} = 0),$$

and therefore, since the harmonic series diverges, that $\sum_{n=0}^{\infty} \mathbb{P}(X_n = 0) = \infty$.

The analysis for the symmetric, nearest neighbor random walk in $\mathbb{Z}^2$ requires an additional ingredient. Namely, the d-dimensional analog of the preceding line of reasoning would lead to $\mathbb{P}(\mathbf{X}_{2n} = \mathbf{0}) \geq (4n+1)^{-d}$, which is inconclusive except when $d = 1$. In order to do better, we need to use the fact that

$$\{\mathbf{X}_n : n \geq 0\} \text{ symmetric} \quad \Longrightarrow \quad \mathbb{E}\big[|\mathbf{X}_n|^2\big] = n. \tag{1.2.8}$$

To prove (1.2.8), note that each coordinate of $\mathbf{B}_n$ is a random variable with mean value 0 and variance $\frac{1}{d}$. Hence, because the $\mathbf{B}_n$'s are mutually independent, the second moment of each coordinate of $\mathbf{X}_n$ is $\frac{n}{d}$.

Knowing (1.2.8), Markov's inequality (7.1.12) says that

$$\mathbb{P}\big(|\mathbf{X}_{2n}| \geq 2\sqrt{n}\big) \leq \frac{1}{4n}\mathbb{E}\big[|\mathbf{X}_{2n}|^2\big] = \frac{1}{2},$$

which allows us to sharpen the preceding argument to give[9]

$$\begin{aligned}\frac{1}{2} &\leq \mathbb{P}\big(|\mathbf{X}_{2n}| < 2\sqrt{n}\big) = \sum_{|\boldsymbol{\ell}|<2\sqrt{n}} \mathbb{P}(\mathbf{X}_{2n} = \boldsymbol{\ell}) \\ &\leq (4\sqrt{n}+1)^d \mathbb{P}(\mathbf{X}_{2n} = \mathbf{0}) \leq 2^{d-1}\big(4n^{\frac{d}{2}} + 1\big)\mathbb{P}(\mathbf{X}_{2n} = \mathbf{0}).\end{aligned}$$

That is, we have now shown that

$$\mathbb{P}(\mathbf{X}_{2n} = \mathbf{0}) \geq 2^{-d}\big(4n^{\frac{d}{2}} + 1\big)^{-1} \tag{1.2.9}$$

for the symmetric, nearest neighbor random walk on $\mathbb{Z}^d$. In particular, when $d = 2$, this proves that *the symmetric, nearest neighbor random walk on $\mathbb{Z}^2$ is recurrent.*

[9]For any $a, b \in [0,\infty)$ and $p \in [1,\infty)$, $(a+b)^p \leq 2^{p-1}(a^p + b^p)$. This can be seen as an application of Jensen's inequality (cf. Exercise 6.6.2), which, in this case, is simply the statement that $x \in [0,\infty) \longmapsto x^p$ is convex.

1.2.4 Transience in $\mathbb{Z}^3$

Although (1.2.9) was sufficient to prove recurrence for the symmetric, nearest neighbor random walk in $\mathbb{Z}^2$, it leaves open the possibility of transience in $\mathbb{Z}^d$ for $d \ge 3$. Thus, in order to nail down the question when $d \ge 3$, we will need to see how good an estimate (1.2.9) really is. In particular, it would suffice to prove that there is an upper bound of the same form.

To get an upper bound which complements the lower bound in (1.2.9), we first do so in the case when $d = 1$. For this purpose, let $0 \le \ell \le n$ be given, and observe that

$$\frac{\mathbb{P}(X_{2n} = 2\ell)}{\mathbb{P}(X_{2n} = 0)} = \frac{(n!)^2}{(n+\ell)!(n-\ell)!} = \frac{n(n-1)\cdots(n-\ell+1)}{(n+\ell)(n+\ell-1)\cdots(n+1)}$$
$$= \prod_{k=0}^{\ell-1}\left(1 - \frac{\ell}{n+\ell-k}\right) \ge \left(1 - \frac{\ell}{n+1}\right)^{\ell}.$$

Now recall that

$$\log(1-x) = -\sum_{m=1}^{\infty} \frac{x^m}{m} \quad \text{for } |x| < 1 \tag{1.2.10}$$

and therefore that $\log(1-x) \ge -\frac{3x}{2}$ for $0 \le x \le \frac{1}{2}$. Hence, the preceding shows that

$$\frac{\mathbb{P}(X_{2n} = 2\ell)}{\mathbb{P}(X_{2n} = 0)} \ge \exp\left(\ell \log \frac{\ell}{n+1}\right) \ge e^{-\frac{3\ell^2}{2(n+1)}}$$

as long as $0 \le \ell \le \frac{n+1}{2}$. Because $\mathbb{P}(X_{2n} = -2\ell) = \mathbb{P}(X_{2n} = 2\ell)$, we can now say that

$$\mathbb{P}(X_{2n} = 0) \le e^{\frac{3}{2}}\mathbb{P}(X_{2n} = 2\ell) \quad \text{for } |\ell| \le \sqrt{n}.$$

But, because $\sum_{\ell \in \mathbb{Z}} \mathbb{P}(X_{2n} = 2\ell) = 1$, this means that $(2\sqrt{n} - 1)\mathbb{P}(X_{2n} = 0) \le e^{\frac{3}{2}}$, and so

$$\mathbb{P}(X_{2n} = 0) \le e^{\frac{3}{2}}(2\sqrt{n} - 1)^{-1}, \quad n \ge 1, \tag{1.2.11}$$

when $\{X_n : n \ge 0\}$ is the symmetric, nearest neighbor random walk on $\mathbb{Z}$.

If, as they most definitely are not, the coordinates of the symmetric, nearest neighbor random walk were independent, then (1.2.11) would yield the sort of upper bound for which we are looking. Thus it is reasonable to examine to what extent we can relate the symmetric, nearest neighbor random walk on $\mathbb{Z}^d$ to d mutually independent symmetric, nearest neighbor random walks $\{X_{i,n} : n \ge 0\}$, $1 \le i \le d$, on $\mathbb{Z}$. To this end, refer to (1.2.1) with $p_\epsilon \equiv \frac{1}{2d}$, and think of choosing $\mathbf{X}_n - \mathbf{X}_{n-1}$ in two steps: first choose the coordinate which is to be non-zero and then choose whether it is to be $+1$ or -1. With this in mind, let $\{I_n : n \ge 0\}$ be a sequence of $\{1, \dots, d\}$-valued, mutually independent, uniformly distributed random variables which are independent of $\{X_{i,n} : 1 \le i \le d \ \& \ n \ge 0\}$, set, for $1 \le i \le d$, $N_{i,0} = 0$ and $N_{i,n} = \sum_{m=1}^{n} \mathbf{1}_{\{i\}}(I_m)$ when $n \ge 1$, and consider the sequence $\{\mathbf{Y}_n : n \ge 0\}$ given by

$$\mathbf{Y}_n = (X_{1,N_{1,n}}, \ldots, X_{d,N_{d,n}}). \tag{1.2.12}$$

Without too much effort, one can check that $\{\mathbf{Y}_n : n \ge 0\}$ satisfies the conditions in (1.2.1) for the symmetric, nearest neighbor random walk on $\mathbb{Z}^d$ and therefore has the same distribution as $\{\mathbf{X}_n : n \ge 0\}$. In particular, by (1.2.11),

$$\begin{aligned}
\mathbb{P}(\mathbf{X}_{2n} = \mathbf{0}) &= \sum_{\mathbf{m}\in\mathbb{N}^d} \mathbb{P}(X_{i,2m_i} = 0 \ \& \ N_{i,2n} = 2m_i \text{ for } 1 \le i \le d) \\
&= \sum_{\substack{\mathbf{m}\in\mathbb{N}^d \\ m_1\wedge\cdots\wedge m_d \ge \frac{n}{d}}} \left(\prod_{i=1}^{d} \mathbb{P}(X_{i,2m_i} = 0)\right) \mathbb{P}(N_{i,2n} = 2m_i \text{ for } 1 \le i \le d) \\
&\quad + \sum_{\substack{\mathbf{m}\in\mathbb{N}^d \\ m_1\wedge\cdots\wedge m_d < \frac{n}{2d}}} \left(\prod_{i=1}^{d} \mathbb{P}(X_{i,2m_i} = 0)\right) \mathbb{P}(N_{i,2n} = 2m_i \text{ for } 1 \le i \le d) \\
&\le e^{\frac{3d}{2}} \left(2\sqrt{\frac{n}{d}} - 1\right)^{-d} + \mathbb{P}\left(N_{i,2n} \le \frac{n}{d} \text{ for some } 1 \le i \le d\right).
\end{aligned}$$

Thus, we will have proved that there is a constant $A(d) < \infty$ such that

$$\mathbb{P}(\mathbf{X}_{2n} = \mathbf{0}) \le A(d) n^{-\frac{d}{2}}, \quad n \ge 1, \tag{1.2.13}$$

once we show that there is a constant $B(d) < \infty$ such that

$$\mathbb{P}(N_{i,2n} \le \tfrac{n}{d} \text{ for some } 1 \le i \le d) \le B(d) n^{-\frac{d}{2}}, \quad n \ge 1. \tag{1.2.14}$$

In particular, this will complete the proof that

$$d \ge 3 \quad \Longrightarrow \quad \sum_{n=0}^{\infty} \mathbb{P}(\mathbf{X}_{2n} = \mathbf{0}) < \infty$$

and therefore that *the symmetric, nearest neighbor random walk in $\mathbb{Z}^d$ is transient when $d \ge 3$.*

To prove (1.2.14), first note that

$$\mathbb{P}\left(N_{i,2n} \le \frac{n}{d} \text{ for some } 1 \le i \le d\right) \le d\,\mathbb{P}\left(N_{1,2n} \le \frac{n}{d}\right).$$

Next, set $Z_m = \mathbf{1}_{\{1\}}(I_m)$, write $N_{1,n} = \sum_1^n Z_m$, and observe that $\{Z_m : m \ge 1\}$ is a sequence of $\{0,1\}$-valued Bernoulli random variables such that $\mathbb{P}(Z_m = 1) =$

$p \equiv \frac{1}{d}$. In particular, for any $\lambda \in \mathbb{R}$,

$$\mathbb{E}\left[\exp\left(\lambda \sum_1^n Z_m\right)\right] = \left(pe^\lambda + q\right)^n,$$

and so

$$\mathbb{E}\left[\exp\left(\lambda\left(np - \sum_1^n Z_m\right)\right)\right] = e^{n\psi(\lambda)} \quad \text{where } \psi(\lambda) \equiv \log\left(pe^{-\lambda q} + qe^{\lambda p}\right).$$

Since $\psi(0) = \psi'(0) = 0$, and

$$\psi''(\lambda) = \frac{pqe^{\lambda(p-q)}}{(qe^{\lambda p} + pe^{-\lambda q})^2} = \frac{pq}{(qx_\lambda + px_\lambda^{-1})^2} \le \frac{1}{4}$$

where $x_\lambda \equiv e^{\frac{1}{2}\lambda(p+q)}$, Taylor's formula allows us to conclude that

$$\mathbb{E}\left[\exp\left(\lambda\left(np - \sum_1^n Z_m\right)\right)\right] \le e^{\frac{n\lambda^2}{8}}, \quad \lambda \in \mathbb{R}. \tag{1.2.15}$$

Starting from (1.2.15), there are many ways to arrive at (1.2.14). For example, for any $\lambda > 0$ and $R > 0$, Markov's inequality (7.1.12) plus (1.2.15) say that

$$\begin{aligned}\mathbb{P}\left(\sum_1^n Z_m \le np - nR\right) &= \mathbb{P}\left(\exp\left(\lambda\left(np - \sum_1^n Z_m\right)\right) \ge e^{n\lambda R}\right) \\ &\le e^{-\lambda nR + \frac{n\lambda^2}{8}},\end{aligned}$$

which, when $\lambda = 4nR$, gives

$$\mathbb{P}\left(\sum_1^n Z_m \le np - nR\right) \le e^{-2nR^2}. \tag{1.2.16}$$

Returning to the notation used earlier and using the remark with which our discussion of (1.2.14) began, one sees from (1.2.16) that

$$\mathbb{P}\left(N_{i,2n} \le \frac{n}{d} \text{ for some } 1 \le i \le d\right) \le de^{-\frac{4n}{d^2}},$$

which is obviously far more than is required by (1.2.14).

1.3 Exercises

Exercise 1.3.1 *Schwarz's inequality* comes in many forms, the most elementary of which is the statement that, for any $\{a_n : n \in \mathbb{Z}\} \subseteq \mathbb{R}$ and $\{b_n : n \in \mathbb{Z}\} \subseteq \mathbb{R}$,

$$\sum_{n\in\mathbb{Z}} |a_n b_n| \le \sqrt{\sum_{n\in\mathbb{Z}} a_n^2}\sqrt{\sum_{n\in\mathbb{Z}} b_n^2}.$$

Moreover, when the right hand side is finite, then

$$\left|\sum_{n\in\mathbb{Z}} a_n b_n\right| = \sqrt{\sum_{n\in\mathbb{Z}} a_n^2}\sqrt{\sum_{n\in\mathbb{Z}} b_n^2}$$

if and only if there is an $\alpha \in \mathbb{R}$ for which either $b_n = \alpha a_n, n \in \mathbb{Z}$, or $a_n = \alpha b_n, n \in \mathbb{Z}$. Here is an outline of one proof of these statements.

(a) Given a real, quadratic polynomial $P(x) = Ax^2 + 2Bx + C$, use the quadratic formula to see that $P \ge 0$ everywhere if and only if $C \ge 0$ and $B^2 \le AC$. Similarly, show that $P > 0$ everywhere if and only if $C > 0$ and $B^2 < AC$.
(b) Begin by noting that there is nothing to do unless $\sum_{n\in\mathbb{Z}}(a_n^2 + b_n^2) < \infty$, in which case $\sum_{n\in\mathbb{Z}} |a_n b_n| < \infty$ as well. Assuming that $\sum_{n\in\mathbb{Z}}(a_n^2 + b_n^2) < \infty$, show that

$$P(x) \equiv \sum_{n\in\mathbb{Z}} (a_n x + b_n)^2 = Ax^2 + 2Bx + C$$

$$\text{where } A = \sum_{n\in\mathbb{Z}} a_n^2,\ B = \sum_{n\in\mathbb{Z}} b_n^2,\ \text{and } C = \sum_{n\in\mathbb{Z}} a_n b_n,$$

and apply (a).

Exercise 1.3.2 Let $\{Y_n : n \ge 1\}$ be a sequence of mutually independent, identically distributed random variables satisfying $\mathbb{E}[|Y_1|] < \infty$. Set $X_n = \sum_{m=1}^n Y_m$ for $n \ge 1$. *The weak law of large numbers* says that

$$\mathbb{P}\left(\left|\frac{X_n}{n} - \mathbb{E}[Y_1]\right| \ge \epsilon\right) \longrightarrow 0 \quad \text{for all } \epsilon > 0.$$

In fact,

$$\lim_{n\to\infty} \mathbb{E}\left[\left|\frac{X_n}{n} - \mathbb{E}[Y_1]\right|\right] = 0, \tag{1.3.3}$$

from which the above follows as an application of Markov's inequality. Here are steps which lead to (1.3.3).

(a) First reduce to the case when $\mathbb{E}[Y_1] = 0$. Next, assume that $\mathbb{E}[Y_1^2] < \infty$, and show that

$$\mathbb{E}\left[\left|\frac{X_n}{n}\right|\right]^2 \le \mathbb{E}\left[\left|\frac{X_n}{n}\right|^2\right] = \frac{\mathbb{E}[Y_1^2]}{n}.$$

Hence the result is proved when Y_1 has a finite second moment.

(b) Given $R > 0$, set $Y_n^{(R)} = Y_n \mathbf{1}_{[0,R)}(|Y_n|) - \mathbb{E}[Y_n, |Y_n| < R]$ and $X_n^{(R)} = \sum_{m=1}^n Y_m^{(R)}$. Note that, for any $R > 0$,

$$\mathbb{E}\left[\left|\frac{X_n}{n}\right|\right] \le \mathbb{E}\left[\left|\frac{X_n^{(R)}}{n}\right|\right] + \mathbb{E}\left[\left|\frac{X_n - X_n^{(R)}}{n}\right|\right]$$

$$\le \sqrt{\mathbb{E}\left[\left(\frac{X_n^{(R)}}{n}\right)^2\right]} + 2\mathbb{E}\big[|Y_1|, |Y_1| \ge R\big] \le \frac{R}{n^{\frac{1}{2}}} + 2\mathbb{E}\big[|Y_1|, |Y_1| \ge R\big],$$

and use this, together with the monotone convergence theorem, to complete the proof of (1.3.3).

Exercise 1.3.4 Refer to Exercise 1.3.2. *The strong law of large numbers* says that the statement in the weak law can be improved to the statement that $\frac{X_n}{n} \longrightarrow \mathbb{E}[Y_1]$ with probability 1. The proof of the strong law when one assumes only that $\mathbb{E}[|Y_1|] < \infty$ is a bit tricky. However, if one is willing to assume that $\mathbb{E}[Y_1^4] < \infty$, then a proof can be based on the same type argument that leads to the weak law.

Let $\{Y_n\}_1^\infty$ be a sequence of mutually independent random variables with the properties that $M = \sup_n \mathbb{E}[|Y_n|^4] < \infty$, and prove that, with probability 1, $\lim_{n\to\infty} \frac{1}{n}\sum_{m=1}^n (Y_m - \mathbb{E}[Y_m]) = 0$. Note that we have not assumed yet that they are identically distributed, but when we add this assumption we get $\lim_{n\to\infty} \frac{1}{n}\sum_{m=1}^n Y_m = \mathbb{E}[Y_1]$ with probability 1.

Here is an outline.

(a) Begin by reducing to the case when $\mathbb{E}[Y_n] = 0$ for all $n \in \mathbb{Z}^+$.

(b) After writing

$$\mathbb{E}\left[\left(\sum_1^n Y_k\right)^4\right] = \sum_{k_1,\dots,k_4=1}^n \mathbb{E}[Y_{k_1} \cdots Y_{k_4}]$$

and noting that the only terms which do not vanish are those for which each index is equal to at least one other index, conclude that

$$\mathbb{E}\left[\left(\sum_1^n Y_k\right)^4\right] = \sum_{k=1}^n \mathbb{E}[Y_k^4] + 6 \sum_{1\le k<\ell\le n} \mathbb{E}[Y_k^2]\mathbb{E}[Y_\ell^2].$$

Hence, since $\mathbb{E}[Y_k^4] - \mathbb{E}[Y_k^2]^2 = \mathrm{Var}(Y_k^2) \ge 0$,

$$\mathbb{E}\left[\left(\sum_1^n Y_k\right)^4\right] \le 3Mn^2. \tag{$*$}$$

(c) Starting from (∗), show that

$$\mathbb{P}\left(\left|\frac{\sum_1^n Y_k}{n}\right| \geq \epsilon\right) \leq \frac{1}{\epsilon^4}\mathbb{E}\left[\left|\frac{\sum_1^n Y_k}{n}\right|^4\right] \leq \frac{3M}{\epsilon^4 n^2} \longrightarrow 0$$

for all $\epsilon > 0$. This is the weak law of large numbers for independent random variables with bounded fourth moments. Of course, the use of four moments here is somewhat ridiculous since the argument using only two moments is easier.

(d) Starting again from (∗) and using (7.1.4), show that

$$\begin{aligned}\mathbb{P}\left(\sup_{n>m}\left|\frac{\sum_1^n Y_k}{n}\right| \geq \epsilon\right) &\leq \sum_{n=m+1}^{\infty} \mathbb{P}\left(\left|\sum_1^n Y_k\right| \geq n\epsilon\right)\\ &\leq \frac{4M}{\epsilon^4}\sum_{n=m+1}^{\infty}\frac{1}{n^2}\\ &\leq \frac{4M}{\epsilon^4 m} \longrightarrow 0 \quad \text{as } m \to \infty \text{ for all } \epsilon > 0.\end{aligned}$$

(e) Use the definition of convergence plus (7.1.4) to show that

$$\begin{aligned}\mathbb{P}\left(\frac{\sum_1^n Y_k}{n} \nrightarrow 0\right) &= \mathbb{P}\left(\bigcup_{N=1}^{\infty}\bigcap_{m=1}^{\infty}\bigcup_{n>m}\left\{\left|\frac{\sum_1^n Y_k}{n}\right| \geq \frac{1}{N}\right\}\right)\\ &\leq \sum_{N=1}^{\infty}\mathbb{P}\left(\bigcap_{m=1}^{\infty}\bigcup_{n>m}\left\{\left|\frac{\sum_1^n Y_k}{n}\right| \geq \frac{1}{N}\right\}\right).\end{aligned}$$

Finally, apply the second line of (7.1.3) plus (d) above to justify

$$\mathbb{P}\left(\bigcap_{m=1}^{\infty}\bigcup_{n>m}\left\{\left|\frac{\sum_1^n Y_k}{n}\right| \geq \frac{1}{N}\right\}\right) = \lim_{m\to\infty}\mathbb{P}\left(\bigcup_{n>m}\left\{\left|\frac{\sum_1^n Y_k}{n}\right| \geq \frac{1}{N}\right\}\right) = 0$$

for each $N \in \mathbb{Z}^+$. Hence, with probability 1, $\frac{1}{n}\sum_1^n Y_k \longrightarrow 0$, which is the strong law of large numbers for independent random variables with bounded fourth moments.

Exercise 1.3.5 Readers who know DeMoivre's proof of the central limit theorem will have realized that the estimate in (1.2.11) is a poor man's substitute for what one can get as a consequence of *Stirling's formula*

$$n! \sim \sqrt{2\pi n}\left(\frac{n}{e}\right)^n \quad \text{as } n \to \infty, \tag{1.3.6}$$

whose meaning is that the ratio of the quantities on the two sides of "∼" tends to 1. Indeed, given (1.3.6), show that

$$\mathbb{P}(X_{2n}=0)\sim\sqrt{\frac{2}{\pi n}}.$$

Next, give a proof of (1.3.6) based on the following line of reasoning.

(a) Let $\tau_1,\dots,\tau_n$ be a mutually independent, unit exponential random variables,[10] and show that for any $0<R\le\sqrt{n}$

$$1-\frac{1}{R^2}\le\mathbb{P}\left(-R\le\frac{\tau_1+\cdots+\tau_n-n}{\sqrt{n}}\le R\right)=\frac{1}{(n-1)!}\int_{-\sqrt{n}R+n}^{\sqrt{n}R+n}t^{n-1}e^{-t}\,dt.$$

(b) Make a change of variables followed by elementary manipulations to show that

$$\begin{aligned}\int_{-\sqrt{n}R+n}^{\sqrt{n}R+n}t^{n-1}e^{-t}\,dt&=\sqrt{n}e^{-n}\int_{-R}^{R}(n+\sqrt{n}\sigma)^{n-1}e^{-\sqrt{n}\sigma}\,d\sigma\\&=n^{n-\frac{1}{2}}e^{-n}\int_{-R}^{R}\left(1+\frac{\sigma}{\sqrt{n}}\right)^{n-1}e^{-\sqrt{n}\sigma}\,d\sigma\\&=n^{n-\frac{1}{2}}e^{-n}\int_{-R}^{R}\exp\left(-\frac{\sigma^2}{2}+E_n(\sigma)\right)d\sigma,\end{aligned}$$

where

$$E_n(\sigma)\equiv(n-1)\log\left(1+\frac{\sigma}{\sqrt{n}}\right)-\sqrt{n}\sigma+\frac{\sigma^2}{2}.$$

(c) As an application of the Taylor's series for $\log(1+x)$ (cf. (1.2.10)), show that $E_n(\sigma)\longrightarrow0$ uniformly for $|\sigma|\le R$ when $n\to\infty$, and combine this with the results in (a) and (b) to arrive at

$$\varlimsup_{n\to\infty}\frac{n^{n+\frac{1}{2}}e^{-n}}{n!}\int_{-R}^{R}e^{-\frac{\sigma^2}{2}}\,d\sigma\le1$$

and

$$\varliminf_{n\to\infty}\frac{n^{n+\frac{1}{2}}e^{-n}}{n!}\int_{-R}^{R}e^{-\frac{\sigma^2}{2}}\,d\sigma\ge1-\frac{1}{R^2}.$$

Because $\int_{-\infty}^{\infty}e^{-\frac{\sigma^2}{2}}\,d\sigma=\sqrt{2\pi}$, it is clear that (1.3.6) follows after one lets $R\nearrow\infty$.

---

[10] A unit exponential random variable is a random variable τ for which $\mathbb{P}(\tau>t)=e^{-t\vee0}$.

Exercise 1.3.7 The argument in Sect. 1.2.3 is quite robust. Indeed, let $\{\mathbf{X}_n : n \ge 0\}$ be any symmetric random walk on $\mathbb{Z}^2$ whose jumps have finite second moment. That is, $\mathbf{X}_0 = \mathbf{0}$, $\{\mathbf{X}_n - \mathbf{X}_{n-1} : n \ge 1\}$ are mutually independent, identically distributed, symmetric ($\mathbf{X}_1$ has the same distribution as $-\mathbf{X}_1$), $\mathbb{Z}^2$-valued random variables with finite second moment. Show that $\{\mathbf{X}_n : n \ge 0\}$ is recurrent in the sense that $\mathbb{P}(\exists n \ge 1\ \mathbf{X}_n = \mathbf{0}) = 1$.

Exercise 1.3.8 Let $\{\mathbf{X}_n : n \ge 0\}$ be a random walk on $\mathbb{Z}^d$: $\mathbf{X}_0 = \mathbf{0}$, $\{\mathbf{X}_n - \mathbf{X}_{n-1} : n \ge 1\}$ are mutually independent, identically distributed, $\mathbb{Z}^d$-valued random variables. Further, for each $1 \le i \le d$, let $(\mathbf{X}_n)_i$ be the ith coordinate of $\mathbf{X}_n$, and assume that

$$\min_{1\le i\le d} \mathbb{P}\big((\mathbf{X}_1)_i \ne 0\big) > 0 \quad \text{but} \quad \mathbb{P}\big(\exists i \ne j\ (\mathbf{X}_1)_i(\mathbf{X}_1)_j \ne 0\big) = 0.$$

If, for some $C < \infty$ and $(\alpha_1, \dots, \alpha_d) \in [0,\infty)^d$ with $\sum_1^d \alpha_i > 1$,

$$\mathbb{P}\big((\mathbf{X}_n)_i = 0\big) \le C n^{-\alpha_i} \quad \text{for all } n \ge 1,$$

show that $\{\mathbf{X}_n : n \ge 0\}$ is transient in the sense that $\mathbb{P}(\exists n \ge 1\ \mathbf{X}_n = \mathbf{0}) < 1$.

Exercise 1.3.9 As in the preceding, let $\{\mathbf{X}_n : n \ge 0\}$ be a random walk on $\mathbb{Z}^d$. Given $\mathbf{k} \in \mathbb{Z}^d$, set

$$T_{\mathbf{k}} = \sum_{n=0}^{\infty} \mathbf{1}_{\{\mathbf{k}\}}(\mathbf{X}_n) \quad \text{and} \quad \zeta^{\{\mathbf{k}\}} = \inf\{n \ge 0 : \mathbf{X}_n = \mathbf{k}\}.$$

Show that

$$\mathbb{E}[T_{\mathbf{k}}] = \mathbb{P}\big(\zeta^{\{\mathbf{k}\}} < \infty\big)\mathbb{E}[T_{\mathbf{0}}] = \frac{\mathbb{P}(\zeta^{\{\mathbf{k}\}} < \infty)}{\mathbb{P}(\rho_{\mathbf{0}} = \infty)}, \tag{1.3.10}$$

where $\rho_{\mathbf{0}} = \inf\{n \ge 1 : \mathbf{X}_n = \mathbf{0}\}$ is the time of first return to $\mathbf{0}$. In particular, if $\{\mathbf{X}_n : n \ge 0\}$ is transient in the sense described in the preceding exercise, show that

$$\mathbb{E}\left[\sum_{m=0}^{\infty} \mathbf{1}_{B(r)}(\mathbf{X}_n)\right] < \infty \quad \text{for all } r \in (0,\infty),$$

where $B(r) = \{\mathbf{k} : |\mathbf{k}| \le r\}$; and from this conclude that $|\mathbf{X}_n| \longrightarrow \infty$ with probability 1. On the other hand, if $\{\mathbf{X}_n : n \ge 0\}$ is recurrent, show that $\mathbf{X}_n = \mathbf{0}$ infinitely often with probability 1. Hence, either $\{\mathbf{X}_n : n \ge 0\}$ is recurrent and $\mathbf{X}_n = \mathbf{0}$ infinitely often with probability 1 or it is transient and $|\mathbf{X}_n| \longrightarrow \infty$ with probability 1.

Exercise 1.3.11 Take $d = 1$ in the preceding, $X_0 = 0$, and $\{X_n - X_{n-1} : n \ge 1\}$ to be mutually independent, identically distributed random variables for which $0 < \mathbb{E}[|X_1|] < \infty$ and $\mathbb{E}[X_1] = 0$. By a slight variation on the argument given in Sect. 1.2.1, we will show here that this random walk is not only recurrent but that

$$\overline{\lim_{n\to\infty}}\ X_n = \infty \quad \text{and} \quad \underline{\lim_{n\to\infty}}\ X_n = -\infty \quad \text{with probability 1.}$$

(a) First show that it suffices to prove that $\sup_n X_n = \infty$ and that $\inf_n X_n = -\infty$. Next, use (1.3.3) to show that

$$\lim_{n\to\infty} \max_{1\le m\le n} \frac{\mathbb{E}[|X_m|]}{n} = 0.$$

(b) For $n \ge 1$, set $T_k^{(n)} = \sum_{m=0}^{n-1} \mathbf{1}_{\{k\}}(X_m)$, show that $\mathbb{E}[T_k^{(n)}] \le \mathbb{E}[T_0^{(n)}]$ for all $k \in \mathbb{Z}$, and use this to arrive at

$$\big(4\mu(n)+1\big)\mathbb{E}\big[T_0^{(n)}\big] \ge \frac{n}{2} \quad \text{where } \mu(n) \equiv \max_{0\le m\le n-1} \mathbb{E}\big[|X_m|\big].$$

Finally, apply part (a) to see that $\mathbb{E}[T_0] = \infty$ and (1.3.10) to get $\mathbb{P}(\rho_0 < \infty) = 1$, which means that $\{X_n : n \ge 0\}$ is recurrent.

(c) To complete the program, proceed as in the derivation of (1.2.2) to pass from (b) to

$$\mathbb{P}\big(\rho_0^{(m)} < \infty\big) = 1 \quad \text{for all } m \ge 1, \tag{$*$}$$

where $\rho_0^{(m)}$ is the time of the mth return to 0. Next, set $\eta_r = \inf\{n \ge 0 : X_n \ge r\}$ for $r \in \mathbb{Z}^+$, show that $\epsilon \equiv \mathbb{P}(\eta_1 > \rho_0) < 1$, and conclude that $\mathbb{P}(\eta_1 > \rho_0^{(m)}) \le \epsilon^m$. Now, combine this with $(*)$ to get $\mathbb{P}(\eta_1 < \infty) = 1$. Finally, argue that

$$\mathbb{P}(\eta_{r+1} < \infty) \ge \mathbb{P}(\eta_r < \infty)\mathbb{P}(\eta_1 < \infty)$$

and therefore that $\mathbb{P}(\eta_r < \infty) = 1$ for all $r \ge 1$. Since this means that, with probability 1, $\sup_n X_n \ge r$ for all $r \ge 1$, it follows that $\sup_n X_n = \infty$ with probability 1. To prove that $\inf_n X_n = -\infty$ with probability 1, simply replace $\{X_n : n \ge 0\}$ by $\{-X_n : n \ge 0\}$.

Exercise 1.3.12[11] Here is an interesting application of one dimensional random walks to elementary *queuing theory*. Queuing theory deals with the distribution of the number of people waiting to be served (i.e., the length of the queue) when, during each time interval, the number of people who arrive and the number of people who are served are random. The queuing model which we will consider here is among the simplest. Namely, we will assume that the queue is initially empty and that during each time interval $[n-1, n)$ the number of people who arrive minus the number who can be served is given by a $\mathbb{Z}$-valued random variable B_n. Further, we assume that the B_n's are mutually independent and identically distributed random variables satisfying $0 < \mathbb{E}[|B_1|] < \infty$. The associated queue is, apart from the fact that there are never a negative number of people waiting, the random walk $\{X_n : n \ge 0\}$ determined by the B_n's: $X_0 = 0$ and $X_n = \sum_{m=1}^{n} B_m$. To take into account the prohibition against having a queue of negative length, the queuing model $\{Q_n : n \ge 0\}$ is given by the prescription

$$Q_0 = 0 \quad \text{and} \quad Q_n = (Q_{n-1} + B_n)^+ \quad \text{for } n \ge 1.$$

[11] So far as I know, this example was invented by Wm. Feller.

(a) Show that

$$Q_n = X_n - \min_{0\le m\le n} X_m = \max_{0\le m\le n} (X_n - X_m),$$

and conclude that, for each $n \ge 0$, the distribution of Q_n is the same as that of $M_n \equiv \max_{0\le m\le n} X_m$.

(b) Set $M_\infty \equiv \lim_{n\to\infty} M_n \in \mathbb{N} \cup \{\infty\}$, and, as a consequence of (a), arrive at

$$\lim_{n\to\infty} \mathbb{P}(Q_n = j) = \mathbb{P}(M_\infty = j) \quad \text{for } j \in \mathbb{N}.$$

(c) Set $\mu \equiv \mathbb{E}[B_1]$. The weak law of large numbers says that, for each $\epsilon > 0$, $\mathbb{P}(|X_n - n\mu| \ge n\epsilon) \longrightarrow 0$ as $n \to \infty$. In particular, when $\mu > 0$, show that $\mathbb{P}(M_\infty = \infty) = 1$. When $\mu = 0$, use Exercise 1.3.11 to reach the same conclusion. Hence, when $\mathbb{E}[B_1] \ge 0$, $\mathbb{P}(Q_n = j) \longrightarrow 0$ for all $j \in \mathbb{N}$. That is, when the expected difference between the number arrivals and the number of people served is non-negative, then $\mathbb{P}(Q_n \ge M) \longrightarrow 1$ for every $M \in \mathbb{Z}^+$.

(d) Now assume that $\mu \equiv \mathbb{E}[B_1] < 0$. Then the strong law of large numbers (cf. Exercise 1.3.4 for the case when B_1 has a finite fourth moment and Theorem 3.3.10 in [8] for the general case) says that $\frac{X_n}{n} \longrightarrow \mu$ with probability 1. In particular, conclude that $M_\infty < \infty$ with probability 1 and therefore that $\sum_{j\in\mathbb{N}} \nu_j = 1$ when $\nu_j \equiv \lim_{n\to\infty} \mathbb{P}(Q_n = j) = \mathbb{P}(M_\infty = j)$.

(e) Specialize to the case when the B_m's are $\{-1, 1\}$-valued Bernoulli random variables with $p \equiv \mathbb{P}(B_1 = 1) \in (0, 1)$, and set $q = 1 - p$. Use the calculations in (1.1.12) to show that

$$\lim_{n\to\infty} \mathbb{P}(Q_n = j) = \begin{cases} 0 & \text{if } p \ge q \\ \frac{q-p}{q}\left(\frac{p}{q}\right)^j & \text{if } p < q. \end{cases}$$

(f) Generalize (e) to the case when $B_m \in \{-1, 0, 1\}$, $p = \mathbb{P}(B_1 = 1)$, and $q = \mathbb{P}(B_1 = -1)$. The idea is that M_∞ in this case has the same distribution as $\sup_n Y_n$, where $\{Y_n : n \ge 0\}$ is the random walk corresponding to $\{-1, 1\}$-valued Bernoulli random variables which are 1 with probability $\frac{p}{p+q}$.

Chapter 2
Doeblin's Theory for Markov Chains

In this chapter we begin in earnest our study of Markov processes. Like the random walks in Chap. 1, the processes with which we will be dealing here take only countably many values and have a discrete (as opposed to continuous) time parameter. In fact, in many ways, these processes are the simplest generalizations of random walks. To be precise, random walks proceed in such a way that the distribution of their increments are independent of everything that has happened before the increment takes place. The processes at which we will be looking now proceed in such a way that the distribution of their new position depends on where they are at the time when they move but not on where they were in the past. A process with this sort of dependence property is said to have the *Markov property* and is called a *Markov chain*.[1]

The set $\mathbb{S}$ in which a process takes its values is called its *state space*, and, as I said, our processes will have state spaces which are either finite or countably infinite. Thus, at least for theoretical purposes, there is no reason for us not to think of $\mathbb{S}$ as the set $\{1, \dots, N\}$ or $\mathbb{Z}^+$, depending on whether $\mathbb{S}$ is finite or countably infinite. On the other hand, always taking $\mathbb{S}$ to be one of these has the disadvantage that it may mask important properties. For example, it would have been a great mistake to describe the nearest neighbor random walk on $\mathbb{Z}^2$ after mapping $\mathbb{Z}^2$ isomorphically onto $\mathbb{Z}^+$.

2.1 Some Generalities

Before getting started, there are a few general facts that we will need to know about Markov chains.

A *Markov chain* on a finite or countably infinite state space $\mathbb{S}$ is a family of $\mathbb{S}$-valued random variables $\{X_n : n \geq 0\}$ with the property that, for all $n \geq 0$ and

[1] The term "chain" is commonly applied to processes with a time discrete parameter.

D.W. Stroock, *An Introduction to Markov Processes*, Graduate Texts in Mathematics 230,
DOI 10.1007/978-3-642-40523-5_2, © Springer-Verlag Berlin Heidelberg 2014

$(i_0, \ldots, i_n, j) \in \mathbb{S}^{n+2}$,

$$\mathbb{P}(X_{n+1} = j \mid X_0 = i_0, \ldots, X_n = i_n) = (\mathbf{P})_{i_n j}, \tag{2.1.1}$$

where **P** is a matrix all of whose entries are non-negative and each of whose rows sums to 1. Equivalently (cf. Sect. 7.4.1)

$$\mathbb{P}(X_{n+1} = j \mid X_0, \ldots, X_n) = (\mathbf{P})_{X_n j}. \tag{2.1.2}$$

It should be clear that (2.1.2) is a mathematically precise expression of the idea that, when a Markov chain jumps, the distribution of where it lands depends only on where it was at the time when it jumped and not on where it was in the past.

2.1.1 Existence of Markov Chains

For obvious reasons, a matrix whose entries are non-negative and each of whose rows sum to 1 is called a *transition probability matrix*: it gives the probability that the Markov chain will move to the state j at time $n+1$ given that it is at state i at time n, independent of where it was prior to time n. Further, it is clear that only a transition probability matrix could appear on the right of (2.1.1). What may not be so immediate is that one can go in the opposite direction. Namely, let $\boldsymbol{\mu}$ be a *probability vector*[2] and **P** a transition probability matrix. Then there exists a Markov chain $\{X_n : n \geq 0\}$ with *initial distribution* $\boldsymbol{\mu}$ and transition probability matrix **P**. That is, $\mathbb{P}(X_0 = i) = (\boldsymbol{\mu})_i$ and (2.1.1) holds.

To prove the preceding existence statement, one can proceed as follows. Begin by assuming, without loss in generality, that $\mathbb{S}$ is either $\{1, \ldots, N\}$ or $\mathbb{Z}^+$. Next, given $i \in \mathbb{S}$, set $\beta(i, 0) = 0$ and $\beta(i, j) = \sum_{k=1}^{j} (\mathbf{P})_{ik}$ for $j \geq 1$, and define $F : \mathbb{S} \times [0, 1) \longrightarrow \mathbb{S}$ so that $F(i, u) = j$ if $\beta(i, j-1) \leq u < \beta(i, j)$. In addition, set $\alpha(0) = 0$ and $\alpha(i) = \sum_{k=1}^{i} (\boldsymbol{\mu})_k$ for $i \geq 1$, and define $f : [0, 1) \longrightarrow \mathbb{S}$ so that $f(u) = i$ if $\alpha(i-1) \leq u < \alpha(i)$. Finally, let $\{U_n : n \geq 0\}$ be a sequence of mutually independent random variables (cf. Theorem 7.3.2) which are uniformly distributed on $[0, 1)$, and set

$$X_n = \begin{cases} f(U_0) & \text{if } n = 0 \\ F(X_{n-1}, U_n) & \text{if } n \geq 1. \end{cases} \tag{2.1.3}$$

We will now show that the sequence $\{X_n : n \geq 0\}$ in (2.1.3) is a Markov chain with the required properties. For this purpose, suppose that $(i_0, \ldots, i_n) \in \mathbb{S}^{n+1}$, and

[2] A probability vector is a row vector whose coordinates are non-negative and sum to 1.

observe that

$$\begin{aligned}
&\mathbb{P}(X_0 = i_0, \dots, X_n = i_n) \\
&\quad = \mathbb{P}\big(U_0 \in \big[\alpha(i_0 - 1), \alpha(i_0)\big) \\
&\quad\quad \& \, U_m \in \big[\beta(i_{m-1}, i_m - 1), \beta(i_{m-1}, i_m)\big) \text{ for } 1 \le m \le n\big) \\
&\quad = \boldsymbol{\mu}_{i_0}(\mathbf{P})_{i_0 i_1} \cdots (\mathbf{P})_{i_{n-1} i_n} = \mathbb{P}(X_0 = i_0, \dots, X_{n-1} = i_{n-1})\mathbf{P}_{i_{n-1}, j}.
\end{aligned}$$

2.1.2 Transition Probabilities & Probability Vectors

Notice that the use of matrix notation here is clever. To wit, if $\boldsymbol{\mu}$ is the row vector with ith entry $(\boldsymbol{\mu})_i = \mathbb{P}(X_0 = i)$, then $\boldsymbol{\mu}$ is called the *initial distribution* of the chain and

$$\big(\boldsymbol{\mu}\mathbf{P}^n\big)_j = \mathbb{P}(X_n = j), \quad n \ge 0 \text{ and } j \in \mathbb{S}, \tag{2.1.4}$$

where we have adopted the convention that $\mathbf{P}^0$ is the identity matrix and $\mathbf{P}^n = \mathbf{P}\mathbf{P}^{n-1}$ for $n \ge 1$.[3] To check (2.1.4), let $n \ge 1$ be given, and note that, by (2.1.1) and induction,

$$\mathbb{P}(X_0 = i_0, \dots, X_{n-1} = i_{n-1}, X_n = j) = (\boldsymbol{\mu})_{i_0}(\mathbf{P})_{i_0 i_1} \cdots (\mathbf{P})_{i_{n-1} j}.$$

Hence (2.1.4) results after one sums with respect to $(i_0, \dots, i_{n-1})$. Obviously, (2.1.4) is the statement that the row vector $\boldsymbol{\mu}\mathbf{P}^n$ *is the distribution of the Markov chain at time n if $\boldsymbol{\mu}$ is its initial distribution* (i.e., its distribution at time 0). Alternatively, $\mathbf{P}^n$ is the n-step transition probability matrix: $(\mathbf{P}^n)_{ij}$ is the conditional probability that $X_{m+n} = j$ given that $X_m = i$.

For future reference, we will introduce here an appropriate way in which to measure the length of row vectors when they are being used to represent measures. Namely, given a row vector $\boldsymbol{\rho}$, we set

$$\|\boldsymbol{\rho}\|_{\mathrm{v}} = \sum_{i \in \mathbb{S}} \big|(\boldsymbol{\rho})_i\big|, \tag{2.1.5}$$

where the subscript "v" is used in recognition that this is the notion of length which corresponds to the *variation norm* on the space of measures. The basic reason for our making this choice of norm is that

$$\|\boldsymbol{\rho}\mathbf{P}\|_{\mathrm{v}} \le \|\boldsymbol{\rho}\|_{\mathrm{v}}, \tag{2.1.6}$$

since, by Theorem 7.1.15,

$$\|\boldsymbol{\rho}\mathbf{P}\|_{\mathrm{v}} = \sum_{j \in \mathbb{S}} \Big|\sum_{i \in \mathbb{S}} (\boldsymbol{\rho})_i (\mathbf{P})_{ij}\Big| \le \sum_{i \in \mathbb{S}} \Big(\sum_{j \in \mathbb{S}} \big|(\boldsymbol{\rho})_i\big| (\mathbf{P})_{ij}\Big) = \|\boldsymbol{\rho}\|_{\mathrm{v}}.$$

[3] The reader should check for itself that $\mathbf{P}^n$ is again a transition probability matrix for all $n \in \mathbb{N}$: all entries are non-negative and each row sums to 1.

Notice that this is a quite different way of measuring the length from the way Euclid would have: he would have used

$$\|\boldsymbol{\rho}\|_2 = \left(\sum_{i\in\mathbb{S}} (\boldsymbol{\rho})_i^2\right)^{\frac{1}{2}}.$$

When $\mathbb{S}$ is finite, these two norms are comparable. Namely,

$$\|\boldsymbol{\rho}\|_2 \le \|\boldsymbol{\rho}\|_{\mathrm{v}} \le \sqrt{\#\mathbb{S}}\|\boldsymbol{\rho}\|_2, \quad \text{where } \#\mathbb{S} \text{ denotes the cardinality of } \mathbb{S}.$$

The first inequality is easily seen by squaring both sides, and the second is an application of Schwarz's inequality (cf. Exercise 1.3.1). However, when $\mathbb{S}$ is infinite, they are not comparable. Nonetheless, $\|\cdot\|_{\mathrm{v}}$ is a good *norm* (i.e., measure of length) in the sense that $\|\boldsymbol{\rho}\|_{\mathrm{v}} = 0$ if and only if $\boldsymbol{\rho} = \mathbf{0}$ and that it satisfies the *triangle inequality*: $\|\boldsymbol{\rho} + \boldsymbol{\rho}'\|_{\mathrm{v}} \le \|\boldsymbol{\rho}\|_{\mathrm{v}} + \|\boldsymbol{\rho}'\|_{\mathrm{v}}$. Furthermore, Cauchy's convergence criterion holds for $\|\cdot\|_{\mathrm{v}}$. That is, if $\{\boldsymbol{\rho}_n\}_1^\infty$ is a sequence in $\mathbb{R}^{\mathbb{S}}$, then there exists $\boldsymbol{\rho} \in \mathbb{R}^{\mathbb{S}}$ for which $\|\boldsymbol{\rho}_n - \boldsymbol{\rho}\|_{\mathrm{v}} \longrightarrow 0$ if and only $\{\boldsymbol{\rho}_n\}_1^\infty$ is *Cauchy convergent* in the sense that

$$\lim_{m\to\infty} \sup_{n>m} \|\boldsymbol{\rho}_n - \boldsymbol{\rho}_m\|_{\mathrm{v}} = 0.$$

As usual, the "only if" direction is an easy application of the triangle inequality:

$$\|\boldsymbol{\rho}_n - \boldsymbol{\rho}_m\|_{\mathrm{v}} \le \|\boldsymbol{\rho}_n - \boldsymbol{\rho}\|_{\mathrm{v}} + \|\boldsymbol{\rho} - \boldsymbol{\rho}_m\|_{\mathrm{v}}.$$

To go the other direction, suppose that $\{\boldsymbol{\rho}_n\}_1^\infty$ is Cauchy convergent, and observe that each coordinate of $\{\boldsymbol{\rho}_n\}_1^\infty$ must be Cauchy convergent as real numbers. Hence, by Cauchy's criterion for real numbers, there exists a $\boldsymbol{\rho}$ to which $\{\boldsymbol{\rho}_n\}_1^\infty$ converges in the sense that each coordinate of the $\boldsymbol{\rho}_n$'s tends to the corresponding coordinate of $\boldsymbol{\rho}$. Thus, by Fatou's lemma, Theorem 7.1.10, as $m \to \infty$,

$$\|\boldsymbol{\rho} - \boldsymbol{\rho}_m\|_{\mathrm{v}} = \sum_{i\in\mathbb{S}} \big|(\boldsymbol{\rho})_i - (\boldsymbol{\rho}_m)_i\big| \le \varliminf_{n\to\infty} \sum_{i\in\mathbb{S}} \big|(\boldsymbol{\rho}_n)_i - (\boldsymbol{\rho}_m)_i\big| \longrightarrow 0.$$

2.1.3 Transition Probabilities and Functions

As we saw in Sect. 2.1.2, the representation of the transition probability as a matrix and the initial distributions as a row vector facilitates the representation of the distribution at later times. In order to understand how to get the analogous benefit when computing expectation values of functions, think of a function f on the state space $\mathbb{S}$ as the column vector $\mathbf{f}$ whose jth coordinate is the value of the function f at j. Clearly, if $\boldsymbol{\mu}$ is the row vector which represents the probability measure μ on $\mathbb{S}$ and $\mathbf{f}$ is the column vector which represents a function f which is either non-negative or bounded, then $\boldsymbol{\mu}\mathbf{f} = \sum_{i\in\mathbb{S}} f(i)\mu(\{i\})$ is the expected value of f with respect to μ.

Similarly, the column vector $\mathbf{P}^n\mathbf{f}$ represents that function whose value at i is the conditional expectation value of $f(X_n)$ given that $X_0 = i$. Indeed,

$$\begin{aligned}\mathbb{E}\big[f(X_n)\,\big|\,X_0 = i\big] &= \sum_{j\in\mathbb{S}} f(j)\mathbb{P}(X_n = j \mid X_0 = i)\\ &= \sum_{j\in\mathbb{S}} \big(\mathbf{P}^n\big)_{ij}(\mathbf{f})_j = \big(\mathbf{P}^n\mathbf{f}\big)_i.\end{aligned}$$

More generally, if f is either a non-negative or bounded function on $\mathbb{S}$ and $\mathbf{f}$ is the column vector which it determines, then, for $0 \le m \le n$,

$$\begin{aligned}&\mathbb{E}\big[f(X_n)\,\big|\,X_0 = i_0,\dots,X_m = i_m\big] = \big(\mathbf{P}^{n-m}\mathbf{f}\big)_{i_m}, \quad \text{or, equivalently,}\\ &\mathbb{E}\big[f(X_n)\,\big|\,X_0,\dots,X_m\big] = \big(\mathbf{P}^{n-m}\mathbf{f}\big)_{X_m}\end{aligned} \tag{2.1.7}$$

since

$$\begin{aligned}&\mathbb{E}\big[f(X_n)\,\big|\,X_0 = i_0,\dots,X_m = i_m\big]\\ &\quad = \sum_{j\in\mathbb{S}} f(j)\mathbb{P}(X_n = j \mid X_0 = i_0,\dots,X_m = i_m)\\ &\quad = \sum_{j\in\mathbb{S}} f(j)\big(\mathbf{P}^{n-m}\big)_{i_m j} = \big(\mathbf{P}^{n-m}\mathbf{f}\big)_{i_m}.\end{aligned}$$

In particular, if $\boldsymbol{\mu}$ is the initial distribution of $\{X_n : n \ge 0\}$, then

$$\mathbb{E}\big[f(X_n)\big] = \boldsymbol{\mu}\mathbf{P}^n\mathbf{f},$$

since $\mathbb{E}[f(X_n)] = \sum_i (\boldsymbol{\mu})_i \mathbb{E}[f(X_n)|X_0 = i]$.

Just as $\|\cdot\|_{\mathrm{v}}$ was the appropriate way to measure the length of row vectors when we were using them to represent measures, the appropriate way to measure the length of column vectors which represent functions is with the *uniform norm* $\|\cdot\|_{\mathrm{u}}$:

$$\|\mathbf{f}\|_{\mathrm{u}} = \sup_{j\in\mathbb{S}} \big|(\mathbf{f})_j\big|. \tag{2.1.8}$$

The reason why $\|\cdot\|_{\mathrm{u}}$ is the norm of choice here is that $|\boldsymbol{\mu}\mathbf{f}| \le \|\boldsymbol{\mu}\|_{\mathrm{v}}\|\mathbf{f}\|_{\mathrm{u}}$, since

$$|\boldsymbol{\mu}\mathbf{f}| \le \sum_{i\in\mathbb{S}} \big|(\boldsymbol{\mu})_i\big|\big|(\mathbf{f})_i\big| \le \|\mathbf{f}\|_{\mathrm{u}} \sum_{i\in\mathbb{S}} \big|(\boldsymbol{\mu})_i\big|.$$

In particular, we have the complement to (2.1.6):

$$\|\mathbf{P}\mathbf{f}\|_{\mathrm{u}} \le \|\mathbf{f}\|_{\mathrm{u}}. \tag{2.1.9}$$

2.1.4 The Markov Property

By definition, if μ is the initial distribution of $\{X_n : n \geq 0\}$, then

$$\mathbb{P}(X_0 = i_0, \dots, X_n = i_n) = (\boldsymbol{\mu})_{i_0}(\mathbf{P})_{i_0 i_1} \cdots (\mathbf{P})_{i_{n-1} i_n}. \tag{2.1.10}$$

Hence, if $m, n \geq 1$ and $F : \mathbb{S}^{n+1} \longrightarrow \mathbb{R}$ is either bounded or non-negative, then

$$\begin{aligned}
&\mathbb{E}\big[F(X_m, \dots, X_{m+n}), \, X_0 = i_0, \dots, X_m = i_m\big] \\
&\quad = \sum_{j_1, \dots, j_n \in \mathbb{S}} F(i_m, j_1, \dots, j_n) \mu_{i_0} (\mathbf{P})_{i_0 i_1} \cdots (\mathbf{P})_{i_{m-1} i_m} (\mathbf{P})_{i_m j_1} \cdots (\mathbf{P})_{j_{1n-1} j_n} \\
&\quad = \mathbb{E}\big[F(X_0, \dots, X_n) \,\big|\, X_0 = i_m\big] \mathbb{P}(X_0 = i_0, \dots, X_m = i_m).
\end{aligned}$$

Equivalently, we have now proved the *Markov property* in the form

$$\begin{aligned}
&\mathbb{E}\big[F(X_m, \dots, X_{m+n}) \,\big|\, X_0 = i_0, \dots, X_m = i_m\big] \\
&\quad = \mathbb{E}\big[F(X_0, \dots, X_n) \,\big|\, X_0 = i_m\big].
\end{aligned} \tag{2.1.11}$$

2.2 Doeblin's Theory

In this section I will introduce an elementary but basic technique, due to Doeblin, which will allow us to study the long time distribution of a Markov chain, particularly ones on a finite state space.

2.2.1 Doeblin's Basic Theorem

For many purposes, what one wants to know about a Markov chain is its distribution after a long time, and, at least when the state space is finite, it is reasonable to think that the distribution of the chain will stabilize. To be more precise, if one is dealing with a chain which can go in a single step from any state i to some state j with positive probability, then that state j is going to visited again and again, and so, after a while, the chain's initial distribution is going to get "forgotten." In other words, we are predicting for such a chain that $\boldsymbol{\mu}\mathbf{P}^n$ will, for sufficiently large n, be nearly independent of $\boldsymbol{\mu}$. In particular, this would mean that $\boldsymbol{\mu}\mathbf{P}^n = (\boldsymbol{\mu}\mathbf{P}^{n-m})\mathbf{P}^m$ is very nearly equal to $\boldsymbol{\mu}\mathbf{P}^m$ when m is large and therefore, by Cauchy's convergence criterion, that $\boldsymbol{\pi} = \lim_{n\to\infty} \boldsymbol{\mu}\mathbf{P}^n$ exists. In addition, if this were the case, then we would have that $\boldsymbol{\pi} = \lim_{n\to\infty} \boldsymbol{\mu}\mathbf{P}^{n+1} = \lim_{n\to\infty}(\boldsymbol{\mu}\mathbf{P}^n)\mathbf{P} = \boldsymbol{\pi}\mathbf{P}$. That is, $\boldsymbol{\pi}$ would have to be a left eigenvector for $\mathbf{P}$ with eigenvalue 1. A probability vector $\boldsymbol{\pi}$ is, for obvious reasons, called a *stationary probability* for the transition probability matrix $\mathbf{P}$ if $\boldsymbol{\pi} = \boldsymbol{\pi}\mathbf{P}$.

Although a state j of the sort in the preceding discussion is most likely to exist when the state space is finite, there are situations in which these musings apply even to infinite state spaces. That is, if, no matter where the chain starts, it has a positive probability of immediately visiting some fixed state, then, as the following theorem shows, it will stabilize.

Theorem 2.2.1 (Doeblin's Theorem) *Let* $\mathbf{P}$ *be a transition probability matrix with the property that, for some state* $j_0 \in \mathbb{S}$ *and* $\epsilon > 0$, $(\mathbf{P})_{i j_0} \geq \epsilon$ *for all* $i \in \mathbb{S}$. *Then* $\mathbf{P}$ *has a unique stationary probability vector* $\boldsymbol{\pi}$, $(\boldsymbol{\pi})_{j_0} \geq \epsilon$, *and, for all initial distributions* $\boldsymbol{\mu}$,

$$\left\|\boldsymbol{\mu}\mathbf{P}^n - \boldsymbol{\pi}\right\|_{\mathrm{v}} \leq (1-\epsilon)^n \|\boldsymbol{\mu} - \boldsymbol{\pi}\|_{\mathrm{v}} \leq 2(1-\epsilon)^n, \quad n \geq 0.$$

Proof The key to the proof lies in the observations that if $\boldsymbol{\rho} \in \mathbb{R}^{\mathbb{S}}$ is a row vector with $\|\boldsymbol{\rho}\|_{\mathrm{v}} < \infty$, then

$$\begin{aligned}
&\sum_{j\in\mathbb{S}} (\boldsymbol{\rho}\mathbf{P})_j = \sum_{i\in\mathbb{S}} (\boldsymbol{\rho})_i, \\
&\text{and} \quad \sum_{i\in\mathbb{S}} (\boldsymbol{\rho})_i = 0 \quad \Longrightarrow \quad \left\|\boldsymbol{\rho}\mathbf{P}^n\right\|_{\mathrm{v}} \leq (1-\epsilon)^n \|\boldsymbol{\rho}\|_{\mathrm{v}} \quad \text{for } n \geq 1.
\end{aligned} \tag{2.2.2}$$

The first of these is trivial, because, by Theorem 7.1.15,

$$\sum_{j\in\mathbb{S}} (\boldsymbol{\rho}\mathbf{P})_j = \sum_{j\in\mathbb{S}}\Big(\sum_{i\in\mathbb{S}} (\boldsymbol{\rho})_i(\mathbf{P})_{ij}\Big) = \sum_{i\in\mathbb{S}}\Big(\sum_{j\in\mathbb{S}} (\boldsymbol{\rho})_i(\mathbf{P})_{ij}\Big) = \sum_{i\in\mathbb{S}} (\boldsymbol{\rho})_i.$$

As for the second, note that, by an easy induction argument, it suffices to check it when $n = 1$. Next, suppose that $\sum_i (\boldsymbol{\rho})_i = 0$, and observe that

$$\begin{aligned}
\left|(\boldsymbol{\rho}\mathbf{P})_j\right| &= \Big|\sum_{i\in\mathbb{S}} (\boldsymbol{\rho})_i(\mathbf{P})_{ij}\Big| \\
&= \Big|\sum_{i\in\mathbb{S}} (\boldsymbol{\rho})_i\big((\mathbf{P})_{ij} - \epsilon\delta_{j,j_0}\big)\Big| \leq \sum_{i\in\mathbb{S}} \left|(\boldsymbol{\rho})_i\right|\big((\mathbf{P})_{ij} - \epsilon\delta_{j,j_0}\big),
\end{aligned}$$

and therefore that

$$\begin{aligned}
\|\boldsymbol{\rho}\mathbf{P}\|_{\mathrm{v}} &\leq \sum_{j\in\mathbb{S}}\Big(\sum_{i\in\mathbb{S}} \left|(\boldsymbol{\rho})_i\right|\big((\mathbf{P})_{ij} - \epsilon\delta_{j,j_0}\big)\Big) \\
&= \sum_{i\in\mathbb{S}} \left|(\boldsymbol{\rho})_i\right|\Big(\sum_{j\in\mathbb{S}}\big((\mathbf{P})_{ij} - \epsilon\delta_{j,j_0}\big)\Big) = (1-\epsilon)\|\boldsymbol{\rho}\|_{\mathrm{v}}.
\end{aligned}$$

Now let $\boldsymbol{\mu}$ be a probability vector, and set $\boldsymbol{\mu}_n = \boldsymbol{\mu}\mathbf{P}^n$. Then, because $\boldsymbol{\mu}_n = \boldsymbol{\mu}_{n-m}\mathbf{P}^m$ and $\sum_i((\boldsymbol{\mu}_{n-m})_i - \boldsymbol{\mu}_i) = 1 - 1 = 0$,

$$\|\boldsymbol{\mu}_n - \boldsymbol{\mu}_m\|_{\mathrm{v}} \leq (1-\epsilon)^m \|\boldsymbol{\mu}_{n-m} - \boldsymbol{\mu}\|_{\mathrm{v}} \leq 2(1-\epsilon)^m$$

for $1 \le m < n$. Hence $\{\boldsymbol{\mu}_n\}_1^\infty$ is Cauchy convergent, and therefore there exists a $\boldsymbol{\pi}$ for which $\|\boldsymbol{\mu}_n - \boldsymbol{\pi}\|_{\mathrm{v}} \longrightarrow 0$. Since each $\boldsymbol{\mu}_n$ is a probability vector, it is clear that $\boldsymbol{\pi}$ must also be a probability vector. In addition, $\boldsymbol{\pi} = \lim_{n\to\infty} \boldsymbol{\mu}\mathbf{P}^{n+1} = \lim_{n\to\infty}(\boldsymbol{\mu}\mathbf{P}^n)\mathbf{P} = \boldsymbol{\pi}\mathbf{P}$, and so $\boldsymbol{\pi}$ is stationary. In particular,

$$(\boldsymbol{\pi})_{j_0} = \sum_{i\in\mathbb{S}} (\boldsymbol{\pi})_i (\mathbf{P})_{ij_0} \ge \epsilon \sum_{i\in\mathbb{S}} (\boldsymbol{\pi})_i = \epsilon.$$

Finally, if $\boldsymbol{\nu}$ is any probability vector, then

$$\left\|\boldsymbol{\nu}\mathbf{P}^m - \boldsymbol{\pi}\right\|_{\mathrm{v}} = \left\|(\boldsymbol{\nu} - \boldsymbol{\pi})\mathbf{P}^m\right\|_{\mathrm{v}} \le (1-\epsilon)^m \|\boldsymbol{\mu} - \boldsymbol{\pi}\|_{\mathrm{v}} \le 2(1-\epsilon)^m,$$

which, of course, proves both the stated convergence result and the uniqueness of $\boldsymbol{\pi}$ as the only stationary probability vector for $\mathbf{P}$. □

The condition in Doeblin's Theorem is called *Doeblin's condition*, and it is instructive to understand what his theorem says in the language of *spectral theory*. Namely, as an operator on the space of bounded functions (a.k.a. column vectors with finite uniform norm), $\mathbf{P}$ has the function $\mathbf{1}$ as a right eigenfunction with eigenvalue 1: $\mathbf{P1} = \mathbf{1}$. Thus, at least if $\mathbb{S}$ is finite, general principles say that there must exist a row vector which is a left eigenvector of $\mathbf{P}$ with eigenvalue 1. Moreover, because 1 and the entries of $\mathbf{P}$ are real, this left eigenvector can be taken to have real components. Thus, from the spectral point of view, it is no surprise that there is a non-zero row vector $\boldsymbol{\mu} \in \mathbb{R}^{\mathbb{S}}$ with the property that $\boldsymbol{\mu}\mathbf{P} = \boldsymbol{\mu}$. On the other hand, standard spectral theory would not predict that $\boldsymbol{\mu}$ can be chosen to have non-negative components, and this is the first place where Doeblin's theorem gives information which is not readily available from standard spectral theory, even when $\mathbb{S}$ is finite. To interpret the estimate in Doeblin's Theorem, let $M_1(\mathbb{S};\mathbb{C})$ denote the space of row vectors $\boldsymbol{\nu} \in \mathbb{C}^{\mathbb{S}}$ with $\|\boldsymbol{\nu}\|_{\mathrm{v}} = 1$. Then,

$$\|\boldsymbol{\nu}\mathbf{P}\|_{\mathrm{v}} \le 1 \quad \text{for all } \boldsymbol{\nu} \in M_1(\mathbb{S};\mathbb{C}),$$

and so

$$\sup\bigl\{|\alpha| : \alpha \in \mathbb{C} \;\&\; \exists\, \boldsymbol{\nu} \in M_1(\mathbb{S};\mathbb{C})\ \boldsymbol{\nu}\mathbf{P} = \alpha\boldsymbol{\nu}\bigr\} \le 1.$$

Moreover, if $\boldsymbol{\nu}\mathbf{P} = \alpha\boldsymbol{\nu}$ for some $\alpha \ne 1$, then $\boldsymbol{\nu}\mathbf{1} = \boldsymbol{\nu}(\mathbf{P1}) = (\boldsymbol{\nu}\mathbf{P})\mathbf{1} = \alpha\boldsymbol{\nu}\mathbf{1}$, and therefore $\boldsymbol{\nu}\mathbf{1} = 0$. Thus, the estimate in (2.2.2) says that all eigenvalues of $\mathbf{P}$ which are different from 1 have absolute value dominated by $1-\epsilon$. That is, the entire spectrum of $\mathbf{P}$ lies in the complex unit disk, 1 is a simple eigenvalue, and all the other eigenvalues lie in the disk of radius $1-\epsilon$. Finally, although general spectral theory fails to predict Doeblin's Theorem, it should be said that there is a spectral theory, the one initiated by Frobenius and developed further by Kakutani, that does cover Doeblin's results. The interested reader should consult Chap. VIII in [2].

2.2.2 A Couple of Extensions

An essentially trivial extension of Theorem 2.2.1 is provided by the observation that, for any $M \geq 1$ and $\epsilon > 0$,[4]

$$\sup_j \inf_i (\mathbf{P}^M)_{ij} \geq \epsilon \implies \|\boldsymbol{\mu}\mathbf{P}^n - \boldsymbol{\pi}\|_{\mathrm{v}} \leq (1-\epsilon)^{\lfloor \frac{n}{M} \rfloor} \|\boldsymbol{\mu} - \boldsymbol{\pi}\|_{\mathrm{v}} \leq 2(1-\epsilon)^{\lfloor \frac{n}{M} \rfloor} \tag{2.2.3}$$

for all probability vectors $\boldsymbol{\mu}$ and a unique stationary probability vector $\boldsymbol{\pi}$. To see this, let $\boldsymbol{\pi}$ be the stationary probability vector for $\mathbf{P}^M$, the one guaranteed by Theorem 2.2.1, and note that, for any probability vector $\boldsymbol{\mu}$, any $m \in \mathbb{N}$, and any $0 \leq r < M$,

$$\|\boldsymbol{\mu}\mathbf{P}^{mM+r} - \boldsymbol{\pi}\|_{\mathrm{v}} = \|(\boldsymbol{\mu}\mathbf{P}^r - \boldsymbol{\pi}\mathbf{P}^r)\mathbf{P}^{mM}\|_{\mathrm{v}} \leq (1-\epsilon)^m \|\boldsymbol{\mu} - \boldsymbol{\pi}\|_{\mathrm{v}} \leq 2(1-\epsilon)^m.$$

Thus (2.2.3) has been proved, and from (2.2.3) the argument needed to show that $\boldsymbol{\pi}$ is the one and only stationary measure for $\mathbf{P}$ is the same as the one given in the proof of Theorem 2.2.1.

The next extension is a little less trivial. In order to appreciate the point that it is addressing, one should keep in mind the following example. Consider the transition probability matrix

$$\mathbf{P} = \begin{pmatrix} 0 & 1 \\ 1 & 0 \end{pmatrix} \quad \text{on } \{1, 2\}.$$

Obviously, this two state chain goes in a single step from one state to the other. Thus, it certainly visits all its states. On the other hand, it does not satisfy the hypothesis in (2.2.3): $(\mathbf{P}^n)_{ij} = 0$ if either $i = j$ and n is odd or if $i \neq j$ and n is even. Thus, it should not be surprising that the conclusion in (2.2.3) fails to hold for this $\mathbf{P}$. Indeed, it is easy to check that although $(\frac{1}{2}, \frac{1}{2})$ is the one and only stationary probability vector for $\mathbf{P}$, $\|(1,0)\mathbf{P}^n - (\frac{1}{2}, \frac{1}{2})\|_{\mathrm{v}} = 1$ for all $n \geq 0$. As we will see later (cf. Sect. 3.1.3), the problems encountered here stem from the fact that $(\mathbf{P}^n)_{ii} > 0$ only if n is even.

In spite of the problems raised by the preceding example, one should expect that the chain corresponding to this $\mathbf{P}$ does equilibrate in some sense. To describe what I have in mind, set

$$\mathbf{A}_n = \frac{1}{n} \sum_{m=0}^{n-1} \mathbf{P}^m. \tag{2.2.4}$$

Although the matrix $\mathbf{A}_n$ is again a transition probability matrix, it is not describing transitions but instead it is giving the average amount of time that the chain will visit

[4] Here and elsewhere, we use $\lfloor s \rfloor$ to denote the *integer part of* s of $s \in \mathbb{R}$. That is, $\lfloor s \rfloor$ is the largest integer dominated by s.

states. To be precise, because

$$(\mathbf{A}_n)_{ij} = \frac{1}{n}\sum_{m=0}^{n-1} \mathbb{P}(X_m = j \mid X_0 = i) = \mathbb{E}\left[\frac{1}{n}\sum_{m=0}^{n-1} \mathbf{1}_{\{j\}}(X_m) \,\middle|\, X_0 = i\right],$$

$(\mathbf{A}_n)_{ij}$ is the expected value of the average time spent at state j during the time interval $[0, n-1]$ given that i was the state from which the chain started. Experience teaches us that data becomes much more forgiving when it is averaged, and the present situation is no exception. Indeed, continuing with the example given above, observe that, for any probability vector $\boldsymbol{\mu}$,

$$\left\|\boldsymbol{\mu}\mathbf{A}_n - \left(\frac{1}{2}, \frac{1}{2}\right)\right\|_{\mathrm{v}} \le \frac{1}{n} \quad \text{for } n \ge 1.$$

What follows is a statement which shows that this sort of conclusion is quite general.

Theorem 2.2.5 *Suppose that* $\mathbf{P}$ *is a transition probability matrix on* $\mathbb{S}$. *If for some* $M \in \mathbb{Z}^+$, $j_0 \in \mathbb{S}$, *and* $\epsilon > 0$, $(\mathbf{A}_M)_{ij_0} \ge \epsilon$ *for all* $i \in \mathbb{S}$, *then there is precisely one stationary probability vector* $\boldsymbol{\pi}$ *for* $\mathbf{P}$, $(\boldsymbol{\pi})_{j_0} \ge \epsilon$, *and*

$$\|\boldsymbol{\mu}\mathbf{A}_n - \boldsymbol{\pi}\|_{\mathrm{v}} \le \frac{M-1}{n\epsilon}$$

for any probability vector $\boldsymbol{\mu}$.

To get started, let $\boldsymbol{\pi}$ be the unique stationary probability that Theorem 2.2.3 guarantees for $\mathbf{A}_M$. Then, because any $\boldsymbol{\mu}$ which is stationary for $\mathbf{P}$ is certainly stationary for $\mathbf{A}_M$, it is clear that $\boldsymbol{\pi}$ is the only candidate for $\mathbf{P}$-stationarity. Moreover, to see that $\boldsymbol{\pi}$ is $\mathbf{P}$-stationary, observe that, because $\mathbf{P}$ commutes with $\mathbf{A}_M$, $(\boldsymbol{\pi}\mathbf{P})\mathbf{A}_M = (\boldsymbol{\pi}\mathbf{A}_M)\mathbf{P} = \boldsymbol{\pi}\mathbf{P}$. Hence, $\boldsymbol{\pi}\mathbf{P}$ is stationary for $\mathbf{A}_M$ and therefore, by uniqueness, must be equal to $\boldsymbol{\pi}$. That is, $\boldsymbol{\pi} = \boldsymbol{\pi}\mathbf{P}$.

In order to prove the asserted convergence result, we will need an elementary property of averaging procedures. Namely, for any probability vector $\boldsymbol{\mu}$,

$$\|\boldsymbol{\mu}\mathbf{A}_n\mathbf{A}_m - \boldsymbol{\mu}\mathbf{A}_n\|_{\mathrm{v}} \le \frac{m-1}{n} \quad \text{for all } m, n \ge 1. \tag{2.2.6}$$

To check this, first note that, by the triangle inequality,

$$\begin{aligned}\|\boldsymbol{\mu}\mathbf{A}_n\mathbf{A}_m - \boldsymbol{\mu}\mathbf{A}_n\|_{\mathrm{v}} &= \frac{1}{m}\left\|\sum_{k=0}^{m-1}\left(\boldsymbol{\mu}\mathbf{A}_n\mathbf{P}^k - \boldsymbol{\mu}\mathbf{A}_n\right)\right\|_{\mathrm{v}} \\ &\le \frac{1}{m}\sum_{k=0}^{m-1}\left\|\boldsymbol{\mu}\mathbf{A}_n\mathbf{P}^k - \boldsymbol{\mu}\mathbf{A}_n\right\|_{\mathrm{v}}.\end{aligned}$$

Second, if $k \geq n$ then $\|\boldsymbol{\mu}\mathbf{A}_n\mathbf{P}^k - \boldsymbol{\mu}\mathbf{A}_n\|_{\mathrm{v}} \leq 2 \leq \frac{2k}{n}$, and if $0 \leq k < n$ then

$$\boldsymbol{\mu}\mathbf{A}_n\mathbf{P}^k - \boldsymbol{\mu}\mathbf{A}_n = \frac{1}{n}\sum_{\ell=0}^{n-1}(\boldsymbol{\mu}\mathbf{P}^{\ell+k} - \boldsymbol{\mu}\mathbf{P}^{\ell}) = \frac{1}{n}\left(\sum_{\ell=k}^{n+k-1}\boldsymbol{\mu}\mathbf{P}^{\ell} - \sum_{\ell=0}^{n-1}\boldsymbol{\mu}\mathbf{P}^{\ell}\right),$$

and so $\|\boldsymbol{\mu}\mathbf{P}^k\mathbf{A}_n - \boldsymbol{\mu}\mathbf{A}_n\|_{\mathrm{v}} \leq \frac{2k}{n}$ for all $n \geq 1$. Hence, after combining this with the first observation, we are lead to

$$\|\boldsymbol{\mu}\mathbf{A}_n\mathbf{A}_m - \boldsymbol{\mu}\mathbf{A}_n\|_{\mathrm{v}} \leq \frac{2}{mn}\sum_{k=0}^{m-1} k = \frac{m-1}{n},$$

which is what we wanted.

To complete the proof of Theorem 2.2.5 from here, assume that $(\mathbf{A}_M)_{ij_0} \geq \epsilon$ for all i, and, as above, let $\boldsymbol{\pi}$ be the unique stationary probability vector for $\mathbf{P}$. Then, $\boldsymbol{\pi}$ is also the unique stationary probability vector for $\mathbf{A}_M$, and so, by the estimate in the second line of (2.2.2) applied to $\mathbf{A}_M$, $\|\boldsymbol{\mu}\mathbf{A}_n\mathbf{A}_M - \boldsymbol{\pi}\|_{\mathrm{v}} = \|(\boldsymbol{\mu}\mathbf{A}_n - \boldsymbol{\pi})\mathbf{A}_M\|_{\mathrm{v}} \leq (1-\epsilon)\|\boldsymbol{\mu}\mathbf{A}_n - \boldsymbol{\pi}\|_{\mathrm{v}}$, which, in conjunction with (2.2.6), leads to

$$\begin{aligned}\|\boldsymbol{\mu}\mathbf{A}_n - \boldsymbol{\pi}\|_{\mathrm{v}} &\leq \|\boldsymbol{\mu}\mathbf{A}_n - \boldsymbol{\mu}\mathbf{A}_n\mathbf{A}_M\|_{\mathrm{v}} + \|\boldsymbol{\mu}\mathbf{A}_n\mathbf{A}_M - \boldsymbol{\pi}\|_{\mathrm{v}} \\ &\leq \frac{M-1}{n} + (1-\epsilon)\|\boldsymbol{\mu}\mathbf{A}_n - \boldsymbol{\pi}\|_{\mathrm{v}}.\end{aligned}$$

Finally, after elementary rearrangement, this gives the required result.

2.3 Elements of Ergodic Theory

In the preceding section we saw that, under suitable conditions, either $\boldsymbol{\mu}\mathbf{P}^n$ or $\boldsymbol{\mu}\mathbf{A}_n$ converge and that the limit is the unique stationary probability vector $\boldsymbol{\pi}$ for $\mathbf{P}$. In the present section, we will provide a more probabilistically oriented interpretation of these results. In particular, we will give a probabilistic interpretation of $\boldsymbol{\pi}$. This will be done again, by entirely different methods, in Chap. 4.

Before going further, it will be useful to have summarized our earlier results in the form (cf. (2.2.3) and remember that $|\boldsymbol{\mu}\mathbf{f}| \leq \|\boldsymbol{\mu}\|_{\mathrm{v}}\|f\|_{\mathrm{u}}$)[5]

$$\sup_j \inf_i (\mathbf{P}^M)_{ij} \geq \epsilon \quad \Longrightarrow \quad \left\|\mathbf{P}^n\mathbf{f} - \boldsymbol{\pi}\mathbf{f}\right\|_{\mathrm{u}} \leq 2(1-\epsilon)^{\lfloor \frac{n}{M} \rfloor}\|\mathbf{f}\|_{\mathrm{u}} \tag{2.3.1}$$

and (cf. Theorem 2.2.5)

$$\sup_j \inf_i (\mathbf{A}_M)_{ij} \geq \epsilon \quad \Longrightarrow \quad \|\mathbf{A}_n\mathbf{f} - \boldsymbol{\pi}\mathbf{f}\|_{\mathrm{u}} \leq \frac{M-1}{n\epsilon}\|\mathbf{f}\|_{\mathrm{u}} \tag{2.3.2}$$

when $\mathbf{f}$ is a bounded column vector.

---

[5] Here, and elsewhere, I abuse notation by using a constant to stand for the associated constant function.

2.3.1 The Mean Ergodic Theorem

Let $\{\mathbf{X}_n : n \geq 0\}$ be a Markov chain with transition probability $\mathbf{P}$. Obviously,

$$\overline{T}_j^{(n)} \equiv \frac{1}{n}\sum_{m=0}^{n-1} \mathbf{1}_{\{j\}}(X_m) \tag{2.3.3}$$

is the average amount of time that the chain spends at j before time n. Thus, if $\boldsymbol{\mu}$ is the initial distribution of the chain (i.e., $(\boldsymbol{\mu})_i = \mathbb{P}(X_0 = i)$), then $(\boldsymbol{\mu}\mathbf{A}_n)_j = \mathbb{E}[\bar{T}_j^{(n)}]$, and so, when it applies, Theorem 2.2.5 implies that $\mathbb{E}[\bar{T}_j^{(n)}] \longrightarrow (\boldsymbol{\pi})_j$ as $n \to \infty$. Here we will be proving that the random variables $\bar{T}_j^{(n)}$ themselves, not just their expected values, tend to $(\boldsymbol{\pi})_j$ as $n \to \infty$. Such results come under the heading of *ergodic theory*. Ergodic theory is the mathematics of the principle, first enunciated by the physicist J.W. Gibbs in connection with the kinetic theory of gases, which asserts that the time-average over a particular trajectory of a dynamical system will approximate the equilibrium state of that system. Unfortunately, in spite of results, like those given here, confirming this principle, even now, nearly 150 years after Gibbs, there are embarrassingly few physically realistic situations in which Gibbs's principle has been mathematically confirmed.

Theorem 2.3.4 (Mean Ergodic Theorem) *Under the hypotheses in Theorem* 2.2.5,

$$\sup_{j\in\mathbb{S}} \mathbb{E}\Big[\big(\overline{T}_j^{(n)} - (\boldsymbol{\pi})_j\big)^2\Big] \leq \frac{2(M-1)}{n\epsilon} \quad \textit{for all } n \geq 1.$$

(See (2.3.10) *below for a more refined, less quantitative version.) More generally, for any bounded function f on $\mathbb{S}$ and all $n \geq 1$:*

$$\mathbb{E}\left[\left(\frac{1}{n}\sum_{m=0}^{n-1} f(X_m) - \boldsymbol{\pi}\mathbf{f}\right)^2\right] \leq \frac{4(M-1)\|\mathbf{f}\|_{\mathrm{u}}^2}{n\epsilon},$$

where $\mathbf{f}$ *denotes the column vector determined by* f.

Proof Let $\bar{\mathbf{f}}$ be the column vector determined by the function $\bar{f} = f - \boldsymbol{\pi}\mathbf{f}$. Obviously,

$$\frac{1}{n}\sum_{m=0}^{n-1} f(X_m) - \boldsymbol{\pi}\mathbf{f} = \frac{1}{n}\sum_{m=0}^{n-1} \bar{f}(X_m),$$

and so

$$\left(\frac{1}{n}\sum_{m=0}^{n-1} f(X_m) - \boldsymbol{\pi}\mathbf{f}\right)^2 = \frac{1}{n^2}\left(\sum_{m=0}^{n-1}\bar{f}(X_m)\right)^2 = \frac{1}{n^2}\sum_{k,\ell=0}^{n-1}\bar{f}(X_k)\bar{f}(X_\ell)$$
$$= \frac{2}{n^2}\sum_{0\le k\le \ell<n}\bar{f}(X_k)\bar{f}(X_\ell) - \frac{1}{n^2}\sum_{k=0}^{n-1}\bar{f}(X_k)^2$$
$$\le \frac{2}{n^2}\sum_{0\le k\le \ell<n}\bar{f}(X_k)\bar{f}(X_\ell).$$

Hence,

$$\mathbb{E}\left[\left(\frac{1}{n}\sum_{m=0}^{n-1} f(X_m) - \boldsymbol{\pi}\mathbf{f}\right)^2\right] \le \frac{2}{n^2}\sum_{k=0}^{n-1}\mathbb{E}\left[\bar{f}(X_k)\sum_{\ell=0}^{n-k-1}\bar{f}(X_{k+\ell})\right]$$
$$= \frac{2}{n^2}\sum_{k=0}^{n-1}\mathbb{E}\left[\bar{f}(X_k)\sum_{\ell=0}^{n-k-1}(\mathbf{P}^\ell\bar{\mathbf{f}})_{X_k}\right]$$
$$= \frac{2}{n^2}\sum_{k=0}^{n-1}(n-k)\mathbb{E}\big[\bar{f}(X_k)(\mathbf{A}_{n-k}\bar{\mathbf{f}})_{X_k}\big].$$

But, by (2.3.2), $\|\mathbf{A}_{n-k}\bar{\mathbf{f}}\|_{\mathrm{u}} \le \frac{M-1}{(n-k)\epsilon}\|\bar{\mathbf{f}}\|_{\mathrm{u}}$, and so, since $\|\bar{\mathbf{f}}\|_{\mathrm{u}} \le 2\|\mathbf{f}\|_{\mathrm{u}}$,

$$(n-k)\mathbb{E}\big[\bar{f}(X_k)(\mathbf{A}_{n-k}\bar{\mathbf{f}})_{X_k}\big] \le \frac{2(M-1)\|\mathbf{f}\|_{\mathrm{u}}^2}{\epsilon}.$$

After plugging this into the preceding, we get the second result. To get the first, simply take $f = \mathbf{1}_{\{j\}}$ and observe that, in this case, $\|\bar{\mathbf{f}}\|_{\mathrm{u}} \le 1$. □

2.3.2 Return Times

As the contents of Sects. 1.1 and 1.2 already indicate, return times ought to play an important role in the analysis of the long time behavior of Markov chains. In particular, if $\rho_j^{(0)} \equiv 0$ and, for $m \ge 1$, the *time of mth return* to j is defined so that $\rho_j^{(m)} = \infty$ if $\rho_j^{(m-1)} = \infty$ or $X_n \ne j$ for every $n > \rho^{(m-1)}$ and $\rho_j^{(m)} = \inf\{n > \rho_j^{(m-1)} : X_n = j\}$ otherwise, then we say that j is *recurrent* if $\mathbb{P}(\rho_j^{(1)} < \infty | X_0 = j) = 1$ and that it is *transient* if $\mathbb{P}(\rho_j^{(1)} < \infty | X_0 = j) < 1$; and we can hope that when j is recurrent, then the history of the chain breaks into epochs which are punctuated by the successive returns to j. In this subsection we will provide evidence which bolsters that hope.

Notice that $\rho_j \equiv \rho_j^{(1)} \geq 1$ and, for $n \geq 1$,

$$\mathbf{1}_{(n,\infty]}(\rho_j) = F_{n,j}(X_0, \dots, X_n) \quad \text{where}$$
$$F_{n,j}(i_0, \dots, i_n) = \begin{cases} 1 & \text{if } i_m \neq j \text{ for } 1 \leq m \leq n, \\ 0 & \text{otherwise.} \end{cases} \tag{2.3.5}$$

In particular, this shows that the event $\{\rho_j > n\}$ is a measurable function of $(X_0, \dots, X_n)$. More generally, because

$$\mathbf{1}_{(n,\infty]}\big(\rho_j^{(m+1)}\big) = \mathbf{1}_{[n,\infty]}\big(\rho_j^{(m)}\big) + \sum_{\ell=1}^{n-1} \mathbf{1}_{\{\ell\}}\big(\rho_j^{(m)}\big) F_{n-\ell,j}(X_\ell, \dots, X_n),$$

an easy inductive argument shows that, for each $m \in \mathbb{N}$ and $n \in \mathbb{N}$, $\{\rho_j^{(m)} > n\}$ is a measurable function of $(X_0, \dots, X_n)$.

Theorem 2.3.6 *For all $m \in \mathbb{Z}^+$ and $(i, j) \in \mathbb{S}^2$,*

$$\mathbb{P}\big(\rho_j^{(m)} < \infty \,\big|\, X_0 = i\big) = \mathbb{P}(\rho_j < \infty \mid X_0 = i)\mathbb{P}(\rho_j < \infty \mid X_0 = j)^{m-1}.$$

In particular, if j is recurrent, then $\mathbb{P}(\rho_j^{(m)} < \infty | X_0 = j) = 1$ for all $m \in \mathbb{N}$. In fact, if j is recurrent, then, conditional on $X_0 = j$, $\{\rho_j^{(m)} - \rho_j^{(m-1)} : m \geq 1\}$ is a sequence of mutually independent random variables each of which has the same distribution as ρ_j.

Proof To prove the first statement, we apply (2.1.11) and the monotone convergence theorem, Theorem 7.1.9, to justify

$$\begin{aligned}
&\mathbb{P}\big(\rho_j^{(m)} < \infty \,\big|\, X_0 = i\big) \\
&= \sum_{n=1}^{\infty} \mathbb{P}\big(\rho_j^{(m-1)} = n \;\&\; \rho_j^{(m)} < \infty \,\big|\, X_0 = i\big) \\
&= \lim_{N \to \infty} \sum_{n=1}^{\infty} \mathbb{P}\big(\rho_j^{(m-1)} = n \;\&\; \rho_j^{(m)} \leq N \,\big|\, X_0 = i\big) \\
&= \sum_{n=1}^{\infty} \lim_{N \to \infty} \mathbb{E}\big[1 - F_{N,j}(X_n, \dots, X_{n+N}),\, \rho_j^{(m-1)} = n \,\big|\, X_0 = i\big] \\
&= \sum_{n=1}^{\infty} \lim_{N \to \infty} \mathbb{E}\big[1 - F_{N,j}(X_0, \dots, X_N) \,\big|\, X_0 = j\big]\mathbb{P}\big(\rho_j^{(m-1)} = n \,\big|\, X_0 = i\big) \\
&= \sum_{n=1}^{\infty} \lim_{N \to \infty} \mathbb{P}(\rho_j \leq N \mid X_0 = j)\mathbb{P}\big(\rho_j^{(m-1)} = n \,\big|\, X_0 = i\big) \\
&= \mathbb{P}(\rho_j < \infty \mid X_0 = j)\mathbb{P}\big(\rho_j^{(m-1)} < \infty \,\big|\, X_0 = i\big).
\end{aligned}$$

Turning to the second statement, note that it suffices for us prove that

$$\mathbb{P}\big(\rho_j^{(m+1)} > n + n_m \,\big|\, X_0 = j, \rho_j^{(1)} = n_1, \ldots, \rho_j^{(m)} = n_m\big)$$
$$= \mathbb{P}(\rho_j > n \mid X_0 = j).$$

But, again by (2.1.11), the expression on the left is equal to

$$\mathbb{E}\big[F_{n,j}(X_{n_m}, \ldots, X_{n_m+n}) \,\big|\, X_0 = j, \rho_j^{(1)} = n_1, \ldots, \rho_j^{(m)} = n_m\big]$$
$$= \mathbb{E}\big[F_{n,j}(X_0, \ldots, X_n) \,\big|\, X_0 = j\big] = \mathbb{P}(\rho_j > n \mid X_0 = j). \qquad \square$$

Reasoning as we did in Sect. 1.2.2, we can derive from the first part of Theorem 2.3.6:

$$\begin{aligned} &\mathbb{E}[T_j \mid X_0 = i] = \delta_{i,j} + \frac{\mathbb{P}(\rho_j < \infty | X_0 = i)}{\mathbb{P}(\rho_j = \infty | X_0 = j)}, \\ &\mathbb{E}[T_j \mid X_0 = j] = \infty \iff \mathbb{P}(T_j = \infty \mid X_0 = j) = 1, \\ &\mathbb{E}[T_j \mid X_0 = j] < \infty \iff \mathbb{P}(T_j < \infty \mid X_0 = j) = 1, \end{aligned} \tag{2.3.7}$$

where $T_j = \sum_{m=0}^{\infty} \mathbf{1}_{\{j\}}(X_m)$ is the total time the chain spends in the state j. Indeed, because

$$\mathbb{P}(T_j > m \mid X_0 = i) = \begin{cases} \mathbb{P}(\rho_j^{(m)} < \infty \mid X_0 = j) & \text{if } i = j \\ \mathbb{P}(\rho_j^{(m+1)} < \infty \mid X_0 = i) & \text{if } i \neq j, \end{cases}$$

all three parts of (2.3.7) follow immediately from the first part of Theorem 2.3.6.

Of course, from (2.3.7) we know that

$$j \text{ is recurrent} \quad \text{if and only if} \quad \mathbb{E}[T_j | X_0 = j] = \infty.$$

In particular, under the conditions in Theorem 2.2.5, this means that j_0 is recurrent since $(\mathbf{A}_n)_{j_0 j_0} \longrightarrow (\boldsymbol{\pi})_{j_0} > 0$ and therefore

$$\mathbb{E}[T_{j_0} \mid X_0 = j_0] = \sum_{m=0}^{\infty} (\mathbf{P}^m)_{j_0 j_0} = \lim_{n\to\infty} n(\mathbf{A}_n)_{j_0 j_0} = \infty.$$

To facilitate the statement of the next result, we will say that j is *accessible* from i and will write $i \to j$ if $(\mathbf{P}^n)_{ij} > 0$ for some $n \geq 0$. Equivalently, $i \to j$ if and only if $i = j$ or $i \neq j$ and $\mathbb{P}(\rho_j < \infty \mid X_0 = i) > 0$.

Theorem 2.3.8 *Assume that* $\inf_k (\mathbf{A}_M)_{ki} \geq \epsilon$ *for some* $M \geq 1$, $i \in \mathbb{S}$, *and* $\epsilon > 0$. *Then* j *is recurrent if and only if* $i \to j$, *in which case* $\inf_k (\mathbf{A}_{M'})_{kj} > 0$ *for some* $M' > 0$. *Moreover, if* k *is recurrent, then* $\sup_{j \in \mathbb{S}} \mathbb{E}[\rho_k^p | X_0 = j] < \infty$ *for all* $p \in (0, \infty)$.

Proof First suppose that $i \nrightarrow j$. Equivalently, $\mathbb{P}(\rho_j = \infty | X_0 = i) = 1$. At the same time, because $(\mathbf{A}_M)_{ji} \geq \epsilon$, there exists an $1 \leq m < M$ such that $(\mathbf{P}^m)_{ji} > 0$, and so (cf. (2.3.5))

$$\begin{aligned}
\mathbb{P}(\rho_j = \infty \mid X_0 = j) &\geq \mathbb{P}(X_n \neq j \text{ for } n \geq m \,\&\, X_m = i \mid X_0 = j) \\
&\geq \lim_{N\to\infty} \mathbb{E}\big[F_{N,j}(X_m, \dots, X_{m+N}),\ X_m = i \mid X_0 = j\big] \\
&= \lim_{N\to\infty} \mathbb{E}\big[F_{N,j}(X_0, \dots, X_N) \,\big|\, X_0 = i\big]\mathbb{P}(X_m = i \mid X_0 = j) \\
&= \mathbb{P}(\rho_j = \infty \mid X_0 = i)\big(\mathbf{P}^m\big)_{ji} > 0.
\end{aligned}$$

Therefore j cannot be recurrent.

We next show that

$$i \rightarrow j \quad \Longrightarrow \quad \inf_k (\mathbf{A}_{M'})_{kj} > 0 \quad \text{for some } M' \geq 1. \tag{$*$}$$

To this end, choose $m \in \mathbb{N}$ so that $(\mathbf{P}^m)_{ij} > 0$. Then, for all $k \in \mathbb{S}$,

$$\begin{aligned}
(\mathbf{A}_{m+M})_{kj} &= \frac{1}{m+M} \sum_{\ell=0}^{M+m-1} \big(\mathbf{P}^\ell\big)_{kj} \geq \frac{1}{m+M} \sum_{\ell=0}^{M-1} \big(\mathbf{P}^\ell\big)_{ki}\big(\mathbf{P}^m\big)_{ij} \\
&= \frac{M}{m+M}(\mathbf{A}_M)_{ki}\big(\mathbf{P}^m\big)_{ij} \geq \frac{M\epsilon}{m+M}\big(\mathbf{P}^m\big)_{ij} > 0.
\end{aligned}$$

In view of ($*$) and what we have already shown, it suffices to show that $\mathbb{E}[\rho_i^p \mid X_0 = j] < \infty$ for all $j \in \mathbb{S}$. For this purpose, set

$$u(n,k) = \mathbb{P}(\rho_i > nM \mid X_0 = k) \quad \text{for } n \in \mathbb{Z}^+ \text{ and } k \in \mathbb{S}.$$

Then, by (2.1.11),

$$\begin{aligned}
u(n+1,k) &= \sum_{j\in\mathbb{S}} \mathbb{P}\big(\rho_i > (n+1)M \,\&\, X_{nM} = j \mid X_0 = k\big) \\
&= \sum_{j\in\mathbb{S}} \mathbb{E}\big[F_{M,i}(X_{nM}, \dots, X_{(n+1)M}),\ \rho_i > nM \,\&\, X_{nM} = j \mid X_0 = k\big] \\
&= \sum_{j\in\mathbb{S}} \mathbb{P}(\rho_i > M \mid X_0 = j)\mathbb{P}(\rho_i > nM \,\&\, X_{nM} = j \mid X_0 = k) \\
&= \sum_{j\in\mathbb{S}} u(1,j)\mathbb{P}(\rho_i > nM \,\&\, X_{nM} = j \mid X_0 = k).
\end{aligned}$$

Hence, $u(n+1,k) \leq U u(n,k)$ where $U \equiv \max_{j\in\mathbb{S}} u(1,j)$. Finally, since $u(1,j) = 1 - \mathbb{P}(\rho_i \leq M | X_0 = j)$ and

$$\mathbb{P}(\rho_i \leq M \mid X_0 = j) \geq \max_{0\leq m<M} \big(\mathbf{P}^m\big)_{ji} \geq (\mathbf{A}_M)_{ji} \geq \epsilon,$$

$U \le 1-\epsilon$. In particular, this means that $u(n+1,k) \le (1-\epsilon)u(n,k)$, and therefore that $\mathbb{P}(\rho_i > nM|X_0 = k) \le (1-\epsilon)^n$, from which

$$\begin{aligned}\mathbb{E}\big[\rho_i^p \,|\, X_0 = k\big] &= \sum_{n=1}^{\infty} n^p \mathbb{P}(\rho_i = n \,|\, X_0 = k)\\ &\le \sum_{m=1}^{\infty} (mM)^p \sum_{n=(m-1)M+1}^{mM} \mathbb{P}(\rho_i = n \,|\, X_0 = k)\\ &\le M^p \sum_{m=1}^{\infty} m^p \mathbb{P}\big(\rho_i > (m-1)M \,\big|\, X_0 = k\big)\\ &\le M^p \sum_{m=1}^{\infty} m^p (1-\epsilon)^{m-1} < \infty\end{aligned}$$

follows immediately. □

2.3.3 *Identification of* π

Under the conditions in Theorem 2.2.5, we know that there is precisely one **P**-stationary probability vector $\boldsymbol{\pi}$. In this section, we will give a probabilistic interpretation of $(\boldsymbol{\pi})_j$. Namely, we will show that

$$\sup_{M\ge 1}\sup_{j\in\mathbb{S}}\inf_{i\in\mathbb{S}}(\mathbf{A}_M)_{ij} > 0$$

$$\implies \quad (\boldsymbol{\pi})_j = \frac{1}{\mathbb{E}[\rho_j|X_0 = j]} \quad (\equiv 0 \text{ if } j \text{ is transient}). \tag{2.3.9}$$

The idea for the proof of (2.3.9) is that, on the one hand, (cf. (2.3.3))

$$\mathbb{E}\big[\overline{T}_j^{(n)} \,|\, X_0 = j\big] = (\mathbf{A}_n)_{jj} \longrightarrow (\boldsymbol{\pi})_j,$$

while, on the other hand,

$$X_0 = j \quad \implies \quad \overline{T}_j^{(\rho_j^{(m)})} = \frac{1}{\rho_j^{(m)}} \sum_{\ell=0}^{\rho_j^{(m)}-1} \mathbf{1}_{\{j\}}(X_\ell) = \frac{m}{\rho_j^{(m)}}.$$

Thus, since, at least when j is recurrent, Theorem 2.3.6 says that $\rho_j^{(m)}$ is the sum of m mutually independent copies of ρ_j, the preceding combined with the weak law of large numbers should lead

$$(\boldsymbol{\pi})_j = \lim_{m\to\infty} \mathbb{E}\big[\overline{T}_j^{(\rho_j^{(m)})} \,|\, X_0 = j\big] = \frac{1}{\mathbb{E}[\rho_j|X_0 = j]}.$$

To carry out the program suggested above, we will actually prove a stronger result. Namely, we will show that, for each $j \in \mathbb{S}$,[6]

$$\mathbb{P}\left(\lim_{n\to\infty} \overline{T}_j^{(n)} = \frac{1}{\mathbb{E}[\rho_j \mid X_0 = j]} \,\Big|\, X_0 = j\right) = 1. \tag{2.3.10}$$

In particular, because $0 \le \overline{T}^{(n)} \le 1$, Lebesgue's dominated convergence theorem, Theorem 7.1.11, says that

$$(\boldsymbol{\pi})_j = \lim_{n\to\infty} (\mathbf{A}_n)_{jj} = \lim_{n\to\infty} \mathbb{E}\big[\overline{T}_j^{(n)} \mid X_0 = j\big] = \frac{1}{\mathbb{E}[\rho_j \mid X_0 = j]}$$

follows from (2.3.10). Thus, we need only prove (2.3.10). To this end, choose j_0, M, and $\epsilon > 0$ so that $(\mathbf{A}_M)_{ij_0} \ge \epsilon$ for all i. If $j_0 \nrightarrow j$, then, by Theorem 2.3.8, j is transient, and so, by (2.3.7), $\mathbb{P}(T_j < \infty \mid X_0 = j) = 1$. Hence, conditional on $X_0 = j$, $\overline{T}_j^{(n)} \le \frac{1}{n} T_j \longrightarrow 0$ with probability 1. At the same time, because j is transient, $\mathbb{P}(\rho_j = \infty \mid X_0 = j) > 0$, and so $\mathbb{E}[\rho_j \mid X_0 = j] = \infty$. Hence, we have proved (2.3.10) in the case when $j_0 \nrightarrow j$.

Next assume that $j_0 \rightarrow j$. Then, again by Theorem 2.3.8, $\mathbb{E}[\rho_j^4 \mid X_0 = j] < \infty$ and, conditional on $X_0 = j$, Theorem 2.3.6 says that the random variables $\rho_j^{(m)} - \rho_j^{(m-1)}$ are mutually independent random variables and have the same distribution as ρ_j. In particular, by the strong law of large numbers (cf. Exercise 1.3.4)

$$\mathbb{P}\left(\lim_{m\to\infty} \frac{\rho_j^{(m)}}{m} = r_j \,\Big|\, X_0 = j\right) = 1 \quad \text{where } r_j \equiv \mathbb{E}[\rho_j \mid X_0 = j].$$

On the other hand, for any $m \ge 1$,

$$\left|\overline{T}_j^{(n)} - \frac{1}{r_j}\right| \le \left|\overline{T}_j^{(n)} - \overline{T}_j^{(\rho_j^{(m)})}\right| + \left|\overline{T}_j^{(\rho_j^{(m)})} - \frac{1}{r_j}\right|,$$

and

$$\begin{aligned}\left|\overline{T}_j^{(n)} - \overline{T}_j^{(\rho_j^{(m)})}\right| &\le \frac{\left|T_j^{(n)} - T_j^{(\rho_j^{(m)})}\right|}{n} + \left|1 - \frac{\rho_j^{(m)}}{n}\right| \overline{T}_j^{(\rho_j^{(m)})} \\ &\le 2\left|1 - \frac{\rho_j^{(m)}}{n}\right| \le 2\left|1 - \frac{mr_j}{n}\right| + \frac{2m}{n}\left|\frac{\rho_j^{(m)}}{m} - r_j\right|\end{aligned}$$

[6] Statements like the one which follows are called *individual ergodic theorems* because they, as distinguished from the first part of Theorem 2.3.4, are about convergence with probability 1 as opposed to convergence in square mean. See Exercise 4.2.10 below for more information.

while, since $\overline{T}_j^{(\rho_j^{(m)})} = \frac{m}{\rho_j^{(m)}} \le 1$,

$$\left|\overline{T}_j^{(\rho_j^{(m)})} - \frac{1}{r_j}\right| \le \frac{1}{r_j}\left|\frac{\rho_j^{(m)}}{m} - r_j\right|.$$

Hence,

$$\left|\overline{T}_j^{(n)} - \frac{1}{r_j}\right| \le 2\left|1 - \frac{mr_j}{n}\right| + \left(\frac{2m}{n} + \frac{1}{r_j}\right)\left|\frac{\rho_j^{(m)}}{m} - r_j\right|.$$

Finally, by taking $m_n = \lfloor \frac{n}{r_j} \rfloor$ we get

$$\left|\overline{T}_j^{(n)} - \frac{1}{r_j}\right| \le \frac{2r_j}{n} + \frac{3}{r_j}\left|\frac{\rho_j^{(m_n)}}{m_n} - r_j\right| \longrightarrow 0 \quad \text{as } n \to \infty.$$

Notice that (2.3.10) is precisely the sort of statement for which Gibbs was looking. That is, it says that, with probability 1, when one observes an individual path, the average time that it spends in each state tends, as one observes for a longer and longer time, to the probability that the equilibrium (i.e., stationary) distribution assigns to that state.

2.4 Exercises

Exercise 2.4.1 In this exercise we will give a probabilistic interpretation of the *adjoint* of a transition probability matrix with respect to a stationary probability. To be precise, suppose that the transition probability matrix $\mathbf{P}$ admits a stationary distribution $\boldsymbol{\pi}$, assume $(\boldsymbol{\pi})_i > 0$ for each $i \in \mathbb{S}$, and determine the matrix $\mathbf{P}^\top$ by $(\mathbf{P}^\top)_{ij} = \frac{(\boldsymbol{\pi})_j}{(\boldsymbol{\pi})_i}(\mathbf{P})_{ji}$.

(a) Show that $\mathbf{P}^\top$ is a transition probability matrix for which $\boldsymbol{\pi}$ is again a stationary probability.
(b) Use $\mathbb{P}$ and $\mathbb{P}^\top$ to denote probabilities computed for the chains determined, respectively, by $\mathbf{P}$ and $\mathbf{P}^\top$ with initial distribution $\boldsymbol{\pi}$, and show that these chains are the *reverse* of one another in the sense that, for each $n \ge 0$ the distribution of $(X_0, \dots, X_n)$ under $\mathbb{P}^\top$ is the same as the distribution of $(X_n, \dots, X_0)$ under $\mathbb{P}$. That is,

$$\mathbb{P}^\top(X_0 = i_0, \dots, X_n = i_n) = \mathbb{P}(X_n = i_0, \dots, X_0 = i_n)$$

for all $n \ge 0$ and $(i_0, \dots, i_n) \in \mathbb{S}^{n+1}$.

Exercise 2.4.2 The Doeblin theory applies particularly well to chains on a finite state. For example, suppose that $\mathbf{P}$ is a transition probability matrix on an N element state space $\mathbb{S}$, and show that there exists an $\epsilon > 0$ such that $(\mathbf{A}_N)_{ij_0} \ge \epsilon$ for all $i \in \mathbb{S}$

if and only if $i \to j_0$ for all $i \in \mathbb{S}$. In particular, if such a j_0 exists, conclude that, for all probability vectors $\boldsymbol{\mu}$,

$$\|\boldsymbol{\mu}\mathbf{A}_n - \boldsymbol{\pi}\|_{\mathrm{v}} \le \frac{2(N-1)}{n\epsilon}, \quad n \ge 1,$$

where $\boldsymbol{\pi}$ is the unique stationary probability vector for $\mathbf{P}$.

Exercise 2.4.3 Here is a version of Doeblin's theorem that sometimes gives a slightly better estimate. Namely, assume that $(\mathbf{P})_{ij} \ge \epsilon_j$ for all (i, j), and set $\epsilon = \sum_j \epsilon_j$. If $\epsilon > 0$, show that the conclusion of Theorem 2.2.1 holds with this ϵ and that $(\boldsymbol{\pi})_i \ge \epsilon_i$ for each $i \in \mathbb{S}$.

Exercise 2.4.4 Assume that $\mathbf{P}$ is a transition probability matrix on the finite state space $\mathbb{S}$, and show that

$$j \in \mathbb{S} \text{ is recurrent} \quad \text{if and only if} \quad \mathbb{E}[\rho_j | X_0 = j] < \infty.$$

Of course, the "if" part is trivial and has nothing to do with the finiteness of the state space.

Exercise 2.4.5 Again assume that $\mathbf{P}$ is a transition probability matrix on the finite state space $\mathbb{S}$. In addition, assume that $\mathbf{P}$ is *doubly stochastic* in the sense that each of its columns as well as each of its rows sums to 1. Under the condition that every state is accessible from every other state, show that $\mathbb{E}[\rho_j | X_0 = j] = \#\mathbb{S}$ for each $j \in \mathbb{S}$.

Exercise 2.4.6 In order to test how good Doeblin's theorem is, consider the case when $\mathbb{S} = \{1, 2\}$ and

$$\mathbf{P} = \begin{pmatrix} 1-\alpha & \alpha \\ \beta & 1-\beta \end{pmatrix} \quad \text{for some } (\alpha, \beta) \in (0, 1).$$

Show that $\boldsymbol{\pi} = (\alpha + \beta)^{-1}(\beta, \alpha)$ is a stationary probability for $\mathbf{P}$, that

$$\|\boldsymbol{\nu}\mathbf{P} - \boldsymbol{\pi}\|_{\mathrm{v}} = |1 - \alpha - \beta| \|\boldsymbol{\mu} - \boldsymbol{\pi}\|_{\mathrm{v}} \quad \text{for all probability vectors } \boldsymbol{\mu}.$$

Hence, in this case, Doeblin's theorem gives the optimal result.

Exercise 2.4.7 One of the earliest examples of Markov processes are the *branching processes* introduced, around the end of the nineteenth century, by Galton and Watson to model demographics. In this model, $\mathbb{S} = \mathbb{N}$, the state $i \in \mathbb{N}$ representing the number of members in the population, and the process evolves so that, at each stage, every individual, independently of all other members of the population, dies and is replaced by a random number of offspring. Thus, 0 is an absorbing state, and, given that there are $i \ge 1$ individuals alive at a given time n, the number of individuals alive at time $n+1$ will be distributed like the sum of i mutually independent, $\mathbb{N}$-valued, identically distributed random variables. To be more precise, if $\boldsymbol{\mu} = (\mu_0, \dots, \mu_k, \dots)$ is the probability vector giving the number of offspring each individual produces, define the m-fold convolution power $\boldsymbol{\mu}^{\star m}$ so that $(\boldsymbol{\mu}^{\star 0})_j = \delta_{0,j}$ and, for $m \geq 1$,

$$(\boldsymbol{\mu}^{\star m})_j = \sum_{i=0}^{j} (\boldsymbol{\mu}^{\star(m-1)})_{j-i} \mu_i.$$

Then the transition probability matrix $\mathbf{P}$ is given by $(\mathbf{P})_{ij} = (\boldsymbol{\mu}^{\star i})_j$.

The first interesting question which one should ask about this model is what it predicts will be the probability of eventual *extinction*. That is, what is $\lim_{n\to\infty} \mathbb{P}(X_n = 0)$? A naïve guess is that eventual extinction should occur or should not occur depending on whether the expected number $\gamma \equiv \sum_{k=0}^{\infty} k\mu_k$ of progeny is strictly less or strictly greater than 1, with the case when the expected number is precisely 1 being more ambiguous. In order to verify this guess and remove trivial special cases, we make the assumptions that $\mu_0 > 0$, $\mu_0 + \mu_1 < 1$, and $\gamma \equiv \sum_{k=0}^{\infty} k\mu_k < \infty$.

(a) Set $f(s) = \sum_{k=0}^{\infty} s^k \mu_k$ for $s \in [0,1]$, and define $f^{\circ n}(s)$ inductively so that $f^{\circ 0}(s) = s$ and $f^{\circ n} = f \circ f^{\circ(n-1)}$ for $n \geq 1$. Show that $\gamma = f'(1)$ and that

$$f^{\circ n}(s)^i = \mathbb{E}\big[s^{X_n} \,\big|\, X_0 = i\big] = \sum_{j=0}^{\infty} s^j (\mathbf{P}^n)_{ij} \quad \text{for } s \in [0,1] \text{ and } i \geq 0.$$

Hint: Begin by showing that $f(s)^i = \sum_{j=0}^{\infty} s^j (\boldsymbol{\mu}^{\star i})_j$.

(b) Observe that $s \in [0,1] \longmapsto f(s) - s$ is a continuous function which is positive at $s = 0$, zero at $s = 1$, and smooth and strictly convex (i.e., $f'' > 0$) on $(0,1)$. Conclude that either $\gamma \leq 1$ and $f(s) > s$ for all $s \in [0,1)$ or $\gamma > 1$ and there is exactly one $\alpha \in (0,1)$ at which $f(\alpha) = \alpha$.

(c) Referring to the preceding, show that

$$\gamma \leq 1 \quad \Longrightarrow \quad \lim_{n\to\infty} \mathbb{E}\big[s^{X_n} \,\big|\, X_0 = i\big] = 1 \quad \text{for all } s \in (0,1]$$

and that

$$\gamma > 1 \quad \Longrightarrow \quad \lim_{n\to\infty} \mathbb{E}\big[s^{X_n} \,\big|\, X_0 = i\big] = \alpha^i \quad \text{for all } s \in (0,1).$$

(d) Based on (c), conclude that $\gamma \leq 1 \implies \mathbb{P}(X_n = 0 | X_0 = i) \longrightarrow 1$ and that $\gamma > 1 \implies \lim_{n\to\infty} \mathbb{P}(X_n = 0 | X_0 = i) = \alpha^i$ and

$$\lim_{n\to\infty} \mathbb{P}(1 \leq X_n \leq L | X_0 = i) = 0 \quad \text{for all } L \geq 1.$$

The last conclusion has the ominous implication that, when the expected number of progeny is larger than 1, then the population either becomes extinct or, what may be worse, grows indefinitely.

Exercise 2.4.8 Continue with the setting and notation in Exercise 2.4.7. We will to show in this exercise that there are significant differences between the cases when $\gamma < 1$ and $\gamma = 1$.

(a) Show that $\mathbb{E}[X_n \mid X_0 = i] = i\gamma^n$. Hence, when $\gamma < 1$, the expected size of the population goes to 0 at an exponential rate. On the other hand, when $\gamma = 1$, the expected size remains constant, this in spite of the fact that as $n \to \infty$ $\mathbb{P}(X_n = 0|X_0 = i) \longrightarrow 1$. Thus, when $\gamma = 1$, we have a typical situation of the sort which demonstrates why Lebesgue had to make the hypotheses he did in his dominated convergence theorem, Theorem 7.1.11. In the present case, the explanation is simple: as $n \to \infty$, with large probability $X_n = 0$ but, nonetheless, with positive probability X_n is enormous.

(b) Let ρ_0 be the time of first return to 0. Show that

$$\mathbb{P}(\rho_0 \le n|X_0 = i) = \mathbb{P}(X_n = 0|X_0 = i) = \left(f^{\circ(n-1)}(\mu_0)\right)^i,$$

and use this to get the estimate

$$\mathbb{P}(\rho_0 > n|X_0 = i) \le i\gamma^{n-1}(1 - \mu_0).$$

In particular, this shows that $\mathbb{E}[\rho_0^k|X_0 = i] < \infty$ for all $k \in \mathbb{Z}^+$ when $\gamma < 1$.

(c) Now assume that $\gamma = 1$. Under the additional condition that $\beta \equiv f''(1) = \sum_{k\ge 2} k(k-1)\mu_k < \infty$, start from $\mathbb{P}(\rho_0 \le n|X_0 = 1) = f^{\circ(n-1)}(\mu_0)$, and show that $\mathbb{E}[\rho_0|X_0 = i] = \infty$ for all $i \ge 1$.

Hint: Begin by showing that

$$1 - f^{\circ n}(\mu_0) \ge \left(\prod_{\ell=m}^{n-1}\left(1 - \beta\left(1 - f^{\circ \ell}(\mu_0)\right)\right)\right)\left(1 - f^{\circ m}(\mu_0)\right)$$

for $n > m$. Next, use this to show that

$$\infty > \mathbb{E}[\rho_0|X_0 = 1] = 1 + \sum_{0}^{\infty}\left(1 - f^{\circ n}(\mu_0)\right)$$

would lead to a contradiction.

(d) Here we show that the conclusion in (c) will, in general, be false without the finiteness condition on the second derivative. To see this, let $\theta \in (0, 1)$ be given, and check that $f(s) \equiv s + \frac{(1-s)^{1+\theta}}{1+\theta} = \sum_{k=0}^{\infty} s^k \mu_k$, for some probability vector $\boldsymbol{\mu} = (\mu_0, \dots, \mu_k, \dots)$ with $\mu_k > 0$ unless $k = 1$. Now use this choice of $\boldsymbol{\mu}$ to see that, when the second derivative condition in (c) fails, $\mathbb{E}[\rho_0|X_0 = 1]$ can be finite even though $\gamma = 1$.

Hint: Set $a_n = 1 - f^{\circ n}(\mu_0)$, note that $a_n - a_{n+1} = \mu_0 a_n^{1+\theta}$, and use this first to see that $\frac{a_{n+1}}{a_n} \longrightarrow 1$ and then that there exist $0 < c_2 < c_2 < \infty$ such that $c_1 \le a_{n+1}^{-\theta} - a_n^{-\theta} \le c_2$ for all $n \ge 1$. Conclude that $\mathbb{P}(\rho_0 > n|X_0 = 1)$ tends to 0 like $n^{-\frac{1}{\theta}}$.

Exercise 2.4.9 The idea underlying this exercise was introduced by J.L. Doob and is called[7] *Doob's h-transformation.* Let **P** is a transition probability matrix on the state space $\mathbb{S}$. Next, let $\emptyset \neq \Gamma \subsetneq \mathbb{S}$ be given, set

$$\rho_\Gamma = \inf\{n \geq 1 : X_n \in \Gamma\},$$

and assume that

$$h(i) \equiv \mathbb{P}(\rho_\Gamma = \infty \mid X_0 = i) > 0 \quad \text{for all } i \in \hat{\mathbb{S}} \equiv \mathbb{S} \setminus \Gamma.$$

(a) Show that $h(i) = \sum_{j \in \hat{\mathbb{S}}} (\mathbf{P})_{ij} h(j)$ for all $i \in \hat{\mathbb{S}}$, and conclude that the matrix $\hat{\mathbf{P}}$ given by $(\hat{\mathbf{P}})_{ij} = \frac{1}{h(i)} (\mathbf{P})_{ij} h(j)$ for $(i, j) \in (\hat{\mathbb{S}})^2$ is a transition probability matrix on $\hat{\mathbb{S}}$.

(b) For all $n \in \mathbb{N}$ and $(j_0, \ldots, j_n) \in (\hat{\mathbb{S}})^{n+1}$, show that, for each $i \in \hat{\mathbb{S}}$,

$$\begin{aligned}\hat{\mathbb{P}}(X_0 = j_0, \ldots, X_n = j_n \mid X_0 = i) \\ = \mathbb{P}(X_0 = j_0, \ldots, X_n = j_n \mid \rho_\Gamma = \infty \,\&\, X_0 = i),\end{aligned}$$

where $\hat{\mathbb{P}}$ is used here to denote probabilities computed for the Markov chain on $\hat{\mathbb{S}}$ whose transition probability matrix is $\hat{\mathbf{P}}$. That is, the Markov chain determined by $\hat{\mathbf{P}}$ is the Markov chain determined by **P** conditioned to never hit Γ.

Exercise 2.4.10 Here is another example of an h-transform. Assume that $j_0 \in \mathbb{S}$ is transient but that $i \rightarrow j_0$ for all $i \in \mathbb{S}$.[8] Set

$$h(j_0) = 1 \quad \text{and} \quad h(i) = \mathbb{P}(\rho_{j_0} < \infty | X_0 = i) \quad \text{for } i \neq j_0.$$

(a) After checking that $h(i) > 0$ for all $i \in \mathbb{S}$, define $\hat{\mathbf{P}}$ so that

$$(\hat{\mathbf{P}})_{ij} = \begin{cases} (\mathbf{P})_{j_0 j} & \text{if } i = j_0 \\ h(i)^{-1} (\mathbf{P})_{ij} h(j) & \text{if } i \neq j_0. \end{cases}$$

Show that $\hat{\mathbf{P}}$ is again a transition probability matrix.

(b) Using $\hat{\mathbb{P}}$ to denote probabilities computed relative to the chain determined by $\hat{\mathbf{P}}$, show that

$$\hat{\mathbb{P}}(\rho_{j_0} > n \mid X_0 = i) = \frac{1}{h(i)} \mathbb{P}(n < \rho_{j_0} < \infty \mid X_0 = i)$$

for all $n \in \mathbb{N}$ and $i \neq j_0$.

(c) Starting from the result in (b), show that j_0 is recurrent for the chain determined by $\hat{\mathbf{P}}$.

---

[7] The "h" comes from the connection with harmonic functions.

[8] By Exercise 2.4.2, this is possible only if $\mathbb{S}$ in infinite.

Chapter 3
Stationary Probabilities

It is important to understand when stationary probabilities exist and how to compute them when they do. In this chapter I will develop some methods for addressing these questions.

3.1 Classification of States

In this section we deal with a topic which was hinted at but not explicitly discussed in Chap. 2. Namely, a transition probability $\mathbf{P}$ determines a relationship structure on the state space. To be precise, given a pair (i, j) of states, recall that we write $i \rightarrow j$ and say that j is accessible from i if, with positive probability, the chain can go from state i to state j. That is, $(\mathbf{P}^n)_{ij} > 0$ for some $n \in \mathbb{N}$. Notice that accessibility is transitive in the sense that

$$i \rightarrow j \quad \text{and} \quad j \rightarrow \ell \quad \Longrightarrow \quad i \rightarrow \ell. \tag{3.1.1}$$

Indeed, if $(\mathbf{P}^m)_{ij} > 0$ and $(\mathbf{P}^n)_{j\ell} > 0$, then

$$\left(\mathbf{P}^{m+n}\right)_{i\ell} = \sum_k \left(\mathbf{P}^m\right)_{ik} \left(\mathbf{P}^n\right)_{k\ell} \geq \left(\mathbf{P}^m\right)_{ij} \left(\mathbf{P}^n\right)_{j\ell} > 0.$$

If i and j are accessible from one another in the sense that $i \rightarrow j$ and $j \rightarrow i$, then we write $i \leftrightarrow j$ and say that i *communicates with* j. It should be clear that "$\leftrightarrow$" is an equivalence relation. To wit, because $(\mathbf{P}^0)_{ii} = 1$, $i \leftrightarrow i$, and it is trivial that $j \leftrightarrow i$ if $i \leftrightarrow j$. Finally, if $i \leftrightarrow j$ and $j \leftrightarrow \ell$, then (3.1.1) makes it obvious that $i \leftrightarrow \ell$. Thus, "$\leftrightarrow$" leads to a partitioning of the state space into equivalence classes made up of communicating states. That is, for each state i, the *communicating equivalence class* $[i]$ of i is the set of states j such that $i \leftrightarrow j$; and, for every pair (i, j), either $[i] = [j]$ or $[i] \cap [j] = \emptyset$. In the case when every state communicates with every other state, we say that the chain is *irreducible*.

D.W. Stroock, *An Introduction to Markov Processes*, Graduate Texts in Mathematics 230,
DOI 10.1007/978-3-642-40523-5_3, © Springer-Verlag Berlin Heidelberg 2014

3.1.1 Classification, Recurrence, and Transience

In this subsection we will show that recurrence and transience are *communicating class properties*. That is, either all members of a communicating equivalence class are recurrent or all members are transient.

Recall (cf. Sect. 2.3.2) that ρ_j is the time of first return to j, and observe that $i \to j$ if and only if $\mathbb{P}(\rho_j < \infty \mid X_0 = 1) > 0$. Also, remember that we say j is recurrent or transient according to whether $\mathbb{P}(\rho_j < \infty | X_0 = j)$ is equal to or strictly less than 1.

Theorem 3.1.2 *Assume that i is recurrent and that $j \neq i$. Then $i \to j$ if and only if $\mathbb{P}(\rho_j < \rho_i | X_0 = i) > 0$. Moreover, if $i \to j$, then $\mathbb{P}(\rho_k < \infty | X_0 = \ell) = 1$ for any $(k, \ell) \in \{i, j\}^2$. In particular, $i \to j$ implies that $i \leftrightarrow j$ and that j is recurrent.*

Proof Given $j \neq i$ and $n \geq 1$, set (cf. (2.3.5))

$$G_n(k_0, \dots, k_n) = \big(F_{n-1,i}(k_0, \dots, k_{n-1}) - F_{n,i}(k_0, \dots, k_n)\big) F_{n,j}(k_0, \dots, k_n).$$

If $\{\rho_j^{(m)} : m \geq 0\}$ are defined as in Sect. 2.3.2, then, by (2.1.11),

$$\begin{aligned}
&\mathbb{P}\big(\rho_i^{(m+1)} < \rho_j \,\big|\, X_0 = i\big) \\
&= \sum_{\ell=1}^{\infty} \mathbb{P}\big(\rho_i^{(m)} = \ell \;\&\; \rho_i^{(m+1)} < \rho_j \,\big|\, X_0 = i\big) \\
&= \sum_{\ell=1}^{\infty} \sum_{n=1}^{\infty} \mathbb{P}\big(\rho_i^{(m)} = \ell,\ \rho_i^{(m+1)} = \ell + n < \rho_j \,\big|\, X_0 = i\big) \\
&= \sum_{\ell,n=1}^{\infty} \mathbb{E}\big[G_n(X_\ell, \dots, X_{\ell+n}),\, \rho_i^{(m)} = \ell < \rho_j \,\big|\, X_0 = i\big] \\
&= \sum_{\ell,n=1}^{\infty} \mathbb{E}\big[G_n(X_0, \dots, X_n) \mid X_0 = i\big] \mathbb{P}\big(\rho_i^{(m)} = \ell < \rho_j \,\big|\, X_0 = i\big) \\
&= \sum_{\ell,n=1}^{\infty} \mathbb{P}(\rho_i = n < \rho_j \mid X_0 = i) \mathbb{P}\big(\rho_i^{(m)} = \ell < \rho_j \,\big|\, X_0 = i\big) \\
&= \mathbb{P}(\rho_i < \rho_j \mid X_0 = i) \mathbb{P}\big(\rho_i^{(m)} < \rho_j \,\big|\, X_0 = i\big),
\end{aligned}$$

and so

$$j \neq i \implies \mathbb{P}\big(\rho_i^{(m)} < \rho_j \,\big|\, X_0 = i\big) = \mathbb{P}(\rho_i < \rho_j \mid X_0 = i)^m. \tag{3.1.3}$$

Now suppose that $i \to j$ but $\mathbb{P}(\rho_j < \rho_i | X_0 = i) = 0$. Then, because

$$\mathbb{P}(\rho_j \neq \rho_i | X_0 = i) \geq \mathbb{P}(\rho_i < \infty | X_0 = i) = 1,$$

$\mathbb{P}(\rho_i < \rho_j|X_0 = i) = 1$. By (3.1.3), this means that $\mathbb{P}(\rho_i^{(m)} < \rho_j|X_0 = i) = 1$ for all $m \geq 1$, which, since $\rho^{(m)} \geq m$, leads to $\mathbb{P}(\rho_j = \infty|X_0 = i) = 1$ and therefore rules out $i \rightarrow j$. Hence, we have already shown that

$$i \rightarrow j \implies \mathbb{P}(\rho_j < \rho_i|X_0 = i) > 0,$$

and the opposite implication needs no comment.

To prove that $i \rightarrow j \implies \mathbb{P}(\rho_i < \infty|X_0 = j) = 1$, first observe that

$$\begin{aligned}
&\mathbb{P}(\rho_j < \rho_i < \infty \,|\, X_0 = i) \\
&\quad = \lim_{n\to\infty} \sum_{m=1}^{\infty} \mathbb{P}(\rho_j = m < \rho_i \leq m + n \,|\, X_0 = i) \\
&\quad = \lim_{n\to\infty} \sum_{m=1}^{\infty} \mathbb{E}\big[1 - F_{n,i}(X_m, \dots, X_{m+n}),\ \rho_j = m < \rho_i \,|\, X_0 = i\big] \\
&\quad = \lim_{n\to\infty} \sum_{m=1}^{\infty} \mathbb{P}(\rho_j = m < \rho_i \,|\, X_0 = i)\mathbb{E}\big[1 - F_{n,i}(X_0, \dots, X_n) \,|\, X_0 = j\big] \\
&\quad = \mathbb{P}(\rho_j < \rho_i \,|\, X_0 = i)\mathbb{P}(\rho_i < \infty \,|\, X_0 = j).
\end{aligned}$$

Thus, after combining this with $\mathbb{P}(\rho_i < \infty|X_0 = i) = 1$, we have

$$\mathbb{P}(\rho_j < \rho_i \,|\, X_0 = i) = \mathbb{P}(\rho_j < \rho_i \,|\, X_0 = i)\mathbb{P}(\rho_i < \infty \,|\, X_0 = j),$$

which, because $\mathbb{P}(\rho_j < \rho_i|X_0 = i) > 0$, is possible only if $\mathbb{P}(\rho_i < \infty|X_0 = j) = 1$. In particular, we have now proved that $j \rightarrow i$ and therefore that $i \leftrightarrow j$.

Similarly,

$$\begin{aligned}
&\mathbb{P}(\rho_j < \infty \,|\, X_0 = i) \\
&\quad = \mathbb{P}(\rho_j < \rho_i \,|\, X_0 = i) + \mathbb{P}(\rho_i < \rho_j < \infty \,|\, X_0 = i) \\
&\quad = \mathbb{P}(\rho_j < \rho_i \,|\, X_0 = i) + \mathbb{P}(\rho_i < \rho_j \,|\, X_0 = i)\mathbb{P}(\rho_j < \infty \,|\, X_0 = i),
\end{aligned}$$

and so, since $\mathbb{P}(\rho_i = \rho_j \,|\, X_0 = i) \leq \mathbb{P}(\rho_i = \infty \,|\, X_0 = i) = 0$,

$$\mathbb{P}(\rho_j < \infty \,|\, X_0 = i)\mathbb{P}(\rho_j < \rho_i \,|\, X_0 = i) = \mathbb{P}(\rho_j < \rho_i \,|\, X_0 = i).$$

Thus, $i \rightarrow j \implies \mathbb{P}(\rho_j < \infty|X_0 = i) = 1$.

Finally,

$$\mathbb{P}(\rho_i < \rho_j < \infty \,|\, X_0 = j) = \mathbb{P}(\rho_j < \infty \,|\, X_0 = i)\mathbb{P}(\rho_i < \rho_j \,|\, X_0 = j).$$

Hence, because we now know that

$$i \rightarrow j \implies \mathbb{P}(\rho_i < \infty|X_0 = j) = 1 = \mathbb{P}(\rho_j < \infty|X_0 = i),$$

we see that $i \to j$ implies

$$\begin{aligned}
&\mathbb{P}(\rho_j < \infty \,|\, X_0 = j) \\
&\quad = \mathbb{P}(\rho_j < \rho_i \,|\, X_0 = j) + \mathbb{P}(\rho_i < \rho_j < \infty \,|\, X_0 = j) \\
&\quad = \mathbb{P}(\rho_j < \rho_i \,|\, X_0 = j) + \mathbb{P}(\rho_j < \infty \,|\, X_0 = i)\mathbb{P}(\rho_i < \rho_j \,|\, X_0 = j) = 1,
\end{aligned}$$

since $\mathbb{P}(\rho_i = \rho_j | X_0 = j) \le \mathbb{P}(\rho_i = \infty | X_0 = j) = 0$. □

As an immediate consequence of Theorem 3.1.2 we have the following corollary.

Corollary 3.1.4 *If $i \leftrightarrow j$, then j is recurrent (transient) if and only if i is. Moreover, if i is recurrent, then $\mathbb{P}(\rho_j < \infty | X_0 = i)$ is either 1 or 0 according whether or not i communicates with j. In particular, if i is recurrent, then $(\mathbf{P}^n)_{ij} = 0$ for all $n \ge 0$ and all j which do not communicate with i.*

When a chain is irreducible, all or none of its states possess any particular communicating class property. Hence, when a chain is irreducible, we will say that it is *recurrent* or *transient* if any one, and therefore all, of its states is.

These considerations allow us to prove the following property of stationary probabilities.

Theorem 3.1.5 *If π is a stationary probability for the transition probability $\mathbf{P}$ on $\mathbb{S}$, then $(\pi)_i = 0$ for all transient $i \in \mathbb{S}$. Furthermore, if i is recurrent and $(\pi)_i > 0$, then $(\pi)_j > 0$ for all j that communicate with i.*

Proof First observe that, for any $n \ge 1$,

$$\begin{aligned}
n(\pi)_i &= \sum_{m=0}^{n-1} \left(\pi \mathbf{P}^m\right)_i = \sum_{j\in\mathbb{S}} (\pi)_j \mathbb{E}\left[\sum_{m=0}^{n-1} \mathbf{1}_{\{i\}}(X_m) \,\middle|\, X_0 = j\right] \\
&\le \sum_{j\in\mathbb{S}} (\pi)_j \mathbb{E}\left[\sum_{m=0}^{\infty} \mathbf{1}_{\{i\}}(X_m) \,\middle|\, X_0 = j\right] \le \mathbb{E}[T_i \,|\, X_0 = i],
\end{aligned}$$

where the final inequality comes from (2.3.7). Hence, if i is transient and therefore $\mathbb{E}[T_i \,|\, X_0 = i] < \infty$, then $(\pi)_i = 0$.

Next suppose that i is recurrent and that $(\pi)_i > 0$. Then, for any j in the communicating class $[i]$ of i, $\mathbf{P}^n_{ij} > 0$ for some $n \ge 0$, and therefore $(\pi)_j = \sum_{k\in\mathbb{S}} (\pi)_k \mathbf{P}^n_{kj} \ge (\pi)_i \mathbf{P}^n_{ij} > 0$. □

3.1.2 Criteria for Recurrence and Transience

There are many tests which can help determine whether a state is recurrent, but no one of them works in all circumstances. In this subsection, we will develop a few of the most common of these tests. Throughout, we will use $\mathbf{u}$ to denote the column vector determined by a function $u : \mathbb{S} \longrightarrow \mathbb{R}$.

We begin with a criterion for transience.

Theorem 3.1.6 *If u is a non-negative function on $\mathbb{S}$ with the property that $(\mathbf{Pu})_i \le (\mathbf{u})_i$ for all $i \in \mathbb{S}$, then $(\mathbf{Pu})_j < (\mathbf{u})_j$ for some $j \in \mathbb{S}$ implies that j is transient.*

Proof Set $\mathbf{f} = \mathbf{u} - \mathbf{Pu}$, and note that, for all $n \ge 1$,

$$\begin{aligned} u(j) \ge (\mathbf{u})_j - \left(\mathbf{P}^n\mathbf{u}\right)_j &= \sum_{m=0}^{n-1}\left(\left(\mathbf{P}^m\mathbf{u}\right)_j - \left(\mathbf{P}^{m+1}\mathbf{u}\right)_j\right) \\ &= \sum_{m=0}^{n-1}\left(\mathbf{P}^m\mathbf{f}\right)_j \ge (\mathbf{f})_j \sum_{m=0}^{n-1}\left(\mathbf{P}^m\right)_{jj}. \end{aligned}$$

Thus $\mathbb{E}[T_j | X_0 = j] = \sum_{m=0}^{\infty}(\mathbf{P}^m)_{jj} \le \frac{u(j)}{(\mathbf{f})_j} < \infty$, which, by (2.3.7), means that j is transient. □

In order to prove our next criterion, we will need the following special case of a general result known as *Doob's stopping time theorem*.

Lemma 3.1.7 *Assume that $u : \mathbb{S} \longrightarrow \mathbb{R}$ is bounded below and that Γ is a non-empty subset of $\mathbb{S}$. If $(\mathbf{Pu})_i \le u(i)$ for all $i \notin \Gamma$ and $\rho_\Gamma \equiv \inf\{n \ge 1 : X_n \in \Gamma\}$, then*

$$\mathbb{E}\left[u(X_{n\wedge\rho_\Gamma}) \mid X_0 = i\right] \le u(i) \quad \text{for all } n \ge 0 \text{ and } i \in \mathbb{S}\setminus\Gamma.$$

Moreover, if the inequality in the hypothesis is replaced by equality, then the inequality in the conclusion can be replaced by equality.

Proof Set $A_n = \{\rho_\Gamma > n\}$. Then, A_n is measurable with respect to $(X_0, \dots, X_n)$, and so, by (2.1.1), for any $i \notin \Gamma$,

$$\begin{aligned} &\mathbb{E}\left[u(X_{(n+1)\wedge\rho_\Gamma}) \mid X_0 = i\right] \\ &\quad = \mathbb{E}\left[u(X_{n\wedge\rho_\Gamma}), A_n\complement \mid X_0 = i\right] + \sum_{k\notin\Gamma}\mathbb{E}\left[u(X_{n+1}), A_n \cap \{X_n = k\} \mid X_0 = i\right] \\ &\quad = \mathbb{E}\left[u(X_{n\wedge\rho_\Gamma}), A_n\complement \mid X_0 = i\right] + \sum_{k\notin\Gamma}\mathbb{E}\left[(\mathbf{Pu})_k, A_n \cap \{X_n = k\} \mid X_0 = i\right] \\ &\quad \le \mathbb{E}\left[u(X_{n\wedge\rho_\Gamma}), A_n\complement \mid X_0 = i\right] + \mathbb{E}\left[u(X_{n\wedge\rho_\Gamma}), A_n \mid X_0 = i\right] \\ &\quad = \mathbb{E}\left[u(X_{n\wedge\rho_\Gamma}) \mid X_0 = i\right]. \end{aligned}$$

Clearly, the same argument works just as well in the case of equality. □

Theorem 3.1.8 *Assume that j is recurrent, and set $C = [j]$. If $u : \mathbb{S} \longrightarrow [0, \infty)$ is a bounded function and either $u(i) = (\mathbf{Pu})_i$ or $u(j) \geq u(i) \geq (\mathbf{Pu})_i$ for all $i \in C \setminus \{j\}$, then u is constant on C. On the other hand, if j is transient, then the function u given by*

$$u(i) = \begin{cases} 1 & \text{if } i = j \\ \mathbb{P}(\rho_j < \infty | X_0 = i) & \text{if } i \neq j \end{cases}$$

is a non-negative, bounded, non-constant solution to $u(i) = (\mathbf{Pu})_i$ for all $i \neq j$.

Proof In proving the first part, we will assume, without loss in generality, that $C = \mathbb{S}$. Now suppose that j is recurrent and that $u(i) = (\mathbf{Pu})_i$ for $i \neq j$. By applying Lemma 3.1.7 with $\Gamma = \{j\}$, we see that, for $i \neq j$,

$$u(i) = u(j)\mathbb{P}(\rho_j \leq n \mid X_0 = i) + \mathbb{E}\big[u(X_n), \rho_j > n \mid X_0 = i\big].$$

Hence, since, by Theorem 3.1.2, $\mathbb{P}(\rho_j < \infty | X_0 = i) = 1$ and u is bounded, we get $u(i) = u(j)$ after letting $n \to \infty$. Next assume that $u(j) \geq u(i) \geq (\mathbf{Pu})_i$ for all $i \neq j$. Then, again by Lemma 3.1.7, we have

$$u(j) \geq u(i) \geq u(j)\mathbb{P}(\rho_j \leq n \mid X_0 = i) + \mathbb{E}\big[u(X_n), \rho_j > n \mid X_0 = i\big],$$

which leads to the required conclusion when $n \to \infty$.

To prove the second part, let u be given by the prescription described, and begin by observing that, because j is transient,

$$1 > \mathbb{P}(\rho_j < \infty \mid X_0 = j) = \mathbf{P}_{jj} + \sum_{i \neq j} \mathbf{P}_{ji} u(i) \geq \mathbf{P}_{jj} + (1 - \mathbf{P}_{jj}) \inf_{i \neq j} u(i).$$

Because j is transient, and therefore $(\mathbf{P})_{jj} < 1$, this proves that $\inf_{i \neq j} u(i) < u(j)$ and therefore that u is non-constant. At the same time, when $i \neq j$, by conditioning on what happens at time 1, we know that

$$u(i) = \mathbb{P}(\rho_j < \infty \mid X_0 = i) = \mathbf{P}_{ij} + \sum_{k \neq j} \mathbf{P}_{ik} \mathbb{P}(\rho_j < \infty \mid X_0 = k) = (\mathbf{Pu})_i.$$

□

Lemma 3.1.9 *If $\mathbf{P}$ is irreducible on $\mathbb{S}$, then, for any finite subset $F \neq \mathbb{S}$, $\mathbb{P}(\rho_{\mathbb{S} \setminus F} < \infty | X_0 = i) = 1$ for all $i \in F$.*

Proof Set $\tau = \rho_{\mathbb{S} \setminus F}$. By irreducibility, $\mathbb{P}(\tau < \infty | X_0 = i) > 0$ for each $i \in F$. Hence, because F is finite, there exists a $\theta \in (0, 1)$ and an $N \geq 1$ such that $\mathbb{P}(\tau > N | X_0 = i) \leq \theta$ for all $i \in F$. But this means that, for each $i \in F$, $\mathbb{P}(\tau > (\ell + 1)N \mid X_0 = i)$ equals

$$
\begin{aligned}
&\sum_{k\in F}\mathbb{P}\big(\tau > (\ell+1)N \ \&\ X_{\ell N}=k \mid X_0=i\big)\\
&\quad=\sum_{k\in F}\mathbb{P}\big(X_n\in F \text{ for } \ell N+1\le n\le(\ell+1)N,\ \tau>\ell N,\ \&\ X_{\ell N}=k\mid X_0=i\big)\\
&\quad=\sum_{k\in F}\mathbb{P}(\tau>N\mid X_0=k)\mathbb{P}(\tau>\ell N \ \&\ X_{\ell N}=k\mid X_0=i)\\
&\quad\le\theta\mathbb{P}(\tau>\ell N\mid X_0=i).
\end{aligned}
$$

Thus, $\mathbb{P}(\tau > \ell N|X_0=i)\le\theta^\ell$, and so $\mathbb{P}(\tau=\infty|X_0=j)=0$ for all $i\in F$. □

Theorem 3.1.10 *Assume that* $\mathbf{P}$ *is irreducible on* $\mathbb{S}$, *and let* $u:\mathbb{S}\longrightarrow[0,\infty)$ *be a function with the property that* $\{k:u(k)\le L\}$ *is finite for each* $L\in(0,\infty)$. *If, for some* $j\in\mathbb{S}$, $(\mathbf{Pu})_i\le u(i)$ *for all* $i\ne j$, *the chain determined by* $\mathbf{P}$ *is recurrent on* $\mathbb{S}$.

Proof If $\mathbb{S}$ is finite, then (cf., for example, Exercise 2.4.2) at least one state is recurrent, and therefore, by irreducibility, all are. Hence, we will assume that $\mathbb{S}$ is infinite.

Given $i\ne j$, set $F_L=\{k:u(k)\le u(i)+u(j)+L\}$ for $L\in\mathbb{N}$, and denote by ρ_L the first return time $\rho_{(\mathbb{S}\setminus F_L)\cup\{j\}}$ to $(\mathbb{S}\setminus F_L)\cup\{j\}$. By Lemma 3.1.6,

$$u(i)\ge\mathbb{E}\big[u(X_{n\wedge\rho_L})\mid X_0=i\big]\ge\big(u(i)+u(j)+L\big)\mathbb{P}(\rho_{\mathbb{S}\setminus F_L}<n\wedge\rho_j\mid X_0=i)$$

for all $n\ge1$. Hence, after letting $n\to\infty$, we conclude that, for all $L\in\mathbb{N}$,

$$
\begin{aligned}
u(i)&\ge\big(u(i)+u(j)+L\big)\mathbb{P}(\rho_{\mathbb{S}\setminus F_L}<\rho_j\mid X_0=i)\\
&\ge\big(u(i)+u(j)+L\big)\mathbb{P}(\rho_j=\infty\mid X_0=i),
\end{aligned}
$$

since, by Lemma 3.1.8, we know that $\mathbb{P}(\rho_{\mathbb{S}\setminus F_L}<\infty|X_0=i)=1$. Thus, after letting $L\to\infty$, we have now shown that $\mathbb{P}(\rho_j<\infty|X_0=i)=1$ for all $i\ne j$. Since

$$\mathbb{P}(\rho_j<\infty\mid X_0=j)=(\mathbf{P})_{jj}+\sum_{i\ne j}\mathbb{P}(\rho_j<\infty\mid X_0=i)(\mathbf{P})_{ji},$$

it follows that $\mathbb{P}(\rho_j<\infty|X_0=j)=1$, which means that j is recurrent. □

Remark The preceding criteria are examples, of which there are many others, that relate recurrence of $j\in\mathbb{S}$ to the existence or non-existence of certain types of functions that satisfy either $\mathbf{Pu}=u$ or $\mathbf{Pu}\le u$ on $\mathbb{S}\setminus\{j\}$. All these criteria can be understood as mathematical implementations of the intuitive idea that

$$u(i)=(\mathbf{Pu})_i\quad\text{or}\quad(\mathbf{Pu})_i\le u(i)\quad\text{for } i\ne j,$$

implies that as long as $X_n\ne j$, $u(X_n)$ will be "nearly constant" or "nearly non-increasing" as n increases. The sense in which these "nearly's" should be interpreted is the subject of *martingale theory*, and our proofs of these criteria would have been simplified had we been able to call on that theory.

3.1.3 Periodicity

Periodicity is another important communicating class property. In order to describe this property, we must recall Euclid's concept of the *greatest common divisor* $\gcd(S)$ of a non-empty subset $S \subseteq \mathbb{Z}$. Namely, we say that $d \in \mathbb{Z}^+$ is a common divisor of S and write $d|S$ if $\frac{s}{d} \in \mathbb{Z}$ for every $s \in S$. Clearly, if $S = \{0\}$, then $d|S$ for every $d \in \mathbb{Z}^+$, and so we take $\gcd(\{0\}) = \infty$. On the other hand, if $S \neq \{0\}$, then no common divisor of S can be larger than $\min\{|s| : s \in S \setminus \{0\}\}$, and so we know that $\gcd(S) < \infty$.

Our interest in this concept comes from the role it plays in the ergodic theory of Markov chains. Namely, as we will see in Chap. 4, it allows us to distinguish between the chains for which powers of the transition probability matrix converge and those for which it is necessary to take averages. More precisely, given a state i, set

$$S(i) = \{n \geq 0 : (\mathbf{P}^n)_{ii} > 0\} \quad \text{and} \quad d(i) = \gcd(S(i)). \tag{3.1.11}$$

Then $d(i)$ is called the *period* of the state i, and, i is said to be *aperiodic* if $d(i) = 1$. In Chap. 4, we will see that averaging is required unless i is aperiodic. However, here we will only take care of a few more mundane matters. In the first place, the period is a communicating class property:

$$i \leftrightarrow j \implies d(i) = d(j). \tag{3.1.12}$$

To see this, assume that $(\mathbf{P}^m)_{ij} > 0$ and $(\mathbf{P}^n)_{ji} > 0$, and let d be a common divisor of $S(i)$. Then for any $k \in S(j)$,

$$(\mathbf{P}^{m+k+n})_{ii} \geq (\mathbf{P}^m)_{ij} (\mathbf{P}^k)_{jj} (\mathbf{P}^n)_{ji} > 0,$$

and so $m+k+n \in S(i)$. Hence $d|\{m+k+n : k \in S(j)\}$. But, because $m+n \in S(i)$, and therefore $d|(m+n)$, this is possible only if d divides $S(j)$, and so we now know that $d(i) \leq d(j)$. After reversing the roles of i and j, one sees that $d(j) \leq d(i)$, which means that $d(i)$ must equal $d(j)$.

We next need the following elementary fact from number theory.

Theorem 3.1.13 *Given $\emptyset \neq S \subseteq \mathbb{Z}$ with $S \neq \{0\}$,*

$$\gcd(S) \leq \min\{|s| : s \in S \setminus \{0\}\}$$

and equality holds if and only if $\{\gcd(S), -\gcd(S)\} \cap S \neq \emptyset$. More generally, there always exists an $M \in \mathbb{Z}^+$, $\{a_m\}_1^M \subseteq \mathbb{Z}$, and $\{s_m\}_1^M \subseteq S$ such that $\gcd(S) = \sum_1^M a_m s_m$. Finally, if $S \subseteq \mathbb{N}$ and $(s_1, s_2) \in S^2 \implies s_1 + s_2 \in S$, then there exists an $M \in \mathbb{Z}^+$ such that

$$\{m \gcd(S) : m \geq M\} = \{s \in S : s \geq M \gcd(S)\}.$$

Proof The first assertion needs no comment. To prove the second assertion, let $\hat{S}$ be the smallest subset of $\mathbb{Z}$ which contains S and has the property that $(s_1, s_2) \in \hat{S}^2 \implies s_1 \pm s_2 \in \hat{S}$. As is easy to check, $\hat{S}$ coincides with the subset of $\mathbb{Z}$ whose elements can be expressed in the form $\sum_1^M a_m s_m$ for some $M \geq 1$, $\{a_m\}_1^M \subseteq \mathbb{Z}$, and $\{s_m\}_1^M \subseteq S$. In particular, this means that $\gcd(S)|\hat{S}$, and so $\gcd(S) \leq \gcd(\hat{S})$. On the other hand, because $S \subseteq \hat{S}$, $\gcd(\hat{S})|S$. Hence, $\gcd(S) = \gcd(\hat{S})$, and so, by the first part, we will be done once we show that $\gcd(\hat{S}) \in \hat{S}$. To this end, let $m = \min\{s \in \mathbb{Z}^+ : s \in \hat{S}\}$. We already know that $\gcd(\hat{S}) \leq m$. Thus, to prove that $\gcd(\hat{S}) \in \hat{S}$, we need only check that $m|\hat{S}$. But, by the Euclidean algorithm, for any $s \in \hat{S}$, we can write $s = am + r$ for some $(a, r) \in \mathbb{Z}^2$ with $0 \leq r < m$. In particular, $r = s - am \in \hat{S}$. Hence, if $r \neq 0$, then r would contradict the condition that m is the smallest positive element of $\hat{S}$.

To prove the final assertion, first note that it suffices to prove that there is an $M \in \mathbb{Z}^+$ such that $\{m \gcd(S) : m \geq M\} \subseteq S$. To this end, begin by checking that, under the stated hypothesis, $\hat{S} = \{s_2 - s_1 : s_1, s_2 \in S \cup \{0\}\}$. Thus $\gcd(S) = s_2 - s_1$ for some $s_1 \in S \cup \{0\}$ and $s_2 \in S \setminus \{0\}$. If $s_1 = 0$, then $m \gcd(S) = m s_2 \in S$ for all $m \in \mathbb{Z}^+$, and so we can take $M = 1$. If $s_1 \neq 0$, choose $a \in \mathbb{Z}^+$ so that $s_1 = a \gcd(S)$. Then, for any $(m, r) \in \mathbb{N}^2$ with $0 \leq r < a$,

$$(a^2 + ma + r)\gcd(S) = ms_1 + rs_2 + (a - r)s_1 = (m + a - r)s_1 + rs_2 \in S.$$

Hence, after another application of the Euclidean algorithm, we see that we can take $M = a^2$. □

As an immediate consequence of Theorem 3.1.13, we see that

$$d(i) < \infty \implies \left(\mathbf{P}^{nd(i)}\right)_{ii} > 0 \quad \text{for all sufficiently large } n \in \mathbb{Z}^+. \tag{3.1.14}$$

In particular,[1]

$$i \text{ is aperiodic} \iff \left(\mathbf{P}^n\right)_{ii} > 0 \quad \text{for all sufficiently large } n \in \mathbb{Z}^+. \tag{3.1.15}$$

We close this subsection with an application of these considerations to stationary probabilities of transition probabilities on a finite state space.

Corollary 3.1.16 *Suppose that* $\mathbf{P}$ *is an transition probability matrix on a finite state space* $\mathbb{S}$. *If there is an aperiodic state* $j_0 \in \mathbb{S}$ *such that* $i \to j_0$ *for every* $i \in \mathbb{S}$, *then there exists an* $M \in \mathbb{Z}^+$ *and an* $\epsilon > 0$ *such that* $(\mathbf{P}^M)_{ij_0} \geq \epsilon$ *for all* $i \in \mathbb{S}$. *In particular, if* π *is the unique stationary probability guaranteed by Theorem 2.2.1, then*

$$\|\mu\mathbf{P}^n - \pi\|_{\mathrm{v}} \leq 2(1-\epsilon)^{\lfloor \frac{n}{M} \rfloor} \quad \textit{for all } n \in \mathbb{Z}^+ \textit{ and initial distributions } \mu.$$

[1]The "if" part of the following statement depends on the existence of infinitely many prime numbers.

Proof Because j_0 is aperiodic, we know that there is an $M_0 \in \mathbb{N}$ such that $(\mathbf{P}^n)_{j_0 j_0} > 0$ for all $n \geq M_0$. Further, because $i \to j_0$, there exists an $m(i) \in \mathbb{Z}^+$ such that $(\mathbf{P}^{m(i)})_{ij_0} > 0$. Hence, $(\mathbf{P}^n)_{ij_0} > 0$ for all $n \geq m(i) + M_0$. Finally, take $M = M_0 + \max_{i \in \mathbb{S}} m(i)$, $\epsilon = \min_{i \in \mathbb{S}} (\mathbf{P}^M)_{ij_0}$, and apply (2.2.3). □

3.2 Computation of Stationary Probabilities

It is important to know how to compute stationary probabilities, and in this section we will develop a procedure that works when the state space is finite.

3.2.1 Preliminary Results

Whether or not the state space is finite, a stationary probability for a transition probability $\mathbf{P}$ is a probability vector $\boldsymbol{\pi}$ that satisfies $\boldsymbol{\pi}\mathbf{P} = \boldsymbol{\pi}$. Equivalently, $\boldsymbol{\pi}(\mathbf{I} - \mathbf{P}) = \mathbf{0}$ and $\boldsymbol{\pi}\mathbf{1} = 1$. If one ignores the necessity of finding a solution with non-negative components, when the state space $\mathbb{S}$ is finite general principles of linear algebra guarantee that one can always solve these equations. Indeed, $(\mathbf{I} - \mathbf{P})\mathbf{1} = \mathbf{0}$ and therefore[2]

$$\det\big((\mathbf{I} - \mathbf{P})^\top\big) = \det\big((\mathbf{I} - \mathbf{P})\big) = 0,$$

which means that there must exist a row vector $\mathbf{v}$ such that $\mathbf{v}(\mathbf{I} - \mathbf{P}) = \mathbf{0}$ with $\mathbf{v}\mathbf{1} = 1$. For us, a better way to see that solutions exist is the following. Clearly, the set of probability vectors on $\mathbb{S}$ is a compact subset of $\mathbb{R}^N$, where N is the cardinality of $\mathbb{S}$. Thus, for any probability vector $\boldsymbol{\mu}$, the sequence $\{\boldsymbol{\mu}\mathbf{A}_M : M \geq 1\}$ admits a subsequence $\{\boldsymbol{\mu}\mathbf{A}_{M_k} : k \geq 1\}$ that converges to a probability vector $\boldsymbol{\pi}$. Moreover,

$$\boldsymbol{\pi}\mathbf{P} = \lim_{k\to\infty} \frac{1}{M_k} \sum_{m=0}^{M_k - 1} \boldsymbol{\mu}\mathbf{P}^{m+1} = \lim_{k\to\infty} \left(\frac{\boldsymbol{\mu}\mathbf{P}^{M_k} - \boldsymbol{\mu}}{M_k} + \boldsymbol{\mu}\mathbf{A}_{M_k} \right) = \boldsymbol{\pi}.$$

Hence $\boldsymbol{\pi}$ is a stationary probability for $\mathbf{P}$.

The following result, which contains a partial converse of Theorem 2.2.5, exploits the preceding construction.

Theorem 3.2.1 *Suppose that $\mathbf{P}$ is a transition probability on a finite state space $\mathbb{S}$. Then there is at least one recurrent $i \in \mathbb{S}$. In addition, if j is transient, then $j \to i$ for some recurrent i. Finally, suppose that $\mathbf{P}$ admits precisely one stationary probability $\boldsymbol{\pi}$. Then all recurrent states communicate with one another and there is*

[2]If $\mathbf{A}$ is a matrix, then $\mathbf{A}^\top$ is its transpose. That is, $\mathbf{A}^\top_{ij} = \mathbf{A}_{ji}$. Thus, if $\mathbf{v}$ is a row vector, then $\mathbf{v}^\top$ is a column vector and $\mathbf{v}\mathbf{A} = \mathbf{A}^\top\mathbf{v}^\top$.

an $M \in \mathbb{Z}^+$ *and an* $\epsilon > 0$ *such that* $(\mathbf{A}_M)_{ji} \geq \epsilon$ *for all* $j \in \mathbb{S}$ *and all recurrent* $i \in \mathbb{S}$. *Thus, when* $\mathbb{S}$ *is finite,* $\mathbf{P}$ *admits precisely one stationary measure if and only if* $\mathbf{A}_M$ *satisfies Doeblin's condition for some* $M \geq 1$.

Proof Begin by noting that, no matter where it starts, the chain can spend only a finite amount of time in the set $\mathcal{T}$ of transient states. Indeed, if $i \in \mathbb{S}$ and $j \in \mathcal{T}$, then (cf. (2.3.7)) $\mathbb{E}[T_j \mid X_0 = i] < \infty$ and therefore, since $\mathcal{T}$ is finite, $\mathbb{E}[\sum_{n=0}^{\infty} \mathbf{1}_{\mathcal{T}}(X_n) \mid X_0 = i] < \infty$. Hence, there must exist at least one recurrent state, and the chain must visit one of them no matter where it starts. In particular, if j is transient, then there is a recurrent i for which $j \to i$.

Now assume that there is only one stationary probability π for $\mathbf{P}$. We will show first that all the recurrent states must communicate with one another. To see this, suppose that i_1 and i_2 were recurrent states that do not communicate. Then $[i_1] \cap [i_2] = \emptyset$ and the restrictions of $\mathbf{P}$ to $[i_1]$ and $[i_2]$ are transition probabilities. Hence, by the preceding, we would be able to construct stationary probabilities π_1 and π_2 for these restrictions. After extending π_k to $\mathbb{S}$ by taking it to be 0 off of $[i_k]$, we would then have two unequal stationary probabilities for $\mathbf{P}$. Thus, $i_1 \leftrightarrow i_2$.

We now know that if i is a recurrent state, then $j \to i$ for all j. In other words, for each $j \in \mathbb{S}$ and recurrent i, $(\mathbf{P}^n)_{ji} > 0$ for some n. Thus, because $\mathbb{S}$ is finite, this completes the proof that M and ϵ exist. As for the concluding assertion, it follows immediately from the result just proved combined with Theorem 2.2.5. □

3.2.2 Computations via Linear Algebra

Our next goal is to find a formula for the stationary probability when there is only one, and the procedure that we will give uses elementary linear algebra. Recall that if $\mathbf{A}$ is a square matrix indexed by an N element set $\mathbb{S}$, then its *eigenvalues* are the roots of its characteristic polynomial $\lambda \in \mathbb{C} \longmapsto \det(\lambda I - \mathbf{A}) \in \mathbb{C}$. An eigenvalue that is different from all the other is said to be a *simple eigenvalue*.

Given an $N \times N$-matrix $\mathbf{A}$, let $\mathrm{cof}(\mathbf{A})$ denote its *cofactor matrix*. That is, $\mathrm{cof}(\mathbf{A})_{ij}$ is $(-1)^{i+j}$ times the determinant of the $(N-1) \times (N-1)$-matrix obtained by removing the ith column and the jth row from $\mathbf{A}$. By Cramer's formula, $\mathbf{A}\mathrm{cof}(\mathbf{A}) = \mathrm{cof}(\mathbf{A})\mathbf{A}$ equals $\det(\mathbf{A})$ times the identity matrix. From this it is easy to see that $\mathrm{cof}(\mathbf{A})_{ji}$ is equal to the derivative of $\det(\mathbf{A})$ with respect to its (i, j)th entry $(\mathbf{A})_{ij}$.

Lemma 3.2.2 *If each row of* $\mathbf{A}$ *sums to* 0, *then* $\mathrm{cof}(\mathbf{A})_{ij} = \mathrm{cof}(\mathbf{A})_{jj}$ *for all* $1 \leq i, j \leq N$ *and*

$$\sum_{i=1}^{N} \mathrm{cof}(\mathbf{A})_{ii} = \Pi_{\mathbf{A}},$$

where $\Pi_{\mathbf{A}}$ *equals the product of the non-zero eigenvalues of* $\mathbf{A}$ *if* 0 *is a simple eigenvalue of* $\mathbf{A}$ *and is equal to* 0 *otherwise.*

Proof Given $i_1 \neq i_2$ and j, let $\mathbf{B}$ be the matrix whose (j, i_1)st entry is 1, whose (j, i_2)th entry is -1, and whose other entries are all 0. Then, for each $t \in \mathbb{R}$, all the rows of $\mathbf{A} + t\mathbf{B}$ sum to 0 and therefore $\det(\mathbf{A} + t\mathbf{B}) = 0$. Hence

$$\mathrm{cof}(\mathbf{A})_{i_2 j} - \mathrm{cof}(\mathbf{A})_{i_1 j} = \frac{d}{dt}\det(\mathbf{A} + t\mathbf{B})\bigg|_{t=0} = 0.$$

To prove the second assertion, consider the polynomial $\lambda \rightsquigarrow \det(\lambda\mathbf{I} + \mathbf{A})$. Because $\det(\mathbf{A}) = 0$, this polynomial vanishes at 0. Furthermore, using Cramer's formula, it is easy to check that $\sum_{i=1}^N \mathrm{cof}(\mathbf{A})_{ii}$ is the coefficient of λ. At the same time, $\det(\lambda\mathbf{I} + \mathbf{A}) = \prod_{k=1}^N (\lambda + \lambda_k)$, where $\lambda_1, \ldots, \lambda_N$ are the eigenvalues of $\mathbf{A}$. Hence, since 0 is one of these eigenvalues, we can take $\lambda_1 = 0$ and thereby see that

$$\prod_{k=2}^N \lambda_k = \lim_{\lambda \to 0} \frac{\det(\lambda\mathbf{I} + \mathbf{A})}{\lambda} = \sum_{i=1}^N \mathrm{cof}(\mathbf{A})_{ii}.$$

□

Theorem 3.2.3 *Assume that each row of* $\mathbf{A}$ *sums to* 0 *and that* 0 *is a simple eigenvalue of* $\mathbf{A}$. *Determine the row vector* $\boldsymbol{\pi}$ *by*

$$(\boldsymbol{\pi})_i = \frac{\mathrm{cof}(\mathbf{A})_{ii}}{\Pi_{\mathbf{A}}}, \quad 1 \le i \le N,$$

where $\Pi_{\mathbf{A}}$ *is the product of the non-zero eigenvalues of* $\mathbf{A}$. *Then* $\boldsymbol{\pi}$ *is the unique row vector* $\mathbf{v} \in \mathbb{C}^N$ *such that* $\mathbf{vA} = 0$ *and* $\sum_{i=1}^N (\mathbf{v})_i = 1$.

Proof Since the null space of $\mathbf{A}^\top$ has the same dimension as that of $\mathbf{A}$ and the latter has dimension 1, we know that there is at most one $\mathbf{v}$. Furthermore, by Lemma 3.2.2, we know that $\sum_{i=1}^N (\boldsymbol{\pi})_i = 1$. Finally, by that same lemma and Cramer's formula,

$$\Pi_{\mathbf{A}} \sum_{i=1}^N (\boldsymbol{\pi})_i a_{ij} = \sum_{i=1}^N \mathrm{cof}(\mathbf{A})_{ji} a_{ij} = \det(\mathbf{A}) = 0 \quad \text{for } 1 \le j \le N.$$

□

In what follows, it will be useful to have introduced the following notation. Given a matrix $\mathbf{A}$ indexed by some finite set $\mathbb{S}$ and $\Delta \subseteq \mathbb{S}$, when $\Delta \neq \mathbb{S}$ let $\mathbf{A}^\Delta$ be the matrix indexed by $\mathbb{S} \setminus \Delta$ obtained by eliminating the rows and columns with indices in Δ and when $\Delta = \mathbb{S}$ define $\mathbf{A}^\Delta = 1$. In this notation, for each $i \in \mathbb{S}$, $\det(\mathbf{A}^{\{i\}})$ equals $\mathrm{cof}(\mathbf{A})_{ii}$, and therefore, if $i \notin \Delta$ and $\det(\mathbf{A}^\Delta) \neq 0$, then Cramer's formula says that

$$\left((\mathbf{A}^\Delta)^{-1}\right)_{ii} = \frac{\det(\mathbf{A}^{\Delta \cup \{i\}})}{\det(\mathbf{A}^\Delta)}.$$

More generally, if $\Delta \subsetneq \Delta'$ and $(i_1, \ldots, i_K)$ is an ordering of the elements of $\Delta' \setminus \Delta$, set $\Delta(1) = \Delta$, $\Delta(k) = \Delta \cup \{i_1, \ldots, i_{k-1}\}$ for $2 \le k \le K$, and assume that

$\det(\mathbf{A}^{\Delta(k)}) \neq 0$ for all $1 \le k \le K$. Then, by induction on K, one sees that

$$\frac{\det(\mathbf{A}^{\Delta'})}{\det(\mathbf{A}^{\Delta})} = \prod_{k=1}^{K} \left((\mathbf{A}^{\Delta(k)})^{-1} \right)_{i_k i_k}. \tag{3.2.4}$$

In conjunction with the next lemma, Theorem 3.2.3 will give us a formula for computing stationary probabilities when there is only one.

Lemma 3.2.5 *Let* $\mathbf{P}$ *be a transition probability on a finite state space* $\mathbb{S}$. *Then all the eigenvalues of* $\mathbf{P}$ *lie inside the closed unit disk* $\mathbb{D}$ *in* $\mathbb{C}$ *centered at* 0. *Also,* $\det((\mathbf{I}-\mathbf{P})^{\{i\}}) \ge 0$ *for all* $i \in \mathbb{S}$. *Finally, there is only one stationary probability* $\boldsymbol{\pi}$ *for* $\mathbf{P}$ *if and only if* 0 *is a simple eigenvalue for* $\mathbf{I}-\mathbf{P}$, *in which case* $(\boldsymbol{\pi})_i > 0$ *if and only if* i *is recurrent.*

Proof Let N be the cardinality of $\mathbb{S}$.

Suppose that λ is an eigenvalue of $\mathbf{P}$. Then there is a non-zero $\mathbf{v} \in \mathbb{C}^N$ satisfying $\|\mathbf{v}\|_{\mathrm{u}} = 1$ and $\mathbf{P}\mathbf{v} = \lambda\mathbf{v}$. Hence $1 \ge \|\mathbf{P}\mathbf{v}\|_{\mathrm{u}} \ge |\lambda|$.

Knowing that the eigenvalues of $\mathbf{P}$ lie in $\mathbb{D}$, we know that $\alpha\mathbf{I}-\mathbf{P}$ is invertible for all $\alpha > 1$. Furthermore, for any $\alpha > 1$ and $\mathbf{v}$, the series $\sum_{m=0}^{\infty} \alpha^{-m-1}\mathbf{P}^m\mathbf{v}$ converges in uniform norm to a limit $\mathbf{v}_\alpha$ which satisfies $(\alpha\mathbf{I}-\mathbf{P})\mathbf{v}_\alpha = \mathbf{v}$. Hence, $\mathbf{v}_\alpha = (\alpha\mathbf{I}-\mathbf{P})^{-1}\mathbf{v}$. In particular, this means that

$$\left((\alpha\mathbf{I}-\mathbf{P})^{-1}\right)_{ii} = \sum_{m=0}^{\infty} \alpha^{-m-1}\left(\mathbf{P}^m\right)_{ii} > 0 \quad \text{for all } \alpha > 1 \text{ and } i \in \mathbb{S}.$$

In addition, because all the real eigenvalues of $\mathbf{I}-\mathbf{P}$ are non-negative and the ones with non-zero imaginary parts must come in conjugate pairs, we have that $\det(\alpha\mathbf{I}-\mathbf{P}) > 0$, and therefore (3.2.4) now implies that

$$\det\left((\alpha\mathbf{I}-\mathbf{P})^{\{i\}}\right) = \left((\alpha\mathbf{I}-\mathbf{P})^{-1}\right)_{ii} \det(\alpha\mathbf{I}-\mathbf{P}) > 0 \quad \text{for all } \alpha > 1 \text{ and } i \in \mathbb{S}.$$

Hence, after letting $\alpha \searrow 1$, we conclude that $\det((\mathbf{I}-\mathbf{P})^{\{i\}}) \ge 0$.

If 0 is a simple eigenvalue of $\mathbf{I}-\mathbf{P}$, then the null space of $\mathbf{I}-\mathbf{P}$ and, therefore of $(\mathbf{I}-\mathbf{P})^\top$, is one dimensional. Hence, there can be only one stationary probability.

Now assume that there is only one stationary probability $\boldsymbol{\pi}$. If 0 were not a simple eigenvalue of $\mathbf{I}-\mathbf{P}$, then there would exist a non-zero column vector $\mathbf{v} \in \mathbb{R}^{\mathbb{S}}$ and an $1 \le n \le N$ such that $\boldsymbol{\pi}\mathbf{v} = 0$, $(\mathbf{I}-\mathbf{P})^{n-1}\mathbf{v} \neq \mathbf{0}$, and $(\mathbf{I}-\mathbf{P})^n\mathbf{v} = \mathbf{0}$. If $n = 1$, then $\mathbf{v} = \mathbf{A}_M\mathbf{v}$ for all $M \ge 1$, and so, by Theorems 3.2.1 and 2.2.5, we would get the contradiction $\mathbf{v} = \boldsymbol{\pi}\mathbf{v}\mathbf{1} = \mathbf{0}$. Next suppose that $n \ge 2$. For $\alpha > 1$, construct $\mathbf{v}_\alpha$ as above. Then $\|\mathbf{v}_\alpha\|_{\mathrm{u}} \le (\alpha-1)^{-1}\|\mathbf{v}\|_{\mathrm{u}}$. At the same time, if $\mathbf{w} = \sum_{m=0}^{n-1}(-1)^m(\alpha-1)^m(\mathbf{I}-\mathbf{P})^m\mathbf{v}$, then, because $(\mathbf{I}-\mathbf{P})^n\mathbf{v} = \mathbf{0}$, $(\alpha\mathbf{I}-\mathbf{P})\mathbf{w} = \mathbf{v}$, and therefore $\mathbf{v}_\alpha = \mathbf{w}$. But this means that

$$(\alpha - 1)^{-1}\|\mathbf{v}\|_{\mathrm{u}} \ge \left\| \sum_{m=0}^{n-1} (-1)^m (\alpha - 1)^{-m-1} (\mathbf{I} - \mathbf{P})^m \mathbf{v} \right\|_{\mathrm{u}}$$

$$\ge (\alpha - 1)^{-n} \left\| (\mathbf{I} - \mathbf{P})^{n-1}\mathbf{v} \right\| - \sum_{m=0}^{n-2} (\alpha - 1)^{-m-1} \left\| (\mathbf{I} - \mathbf{P})^m \mathbf{v} \right\|_{\mathrm{v}}$$

for all $\alpha > 1$. Hence, after multiplying both sides by $(\alpha - 1)^n$ and letting $\alpha \searrow 1$, we would get the contradiction that $(\mathbf{I} - \mathbf{P})^{n-1}\mathbf{v} = \mathbf{0}$.

Turning to the final assertion, Theorem 3.1.5 says that $(\boldsymbol{\pi})_i = 0$ if i is transient. Hence, all that remains is to check that $(\boldsymbol{\pi})_i > 0$ if i is recurrent. But we know that $(\boldsymbol{\pi})_i > 0$ for some recurrent i, and, by Theorem 3.2.1, we know that all recurrent states communicate with one another. Hence, the desired conclusion follows from Theorem 3.1.5. □

Theorem 3.2.6 *Let* $\mathbf{P}$ *be a transition probability on a finite state space* $\mathbb{S}$, *and assume that* $\boldsymbol{\pi}$ *is the one and only stationary probability for* $\mathbf{P}$. *Then* 0 *is a simple eigenvalue of* $\mathbf{I} - \mathbf{P}$ *and, if* $\Pi_{\mathbf{I}-\mathbf{P}}$ *is the product of the other eigenvalues of* $\mathbf{I} - \mathbf{P}$, *then* $\Pi_{\mathbf{I}-\mathbf{P}} > 0$ *and*

$$(\boldsymbol{\pi})_i = \frac{\det((\mathbf{I} - \mathbf{P})^{\{i\}})}{\Pi_{\mathbf{I}-\mathbf{P}}} = \frac{\det((\mathbf{I} - \mathbf{P})^{\{i\}})}{\sum_{j\in\mathbb{S}} \det((\mathbf{I} - \mathbf{P})^{\{j\}})} \quad \text{for all } i \in \mathbb{S}. \tag{3.2.7}$$

In particular, i *is transient if and only if* $\det((\mathbf{I} - \mathbf{P})^{\{i\}}) = 0$ *and* i *is recurrent if and only if* $\det((\mathbf{I} - \mathbf{P})^{\{i\}}) > 0$.

Proof To prove (3.2.7), take $\mathbf{A} = \mathbf{I} - \mathbf{P}$. Clearly the rows of $\mathbf{A}$ each sum to 0, and, by Lemma 3.2.5, 0 is a simple eigenvalue of $\mathbf{A}$. In addition, again by Lemma 3.2.5, the $\boldsymbol{\pi}$ in (3.2.7) is a probability vector and, by Theorem 3.2.3, it is the one and only stationary probability for $\mathbf{P}$. □

The following theorem can sometimes simplify calculations.

Corollary 3.2.8 *Let* $\mathbf{P}$ *and* $\boldsymbol{\pi}$ *be as in Theorem* 3.2.6, *and denote by* $\mathcal{T}$ *the set of transient states. Then* $(\boldsymbol{\pi})_i = 0$ *for* $i \in \mathcal{T}$ *and*

$$(\boldsymbol{\pi})_i = \frac{\det((\mathbf{I} - \mathbf{P})^{\mathcal{T}\cup\{i\}})}{\Pi_{(\mathbf{I}-\mathbf{P})^{\mathcal{T}}}} = \frac{\det((\mathbf{I} - \mathbf{P})^{\mathcal{T}\cup\{i\}})}{\sum_{j\in\mathbb{S}} \det((\mathbf{I} - \mathbf{P})^{\mathcal{T}\cup\{j\}})} \quad \text{for } i \in \mathbb{S}\setminus\mathcal{T}.$$

Proof First notice that $\mathbf{P}^{\mathcal{T}}$ is a transition probability on $\mathbb{S}\setminus\mathcal{T}$ and that $\boldsymbol{\pi}^{\mathcal{T}} = \boldsymbol{\pi}^{\mathcal{T}}\mathbf{P}^{\mathcal{T}}$ if $\boldsymbol{\pi}^{\mathcal{T}}$ is the restriction of $\boldsymbol{\pi}$ to $\mathbb{S}\setminus\mathcal{T}$. Conversely, observe that if $\boldsymbol{\mu}^{\mathcal{T}}$ is a stationary probability for $\mathbf{P}^{\mathcal{T}}$ and $\boldsymbol{\mu}$ is the extention of $\boldsymbol{\mu}^{\mathcal{T}}$ that is 0 on $\mathcal{T}$, then $\boldsymbol{\mu}$ is a stationary probability for $\mathbf{P}$. Hence, $\boldsymbol{\pi}^{\mathcal{T}}$ is the unique stationary probability for $\mathbf{P}^{\mathcal{T}}$ and therefore, by (3.2.7) applied to $\mathbf{P}^{\mathcal{T}}$, the assertion is proved. □

Another interesting application of the ideas here is contained in the following theorem.

Theorem 3.2.9 *Let* $\mathbf{P}$ *and* $\mathbb{S}$ *be as in Theorem* 3.2.6. *If* i *is recurrent and* $i \in \Delta \subsetneq \mathbb{S}$, *then* $(\mathbf{I}-\mathbf{P})^{\Delta}$ *is invertible and*

$$\left(\left((\mathbf{I}-\mathbf{P})^{\Delta}\right)^{-1}\right)_{jk} = \mathbb{E}\left[\sum_{m=0}^{\zeta^{\Delta}-1} \mathbf{1}_{\{k\}}(X_m) \,\middle|\, X_0 = j\right] \quad \text{for } j,k \in \mathbb{S}\setminus\Delta, \tag{3.2.10}$$

where $\zeta^{\Delta} \equiv \inf\{n \geq 0 : X_n \in \Delta\}$ *is the time of the first visit by* $\{X_n : n \geq 0\}$ *to* Δ. *In particular,*

$$\mathbb{E}\left[\sum_{m=0}^{\zeta^{\Delta}-1} \mathbf{1}_{\{j\}}(X_m) \,\middle|\, X_0 = j\right] = \frac{\det((\mathbf{I}-\mathbf{P})^{\Delta\cup\{j\}})}{\det((\mathbf{I}-\mathbf{P})^{\Delta})} \quad \text{for } j \in \mathbb{S}\setminus\Delta. \tag{3.2.11}$$

(*See Exercise* 4.2.8 *for an interesting application of* (3.2.11).)

Proof First observe that

$$\begin{aligned}&\mathbb{P}\left(X_m = j_m \text{ for } 0 \leq m \leq n \;\&\; \zeta^{\Delta} > n \,\middle|\, X_0 = j_0\right)\\ &\quad = \mathbf{1}_{\Delta\complement}(j_0) \prod_{m=1}^{n} \mathbf{1}_{\Delta\complement}(j_m)(\mathbf{P})_{j_{m-1}j_m}\end{aligned}$$

and therefore that

$$\mathbb{P}\left(X_n = k \;\&\; \zeta^{\Delta} > n \,\middle|\, X_0 = j\right) = \left(\mathbf{P}^{\Delta}\right)^n_{jk} \quad \text{for } j,k \in \mathbb{S}\setminus\Delta.$$

From this it follows that

$$\mathbb{E}\left[\sum_{m=0}^{n\wedge\zeta^{\Delta}-1} \mathbf{1}_{\{k\}}(X_m) \,\middle|\, X_0 = j\right] = \sum_{m=0}^{n-1} \left(\mathbf{P}^{\Delta}\right)^m_{jk} \quad \text{for all } n \geq 1. \tag{$*$}$$

Hence, since $\mathbb{E}[\rho_i \mid X_0 = j] \geq \mathbb{E}[\zeta^{\Delta} \mid X_0 = j]$ and, by Theorems 2.3.8 and 3.2.1, we know that $\mathbb{E}[\rho_i \mid X_0 = j] < \infty$, we have that

$$\sum_{k\in\mathbb{S}} \sum_{m=0}^{\infty} \left(\mathbf{P}^{\Delta}\right)_{jk} = \sum_{k\in\mathbb{S}} \mathbb{E}\left[\sum_{m=0}^{\zeta^{\Delta}-1} \mathbf{1}_{\{k\}}(X_m) \,\middle|\, X_0 = j\right] = \mathbb{E}\left[\zeta^{\Delta} \,\middle|\, X_0 = j\right] < \infty$$

for all $j \in \mathbb{S}$. In particular, if $\mathbf{v} \in \mathbb{R}^{\mathbb{S}\setminus\Delta}$, then $\sum_{m=0}^{\infty}(\mathbf{P}^{\Delta})^m\mathbf{v}$ converges to a $\mathbf{w} \in \mathbb{R}^{\mathbb{S}}$, and it is easy to see that $(\mathbf{I}-\mathbf{P})^{\Delta}\mathbf{w} = \mathbf{v}$. This proves that $(\mathbf{I}-\mathbf{P})^{\Delta}$ is invertible and, after letting $n \to \infty$ in ($*$), that (3.2.10) holds. Furthermore, (3.2.11) is just (3.2.10) followed by an application of (3.2.4). □

3.3 Wilson's Algorithm and Kirchhoff's Formula

A *graph* Γ is an ordered pair $(V, \mathcal{E})$, where V is a finite set and $\mathcal{E}$ is a collection of unordered pairs of elements from V. An element of V is called a *vertex* and an element of $\mathcal{E}$ is called an *edge*. Without further notice, we will be considering graphs in which no vertex has an edge connecting it to itself and no pair of vertices has more than one edge connecting them. When an edge exists between vertices v and w, we will write $v \sim w$.

A *directed path* is an ordered n-tuple $(v_1, \dots, v_n) \in V^n$ such that $v_m \sim v_{m-1}$ for each $2 \le m \le n$, and a graph is said to be *connected* if every pair of distinct vertices has a directed path connecting them. A directed path $(v_1, \dots, v_n)$ is said to be a *loop* if $v_1 = v_n$, and a graph is called a *tree* if it is connected and no directed path in it is a loop. Equivalently, a graph is a tree if every pair of distinct vertices has precisely one directed path running from one to the other.

Given a graph $\Gamma = (V, \mathcal{E})$, a *subgraph* is a graph $\Gamma' = (V', \mathcal{E}')$, where $V' \subseteq V$ and $\mathcal{E}'$ is a subset of $\{\{v, w\} \in \mathcal{E} : v, w \in V'\}$. Finally, given a connected graph, a *spanning tree* is a subgraph that is a tree in which all the vertices of the graph are present. A challenging combinatorial problem is that of determining how many spanning trees a graph has. When a graph with N elements is *complete*, in the sense that there is an edge between every pair of distinct vertices, G. Cayley showed that there are N^{N-2} spanning trees, although the problem had been solved in the general case somewhat earlier by G. Kirchhoff.

A related problem is that of finding an algorithm for constructing the uniform measure on the set of all spanning trees. Such an algorithm based on Markov chains was discovered by D. Wilson. In this section, I will present Wilson's algorithm and from it derive Cayley's and Kirchhoff's results.

3.3.1 Spanning Trees and Wilson Runs

In order to explain Wilson's algorithm, I must first give a method for labeling the spanning tree in terms of what I will call *Wilson runs*.

A Wilson run is an L-tuple $(P_1, \dots, P_L)$ where:

(1) for each $1 \le \ell \le L$, $P_\ell = (w_{1,\ell}, \dots, w_{K_\ell,\ell})$ is loop free (i.e., $w_{k,\ell} \ne w_{k',\ell}$ if $k \ne k'$) directed path of length $K_\ell \ge 2$,
(2) $V = \{w_{k,\ell} : 1 \le \ell \le L \;\&\; 1 \le k \le K_\ell\}$,
(3) if $2 \le \ell \le L$, then

$$\{w_{k,\ell} : 1 \le k \le K_\ell\} \cap \{w_{k,j} : 1 \le j < \ell \;\&\; 1 \le k \le K_j\} = \{w_{K_\ell,\ell}\}.$$

Each Wilson run determines a unique spanning tree. Namely, the spanning tree whose edges are $\{\{w_{k,\ell}, w_{k+1,\ell}\} : 1 \le \ell \le L \;\&\; 1 \le k < K_\ell\}$.

Although a Wilson run determines a unique spanning tree, there are many Wilson runs that determine the same spanning tree. For example, if Γ is a complete graph with three vertices v_1, v_2, v_3 and T is the spanning tree with edges

$\{\{v_1, v_2\}, \{v_2, v_3\}\}$, then both (v_1, v_2, v_3) and $((v_1, v_2), (v_3, v_2))$ are Wilson runs that determine T. Nonetheless, one can make the correspondence one-to-one by specifying an ordering $(v_1, \dots, v_N)$ of the vertices in V and insisting that the initial segment in the Wilson run start at v_1 and end at v_2 and that later segments always start at the first v_m that is not in any of the earlier segments. To be precise, given a spanning tree T, take P_1 to be the directed path in T that begins at v_1 and ends at v_2. Because T is a tree, there is only one such path. If all the vertices in V appear in P_1, stop. Otherwise, take P_2 to be the directed path in T that starts at the first v_m not in P_1 and ends when it first hits P_1. Again, because T is a tree, P_2 is uniquely determined. More generally, if one has constructed $(P_1, \dots, P_\ell)$ and there is a vertex that is not in[3] $\bigcup_{j=1}^{\ell} P_j$, take $P_{\ell+1}$ to be the unique directed path in T that connects the first v_m that is not in $\bigcup_{j=1}^{\ell} P_j$ to $\bigcup_{j=1}^{\ell} P_j$. After $L \le N-1$ steps, one will have produced a Wilson run $(P_1, \dots, P_L)$ that determines T, and this Wilson run will be the only one that respects the ordering $(v_1, \dots, v_N)$.

Finally, I must describe the procedure that will be used to erase loops from a directed path. For this purpose, suppose that $J \ge 2$ and that $(X_1, \dots, X_J)$ is a directed path in the graph with the property that $X_J \notin \{X_0, \dots, X_{J-1}\}$. We will say that $(w_1, \dots, w_K)$ is the *loop erasure* of $(X_0, \dots, X_J)$ if it is obtained by sequentially erasing the loops from $(X_1, \dots, X_J)$. To be precise, $w_1 = X_0$, $w_K = X_J$, and, for $2 \le k \le K$, $w_k = X_{\ell_{k-1}+1}$, where $\ell_1 < \dots < \ell_{K-1}$ are determined by $\ell_1 = \max\{j > 0 : X_j = w_1\}$ and, for $2 \le k \le K$, $\ell_k = \max\{j > \ell_{k-1} : X_j = w_k\}$. Clearly, $(w_1, \dots, w_K)$ is a loop free directed path.

3.3.2 Wilson's Algorithm

Let $\Gamma = (V, \mathcal{E})$ be a connected graph with $N \ge 2$ vertices. For each $v \in V$, the *degree* of v is the number d_v of $w \in V$ such that $\{v, w\} \in \mathcal{E}$. Because Γ is connected, $d_v \ge 1$ for all $v \in V$. Next, let $\mathcal{A}$ be the *adjacency matrix*. That is, $\mathcal{A}_{vw} = 1$ if $\{v, w\} \in \mathcal{E}$ and $\mathcal{A}_{vw} = 0$ otherwise. Because of the running assumptions that I made about our graphs, the diagonal entries of $\mathcal{A}$ are 0 and $\sum_{w \in V} \mathcal{A}_{vw} = d_v = \sum_w \mathcal{A}_{wv}$ for each $v \in V$. Further, because Γ is connected, $\sum_{n=0}^{N} (\mathcal{A}^n)_{vw} \ge 1$ for all $(v, w) \in V^2$.

Now let $\mathcal{D}$ be the diagonal matrix $\mathcal{D}_{vw} = \delta_{v,w} d_v$, and set $\mathbf{P} = \mathcal{D}^{-1}\mathcal{A}$. Then $\mathbf{P}$ is a transition probability, and the paths $\{X_n : n \ge 0\}$ followed by a Markov process with transition probability $\mathbf{P}$ will have the property that $X_n \sim X_{n-1}$ for all $n \ge 1$. In addition, $\sum_{n=0}^{N} (\mathbf{P}^n)_{vw} \ge \epsilon$ for all $(v, w) \in V^2$ and some $\epsilon > 0$. Hence, by Theorem 2.2.5, there is precisely one stationary probability $\boldsymbol{\pi}$ for $\mathbf{P}$, $(\boldsymbol{\pi})_v \ge \epsilon$ for all v, and therefore every v is recurrent.

Wilson's algorithm is the following. Choose and fix an ordering $(v_1, \dots, v_N)$ of the vertices, and start the Markov process $\{X_n : n \ge 0\}$ with transition probability $\mathbf{P}$

---

[3] Although the P_ℓ's are tuples and not sets, the following union is the set consisting of the entries of the tuples over which it is taken.

at v_1. Set $\Delta_1(\omega) = \{v_2\}$ and, assuming that (cf. Theorem 3.2.9) $\zeta^{\Delta_1}(\omega) < \infty$, and take $P_1(\omega)$ to be the loop erasure of $(X_0(\omega), \dots, X_{\zeta^{\Delta_1}}(\omega))$. If $P_1(\omega)$ contains all the vertices in V, the algorithm terminates at time $\zeta^{\Delta_1}(\omega)$. Otherwise, take $\Delta_2(\omega) = P_1(\omega)$, start the Markov chain at the first v_m that is not in $\Delta_2(\omega)$, and, assuming that $\zeta^{\Delta_2}(\omega) < \infty$, take $P_2(\omega)$ to be the loop erasure of $(X_0(\omega), \dots, X_{\zeta^{\Delta_2}}(\omega))$. More generally, assuming that $\zeta^{\Delta_{\ell-1}}(\omega) < \infty$ and that $\Delta_\ell(\omega) \equiv \bigcup_{j=1}^{\ell-1} P_j(\omega) \neq V$, one starts the chain at the first $v_m \notin \Delta_\ell(\omega)$ and takes $P_\ell(\omega)$ to be the loop erasure of $(X_0(\omega), \dots, X_{\zeta^{\Delta_\ell}}(\omega))$. Assuming that $\zeta^{\Delta_\ell} < \infty$ for all ℓ, one continues in this way until one generates a Wilson run $(P_1(\omega), \dots, P_L(\omega))$ that respects the ordering $(v_1, \dots, v_N)$. Of course, one has to worry about the fact that one or more of the ζ^{Δ_ℓ}'s may be infinite. However, by Theorem 2.3.8, with probability 1, none of them will be. Thus, with probability 1, Wilson's algorithm will generate a Wilson run that respects the ordering $(v_1, \dots, v_N)$.

The problem now is to compute the probability $\mathcal{P}(T)$ that Wilson's algorithm generates the Wilson run corresponding to a specific spanning tree T. To solve this problem, for a given $\emptyset \neq \Delta \subsetneq V$ and a loop free directed path $P = (w_1, \dots, w_K) \in (V \setminus \Delta)^{K-1} \times \Delta$, let $\mathcal{P}^\Delta(P)$ be the probability that, when Markov chain is started at w_1, $\zeta^\Delta < \infty$ and the loop erasure of $(X_0, \dots, X_{\zeta^\Delta})$ is P. If the Wilson run determining T is $(P_1, \dots, P_L)$, then

$$\mathcal{P}(T) = \prod_{\ell=1}^{L} \mathcal{P}^{\Delta_\ell}(P_\ell), \tag{3.3.1}$$

where $\Delta_1 = \{v_2\}$ and $\Delta_\ell = \bigcup_{j=1}^{\ell-1} P_j$ for $2 \leq \ell \leq L$. Thus, what we need to learn is how to compute $\mathcal{P}^\Delta(P)$, and for this purpose I will use a strategy that was introduced by Greg Lawler.

Lemma 3.3.2 *Let $\emptyset \neq \Delta \subsetneq V$ and a loop free directed path*

$$P = (w_1, \dots, w_K) \in (V \setminus \Delta)^{K-1} \times \Delta$$

be given. Set $\Delta(1) = \Delta$ and $\Delta(k) = \Delta \cup \{w_1, \dots, w_{k-1}\}$ for $2 \leq k \leq K$. Then

$$\mathcal{P}^\Delta(P) = \prod_{k=1}^{K-1} \frac{(((\mathbf{I} - \mathbf{P})^{\Delta(k)})^{-1})_{w_k, w_k}}{d_{w_k}} = \frac{\det((\mathcal{D} - \mathcal{A})^{\Delta \cup P})}{\det((\mathcal{D} - \mathcal{A})^{\Delta})}. \tag{3.3.3}$$

Proof Let $B = \{\omega : \zeta^\Delta(\omega) < \infty \ \&\ P(\omega) = P\}$, and, for

$$\mathbf{m} = (m_1, \dots, m_{K-1}) \in \mathbb{N}^{K-1},$$

let $B^{\mathbf{m}}$ be the set of $\omega \in B$ such that, for each $1 \leq k < K$, precisely m_k loops are erased at step k in the passage from $(X_0(\omega), \dots, X_{\zeta^{\Delta}(\omega)}(\omega))$ to P. Clearly

$$\mathbb{P}(B \mid X_0 = w_1) = \sum_{\mathbf{m} \in \mathbb{N}^{K-1}} \mathbb{P}\big(B^{\mathbf{m}} \,\big|\, X_0 = w_1\big).$$

To compute $\mathbb{P}(B^{\mathbf{m}} \mid X_0 = w_1)$, let $\sigma_1(\omega)$ be the time of the (m_1+1)th visit to w_1 and $\zeta_1(\omega)$ the time of the first visit to $\Delta(1)$ by the path $\{X_n(\omega) : n \geq 0\}$. Proceeding by induction, for $2 \leq k < K$, take $\sigma_k(\omega) = \zeta_k(\omega) = \infty$ if $\sigma_{k-1}(\omega) \geq \zeta_{k-1}(\omega)$ and otherwise let $\sigma_k(\omega)$ be the time of the (m_k+1)th visit to w_k and $\zeta_k(\omega)$ the time of the first visit to $\Delta(k)$ by $\{X_n(\omega) : n > \sigma_{k-1}(\omega)\}$. Then

$$B^{\mathbf{m}} = \{\zeta^{\Delta} < \infty\} \cap \{\sigma_k < \zeta_k < \infty \;\&\; X_{\sigma_k+1} = w_{k+1} \text{ for } 1 \leq k < K\}.$$

We will now use an induction procedure to compute $\mathbb{P}(B^{\mathbf{m}} \mid X_0 = w_1)$. For $1 \leq k < K$, set

$$B_k^{\mathbf{m}} = \{\zeta^{\Delta} < \infty\} \cap \{\sigma_j < \zeta_j < \infty \;\&\; X_{\sigma_j+1} = w_{j+1} \text{ for } 1 \leq j \leq k\}.$$

Obviously, $B^{\mathbf{m}} = B_{K-1}^{\mathbf{m}}$. By the Markov property,

$$\begin{aligned}
&\mathbb{P}\big(B_1^{\mathbf{m}} \,\big|\, X_0 = w_1\big) \\
&\quad = \sum_{n=0}^{\infty} \mathbb{P}\big(\sigma_1 = n < \zeta^{\Delta(1)} \;\&\; X_{n+1} = w_2 \,\big|\, X_0 = w_1\big) \\
&\quad = \sum_{n=0}^{\infty} \frac{\mathbb{P}(\rho_{w_1}^{(m_1)} = n < \zeta^{\Delta(1)} \mid X_0 = w_1)}{d_{w_1}} = \frac{\mathbb{P}(\rho_{w_1}^{(m_1)} < \zeta^{\Delta(1)} \mid X_0 = w_1)}{d_{w_1}},
\end{aligned}$$

where, for any $m \in \mathbb{N}$ and $v \in V$, $\rho_v^{(m)}(\omega)$ is the time of the mth return to v by $\{X_n(\omega) : n \geq 0\}$. Similarly, for $2 \leq k < K$,

$$\begin{aligned}
&\mathbb{P}\big(B_k^{\mathbf{m}} \,\big|\, X_0 = w_1\big) \\
&\quad = \mathbb{P}\big(B_{k-1}^{\mathbf{m}} \cap \{\sigma_k < \zeta_k \;\&\; X_{\sigma_k+1} = w_{k+1}\} \,\big|\, X_0 = w_1\big) \\
&\quad = \mathbb{P}\big(B_{k-1}^{\mathbf{m}}\big) \frac{\mathbb{P}(\rho_{w_k}^{(m_k)} < \zeta^{\Delta(k)} \mid X_0 = w_k)}{d_{w_k}}.
\end{aligned}$$

Since $\rho_v^{(m)} < \zeta^{\Delta(k)} \iff \sum_{n=0}^{\zeta^{\Delta(k)}-1} \mathbf{1}_{\{v\}}(X_n) > m$,

$$\sum_{m_k=0}^{\infty} \mathbb{P}\big(\rho_{w_k}^{(m_k)} < \zeta^{\Delta(k)} \,\big|\, X_0 = w_k\big) = \mathbb{E}\left[\sum_{n=0}^{\zeta^{\Delta(k)}-1} \mathbf{1}_{\{w_k\}}(X_n) \,\middle|\, X_0 = w_k\right],$$

which, by Theorem 3.2.9, is equal to $(((\mathbf{I}-\mathbf{P})^{\Delta(k)})^{-1})_{w_k w_k}$, and so

$$\sum_{\mathbf{m} \in \mathbb{N}^{K-1}} \mathbb{P}(B^{\mathbf{m}}) = \prod_{k=1}^{K-1} \frac{(((\mathbf{I}-\mathbf{P})^{\Delta(k)})^{-1})_{w_k w_k}}{d_{w_k}},$$

which is the first equality in (3.3.3). To prove the second one, note that, because $\mathcal{D} - \mathcal{A} = \mathcal{D}(\mathbf{I} - \mathbf{P})$,

$$\frac{(((\mathbf{I} - \mathbf{P})^{\Delta(k)})^{-1})_{w_k w_k}}{d_{w_k}} = \left(((\mathcal{D} - \mathcal{A})^{\Delta(k)})^{-1}\right)_{w_k w_k},$$

and apply (3.2.4). □

By combining this lemma with (3.3.1), we arrive at the following statement of Wilson's theorem.

Theorem 3.3.4 (Wilson) *For any spanning tree T and any vertex $v \in V$,*

$$\mathcal{P}(T) = \frac{1}{\det((\mathcal{D} - \mathcal{A})^{\{v\}})}.$$

In particular, $\mathcal{P}(T)$ is the same for all spanning trees T.

Proof Choose an ordering $(v_1, \dots, v_N)$ with $v_2 = v$, let $(P_1, \dots, P_L)$ be the Wilson run determined by T that respects this ordering, define Δ_ℓ as in (3.3.1) for $1 \le \ell \le L$, and set $\Delta_{L+1} = V$. Then, by Lemma 3.3.2 and (3.3.1),

$$\mathcal{P}(T) = \prod_{\ell=1}^{L} \frac{\det((\mathcal{D} - \mathcal{A})^{\Delta_{\ell+1}})}{\det((\mathcal{D} - \mathcal{A})^{\Delta_\ell})} = \frac{1}{\det((\mathcal{D} - \mathcal{A})^{\{v\}})}.$$ □

3.3.3 Kirchhoff's Matrix Tree Theorem

Let everything be as in the preceding subsection.

Obviously each row and column of $\mathcal{D} - \mathcal{A}$ sums to 0. In addition, because 0 is a simple eigenvalue for $\mathbf{I} - \mathbf{P}$ and therefore for $\mathcal{D} - \mathcal{A}$, Theorem 3.2.3 applies. Further, because the columns of $\mathcal{D} - \mathcal{A}$ sum to 0, $N^{-1}\mathbf{1}$ is a left eigenvector, and therefore Theorem 3.2.3 says that

$$\det\left((\mathcal{D} - \mathcal{A})^{\{v\}}\right) = \frac{\Pi_{\mathcal{D}-\mathcal{A}}}{N}.$$

Hence, by Theorem 3.3.4,

$$\mathcal{P}(T) = \frac{1}{\det((\mathcal{D} - \mathcal{A})^{\{v\}})} = \frac{N}{\Pi_{\mathcal{D}-\mathcal{A}}} \quad \text{for any spanning tree } T. \tag{3.3.5}$$

Since, with probability 1, Wilson's algorithm produces a spanning tree, it follows that, for all $v \in V$,

$$\text{the number of spanning trees} = \det\left((\mathcal{D} - \mathcal{A})^{\{v\}}\right) = \frac{\Pi_{\mathcal{D}-\mathcal{A}}}{N}, \tag{3.3.6}$$

which is one statement of Kirchhoff's renowned *matrix tree theorem*.

To derive Cayley's formula from (3.3.6), assume that the graph is complete, and let $(v_1, \dots, v_N)$ be an ordering of its vertices. Then $(\mathcal{D} - \mathcal{A})^{\{v_N\}}$ is equal to the $(N-1) \times (N-1)$-matrix

$$\begin{pmatrix} N-1 & -1 & -1 & \dots & -1 & -1 \\ -1 & N-1 & -1 & \dots & -1 & -1 \\ \cdots & \cdots & \cdots & \cdots & \cdots & \cdots \\ \cdots & \cdots & \cdots & \cdots & \cdots & \cdots \\ -1 & -1 & -1 & \dots & N-1 & -1 \\ -1 & -1 & -1 & \dots & -1 & N-1 \end{pmatrix}.$$

By subtracting the last column from each of the columns 2 through $N-2$, one sees that the determinant of $(\mathcal{D} - \mathcal{A})^{\{v_N\}}$ is the same as that of

$$\begin{pmatrix} N-1 & 0 & 0 & \dots & 0 & -1 \\ -1 & N & 0 & \dots & 0 & -1 \\ \cdots & \cdots & \cdots & \cdots & \cdots & \cdots \\ \cdots & \cdots & \cdots & \cdots & \cdots & \cdots \\ -1 & 0 & 0 & \dots & N & -1 \\ -1 & -N & -N & \dots & -N & N-1 \end{pmatrix},$$

and by adding the sum of the columns 1 through $N-2$ times $(N-1)^{-1}$ to the last column, one sees that this has the same determinant as

$$\begin{pmatrix} N-1 & 0 & 0 & \dots & 0 & 0 \\ -1 & N & 0 & \dots & 0 & 0 \\ \cdots & \cdots & \cdots & \cdots & \cdots & \cdots \\ \cdots & \cdots & \cdots & \cdots & \cdots & \cdots \\ -1 & 0 & 0 & \dots & N & 0 \\ -1 & -N & -N & \dots & -N & \frac{N}{N-1} \end{pmatrix}.$$

Hence there are N^{N-2} spanning trees in a complete graph with N vertices.

3.4 Exercises

Exercise 3.4.1 In Sect. 1.2.4 we worked quite hard to prove that the nearest neighbor, symmetric random walk on $\mathbb{Z}^3$ is transient, and one might hope that the criteria provided in Sect. 3.1.2 would allow us to avoid working so hard. However, even if one knows which function u ought to be plugged into Theorem 3.1.6 or 3.1.8, the computation to show that it works is rather delicate.

Let $\mathbf{P}$ be the transition probability for the simple, symmetric random walk in $\mathbb{Z}^3$. That is,

$$(\mathbf{P})_{\mathbf{k}\boldsymbol{\ell}} = \begin{cases} \frac{1}{6} & \text{if } \sum_{i=1}^{3} |(\mathbf{k})_i - (\boldsymbol{\ell})_i| = 1 \\ 0 & \text{otherwise.} \end{cases}$$

Show that if $\alpha \geq 1$ is sufficiently large and

$$u(\mathbf{k}) = \left(\alpha^2 + \sum_{i=1}^{3} (\mathbf{k})_i^2\right)^{-\frac{1}{2}} \quad \text{for } \mathbf{k} \in \mathbb{Z}^3,$$

then $(\mathbf{Pu})_\mathbf{k} \leq u(\mathbf{k}) \leq u(\mathbf{0})$ when $\mathbf{u}$ is the column vector determined by the function u. By Theorem 3.1.8, this proves that $\mathbf{0}$ is transient.

What follows are some hints.

(a) Let $\mathbf{k} \in \mathbb{Z}^3$ be given, and set

$$M = 1 + \alpha^2 + \sum_{i=1}^{3} (\mathbf{k})_i^2 \quad \text{and} \quad x_i = \frac{(\mathbf{k})_i}{M} \quad \text{for } 1 \leq i \leq 3.$$

Show that $(\mathbf{Pu})_\mathbf{k} \leq u(\mathbf{k})$ if and only if

$$\left(1 - \frac{1}{M}\right)^{-\frac{1}{2}} \geq \frac{1}{3} \sum_{i=1}^{3} \frac{(1 + 2x_i)^{\frac{1}{2}} + (1 - 2x_i)^{\frac{1}{2}}}{2(1 - 4x_i^2)^{\frac{1}{2}}}.$$

(b) Show that $(1 - \frac{1}{M})^{-\frac{1}{2}} \geq 1 + \frac{1}{2M}$ and that

$$\frac{(1+\xi)^{\frac{1}{2}} + (1-\xi)^{\frac{1}{2}}}{2} \leq 1 - \frac{\xi^2}{8} \quad \text{for } |\xi| < 1,$$

and conclude that $(\mathbf{Pu})_\mathbf{k} \leq u(\mathbf{k})$ if

$$1 + \frac{1}{2M} \geq \frac{1}{3} \sum_{i=1}^{3} \frac{1}{(1 - 4x_i^2)^{\frac{1}{2}}} - \frac{|\mathbf{x}|^2}{6},$$

where $|\mathbf{x}| = \sqrt{\sum_1^3 x_i^2}$ is the Euclidean length of $\mathbf{x} = (x_1, x_2, x_3)$.

(c) Show that there is a constant $C < \infty$ such that, as long as $\alpha \geq 1$,

$$\frac{1}{3} \sum_{i=1}^{3} \frac{1}{(1 - 4x_i^2)^{\frac{1}{2}}} \leq 1 + \frac{2|\mathbf{x}|^2}{3} + C|\mathbf{x}|^4,$$

and put this together with the preceding to conclude that we can take any $\alpha \geq 1$ with $\alpha^2 + 1 \geq 2C$.

An analogous computation shows that, for each $d \geq 3$ and all sufficiently large $\alpha > 0$, the function $(\alpha^2 + \sum_1^d (\mathbf{k})_i^2)^{-\frac{d-2}{2}}$ can be used to prove that the nearest neighbor, symmetric random walk is transient in $\mathbb{Z}^d$. The reason why one does not get a contradiction when $d = 2$ and the walk is recurrent is that the non-constant, non-negative functions which satisfies $\mathbf{P}u \leq u$ when $d = 2$ are of the form $\log(\alpha^2 + (\mathbf{k})_1^2 + (\mathbf{k})_2^2)$ and therefore do not achieve their maximum value at $\mathbf{0}$.

Exercise 3.4.2 Let $\mathbf{P} = ((p_{i,j}))_{1\le i,j\le 3}$ be a transition probability on $\{1,2,3\}$, and set

$$D(i,j,k) = p_{ji}(1-p_{kk}) + p_{jk}p_{ki}.$$

Show that

$$\det\big((\mathbf{I}-\mathbf{P})^{\{1\}}\big) = D(1,2,3), \qquad \det\big((\mathbf{I}-\mathbf{P})^{\{2\}}\big) = D(2,3,1),$$
$$\text{and} \quad \det\big((\mathbf{I}-\mathbf{P})^{\{3\}}\big) = D(3,1,2),$$

and conclude that $\mathbf{P}$ has a unique stationary distribution π if and only if

$$\Pi \equiv D(1,2,3) + D(2,3,1) + D(3,1,2) > 0,$$

in which case,

$$(\pi)_1 = \frac{D(1,2,3)}{\Pi}, \qquad (\pi)_2 = \frac{D(2,3,1)}{\Pi}, \qquad \text{and} \quad (\pi)_3 = \frac{D(3,1,2)}{\Pi}.$$

Exercise 3.4.3 Given a transition probability $\mathbf{P}$ on a finite state space $\mathbb{S}$, show that, for each $\Delta \subsetneq \mathbb{S}$,

$$0 \le \det\big((\mathbf{I}-\mathbf{P})^{\Delta}\big) \le \det\big((\mathbf{I}-\mathbf{P})^{\Delta\cup\{i\}}\big) \quad \text{for all } i \in \mathbb{S}\setminus\Delta.$$

Conclude that if $\Delta \subseteq \Delta' \subseteq \mathbb{S}$, then

$$0 \le \det\big((\mathbf{I}-\mathbf{P})^{\Delta}\big) \le \det\big((\mathbf{I}-\mathbf{P})^{\Delta'}\big).$$

Hint: Show that for each $\epsilon > 0$ there is a $\mathbf{P}^{\epsilon}$ that satisfies Doeblin's condition and whose entries differ from those of $\mathbf{P}$ by less that ϵ. In this way, reduce the problem to transition probabilities that satisfies Doeblin's condition.

Exercise 3.4.4 Let $\mathbf{P}$ the transition probability introduced in Sect. 3.2.2, and show that the unique stationary probability π for $\mathbf{P}$ is given by $(\pi)_v = \frac{d_v}{D}$, where $D = \sum_{w\in V} d_w$.

Chapter 4
More About the Ergodic Properties of Markov Chains

In Chap. 2 all of our considerations centered around one form or another of the Doeblin condition which says that there is a state which can be reached from any other state at a uniformly fast rate. Although there are lots of chains on an infinite state space which satisfy his condition, most do not. In fact, many chains on an infinite state space will not even admit a stationary probability distribution because the availability of infinitely many states means that there is enough room for the chain to "get lost and disappear." There are two ways in which this can happen. Namely, the chain can disappear because, like the a nearest neighbor, non-symmetric random walk in $\mathbb{Z}$ (cf. (1.1.13)) or even the symmetric one in $\mathbb{Z}^3$ (cf. Sect. 1.2.4), it may have no recurrent states and, as a consequence, will spend a finite amount of time in any given state. A more subtle way for the chain to disappear is for it to be recurrent but not sufficiently recurrent for there to exist a stationary distribution. Such an example is the symmetric, nearest neighbor random walk in $\mathbb{Z}$ which is recurrent, but just barely so. In particular, although this random walk returns infinitely often to the place where is begins, it does so at too sparse a set of times. More precisely, by (1.2.7) and (1.2.13), if $\mathbf{P}$ is the transition probability matrix for the symmetric, nearest neighbor random walk on $\mathbb{Z}$, then

$$\left(\mathbf{P}^{2n}\right)_{ij} \le \left(\mathbf{P}^{2n}\right)_{ii} = \mathbb{P}(X_{2n} = 0) \le A(1)n^{-\frac{1}{2}} \longrightarrow 0,$$

and so, if $\boldsymbol{\mu}$ were a probability vector which was stationary for $\mathbf{P}$, then, by Lebesgue's dominated convergence theorem, Theorem 7.1.11, we would have the contradiction

$$(\boldsymbol{\mu})_j = \lim_{n\to\infty} \sum_{i\in\mathbb{Z}} (\boldsymbol{\mu})_i \left(\mathbf{P}^{2n}\right)_{ij} = 0 \quad \text{for all } j \in \mathbb{Z}.$$

In this chapter we will see that ergodic properties can exist in the absence of Doeblin's condition. However, as we will see, what survives does so in a weaker form. See part (c) of Exercise 4.2.5 for a comparison of the results here with the earlier ones.

D.W. Stroock, *An Introduction to Markov Processes*, Graduate Texts in Mathematics 230, DOI 10.1007/978-3-642-40523-5_4, © Springer-Verlag Berlin Heidelberg 2014

4.1 Ergodic Theory Without Doeblin

In this section we will see to what extent the results obtained in Chap. 2 for Markov chains which satisfy Doeblin's condition can be reproduced without Doeblin's condition. The progression which we will adopt here runs in the opposite direction to that in Chap. 2. That is, here we will start with the most general but weakest form of the result and afterwards will see what can be done to refine it.

4.1.1 Convergence of Matrices

Because we will be looking at power series involving matrices that may have infinitely many entries, it will be important for us to be precise about what is the class of matrices with which we are dealing and in what sense our series are converging. For our purposes, the most natural class will be of matrices $\mathbf{M}$ for which

$$\|\mathbf{M}\|_{\mathrm{u,v}} \equiv \sup_{i\in\mathbb{S}} \sum_{j\in\mathbb{S}} \left|(\mathbf{M})_{ij}\right| \tag{4.1.1}$$

is finite, and the set of all such matrices will be denoted by $M_{\mathrm{u,v}}(\mathbb{S})$. An easy calculation shows that $M_{\mathrm{u,v}}(\mathbb{S})$ is a vector space over $\mathbb{R}$ and that $\|\cdot\|_{\mathrm{u,v}}$ is a good norm on $M_{\mathrm{u,v}}(\mathbb{S})$. That is,

$$\begin{aligned} \|\mathbf{M}\|_{\mathrm{u,v}} &= 0 \quad \text{if and only if} \quad \mathbf{M}=\mathbf{0}, \\ \|\alpha\mathbf{M}\|_{\mathrm{u,v}} &= |\alpha|\,\|\mathbf{M}\|_{\mathrm{u,v}} \quad \text{for } \alpha\in\mathbb{R}, \end{aligned}$$

and

$$\left\|\mathbf{M}+\mathbf{M}'\right\|_{\mathrm{u,v}} \le \|\mathbf{M}\|_{\mathrm{u,v}} + \left\|\mathbf{M}'\right\|_{\mathrm{u,v}}.$$

Slightly less obvious is the fact that

$$\left(\mathbf{M},\mathbf{M}'\right)\in M_{\mathrm{u,v}}(\mathbb{S})^2 \quad\Longrightarrow\quad \begin{aligned} &\mathbf{M}\mathbf{M}' \text{ exists} \quad \text{and} \\ &\left\|\mathbf{M}\mathbf{M}'\right\|_{\mathrm{u,v}} \le \|\mathbf{M}\|_{\mathrm{u,v}}\left\|\mathbf{M}'\right\|_{\mathrm{u,v}}. \end{aligned} \tag{4.1.2}$$

To see this, observe first that, since

$$\sum_k \left|(\mathbf{M})_{ik}\right|\left|\left(\mathbf{M}'\right)_{kj}\right| \le \left(\sum_k \left|(\mathbf{M})_{ik}\right|\right)\sup_k\left|\left(\mathbf{M}'\right)_{kj}\right| < \infty,$$

the sum in

$$\left(\mathbf{M}\mathbf{M}'\right)_{ij} = \sum_k (\mathbf{M})_{ik}\left(\mathbf{M}'\right)_{kj}$$

is absolutely convergent. In addition, for each i,

$$\sum_j \left|(\mathbf{M}\mathbf{M}')_{ij}\right| \le \sum_j \sum_k |(\mathbf{M})_{ik}|\left|(\mathbf{M}')_{kj}\right|$$

$$= \sum_k |(\mathbf{M})_{ik}| \left(\sum_j \left|(\mathbf{M}')_{kj}\right|\right) \le \left(\sum_k |(\mathbf{M})_{ik}|\right) \|\mathbf{M}'\|_{\mathrm{u,v}},$$

and so the inequality in (4.1.2) follows.

We will show next that the metric that $\|\cdot\|_{\mathrm{u,v}}$ determines on $M_{\mathrm{u,v}}(\mathbb{S})$ is complete. In order to do this, it will be useful to know that if $\{\mathbf{M}_n\}_0^\infty \subseteq M_{\mathrm{u,v}}(\mathbb{S})$, then

$$(\mathbf{M})_{ij} = \lim_{n\to\infty} (\mathbf{M}_n)_{ij} \quad \text{for each } (i, j)$$
$$\implies \quad \|\mathbf{M}\|_{\mathrm{u,v}} \le \varliminf_{n\to\infty} \|\mathbf{M}_n\|_{\mathrm{u,v}}. \tag{4.1.3}$$

By Fatou's lemma, Theorem 7.1.10,

$$\sum_{j\in\mathbb{S}} |(\mathbf{M})_{ij}| \le \varliminf_{n\to\infty} \sum_{j\in\mathbb{S}} |(\mathbf{M}_n)_{ij}| \quad \text{for each } i \in \mathbb{S},$$

and so (4.1.3) is proved.

Knowing (4.1.3), the proof of completeness goes as follows. Assume that $\{\mathbf{M}_n\}_0^\infty \subseteq M_{\mathrm{u,v}}(\mathbb{S})$ is $\|\cdot\|_{\mathrm{u,v}}$-Cauchy convergent:

$$\lim_{m\to\infty} \sup_{n>m} \|\mathbf{M}_n - \mathbf{M}_m\|_{\mathrm{u,v}} = 0.$$

Obviously, for each $(i, j) \in \mathbb{S}^2$, $\{(\mathbf{M}_n)_{ij}\}_0^\infty$ is Cauchy convergent as a sequence in $\mathbb{R}$. Hence, there is a matrix $\mathbf{M}$ such that, for each (i, j), $(\mathbf{M})_{ij} = \lim_{n\to\infty}(\mathbf{M}_n)_{ij}$. Furthermore, by (4.1.3),

$$\|\mathbf{M} - \mathbf{M}_m\|_{\mathrm{u,v}} \le \varliminf_{n\to\infty} \|\mathbf{M}_n - \mathbf{M}_m\|_{\mathrm{u,v}} \le \sup_{n>m} \|\mathbf{M}_n - \mathbf{M}_m\|_{\mathrm{u,v}}.$$

Hence, $\|\mathbf{M} - \mathbf{M}_m\|_{\mathrm{u,v}} \longrightarrow 0$.

4.1.2 Abel Convergence

As I said in the introduction, we will begin with the weakest form of the convergence results at which we are aiming. That is, rather than attempting to prove the convergence of $(\mathbf{P}^n)_{ij}$ or even the Césaro means $\frac{1}{n}\sum_{m=0}^{n-1}(\mathbf{P}^m)_{ij}$ as $n \to \infty$, we will begin by studying the Abel sums $(1-s)\sum_{m=0}^{\infty} s^m (\mathbf{P}^m)_{ij}$ as $s \nearrow 1$.

I will say that a bounded sequence $\{x_n\}_0^\infty \subseteq \mathbb{R}$ is *Abel convergent* to x if

$$\lim_{s\nearrow 1}(1-s)\sum_{n=1}^{\infty} s^n x_n = x.$$

It should be clear that Abel convergence is weaker than (i.e., is implied by) ordinary convergence.[1] Indeed, if $x_n \longrightarrow x$, then, since $(1-s)\sum_0^\infty s^n = 1$, for any N:

$$\begin{aligned}\left|x-(1-s)\sum_{n=0}^{\infty} s^n x_n\right| &= (1-s)\left|\sum_{n=0}^{\infty} s^n (x-x_n)\right| \\ &\le (1-s)\sum_{n=0}^{\infty} s^n |x-x_n| \\ &\le N(1-s)\sup_{n\in\mathbb{N}}|x-x_n| + \sup_{n\ge N}|x-x_n|.\end{aligned}$$

Hence,

$$\varlimsup_{s\nearrow 1}\left|x-(1-s)\sum_{1}^{\infty} s^n x_n\right| \le \lim_{N\to\infty}\sup_{n\ge N}|x-x_n| = 0$$

if $x_n \longrightarrow x$. On the other hand, although $\{(-1)^n\}_1^\infty$ fails to converge to anything,

$$(1-s)\sum_{n=0}^{\infty} s^n(-1)^n = (1-s)\frac{1}{1+s} \longrightarrow 0 \quad \text{as } s\nearrow 1.$$

That is, Abel convergence does not in general imply ordinary convergence.

With the preceding in mind, we set

$$\mathbf{R}(s) = (1-s)\sum_{n=0}^{\infty} s^n \mathbf{P}^n \quad \text{for } s\in[0,1). \tag{4.1.4}$$

However, before accepting this definition, it is necessary to check that the above series converges. To this end, first note that, because $\mathbf{P}^n$ is a transition probability matrix for each $n\ge 0$, $\|\mathbf{P}^n\|_{\mathrm{u,v}} = 1$. Hence, for $0\le m<n$,

$$\left\|(1-s)\sum_{\ell=0}^{n} s^\ell \mathbf{P}^\ell - (1-s)\sum_{\ell=0}^{m} s^\ell \mathbf{P}^\ell\right\|_{\mathrm{u,v}} \le (1-s)\sum_{\ell=m+1}^{n} s^\ell \left\|\mathbf{P}^\ell\right\|_{\mathrm{u,v}} \le s^m,$$

and so, by the completeness proved above, the series in (4.1.4) converges with respect to the $\|\cdot\|_{\mathrm{u,v}}$-norm.

[1]In Exercise 4.2.1 below, it is shown that Abel convergence is also weaker than Césaro convergence.

Our goal here is to prove that

$$\lim_{s\nearrow 1}\big(\mathbf{R}(s)\big)_{ij}=\pi_{ij}\equiv\begin{cases}(\mathbb{E}[\rho_j|X_0=j])^{-1} & \text{if } i=j\\ \mathbb{P}(\rho_j<\infty|X_0=i)\pi_{jj} & \text{if } i\neq j,\end{cases} \tag{4.1.5}$$

and the key to our doing so lies in the *renewal equation*

$$\begin{gathered}\big(\mathbf{P}^n\big)_{ij}=\sum_{m=1}^{n}f(m)_{ij}\big(\mathbf{P}^{n-m}\big)_{jj}\quad \text{for } n\geq 1,\\ \text{where } f(m)_{ij}\equiv\mathbb{P}(\rho_j=m\,|\,X_0=i),\end{gathered} \tag{4.1.6}$$

which is an elementary application of (2.1.11):

$$\begin{aligned}\big(\mathbf{P}^n\big)_{ij}&=\sum_{m=1}^{n}\mathbb{P}(X_n=j\ \&\ \rho_j=m\,|\,X_0=i)\\ &=\sum_{m=1}^{n}\mathbb{P}(X_{n-m}=j\,|\,X_0=j)\mathbb{P}(\rho_j=m\,|\,X_0=i).\end{aligned}$$

Next, for $s\in[0,1)$, set

$$\hat{f}(s)_{ij}\equiv\sum_{m=1}^{\infty}s^m f(m)_{ij}=\mathbb{E}\big[s^{\rho_j}\,\big|\,X_0=i\big],$$

and, starting from (4.1.6), conclude that

$$\begin{aligned}\big(\mathbf{R}(s)\big)_{ij}&=(1-s)\delta_{i,j}+(1-s)\sum_{n=1}^{\infty}s^n\left(\sum_{m=1}^{n}f(m)_{ij}\big(\mathbf{P}^{n-m}\big)_{jj}\right)\\ &=(1-s)\delta_{i,j}+(1-s)\sum_{m=1}^{\infty}s^m f(m)_{ij}\left(\sum_{n=m}^{\infty}s^{n-m}\big(\mathbf{P}^{n-m}\big)_{jj}\right)\\ &=(1-s)\delta_{i,j}+\hat{f}(s)_{ij}\big(\mathbf{R}(s)\big)_{jj}.\end{aligned}$$

That is,

$$\big(\mathbf{R}(s)\big)_{ij}=(1-s)\delta_{i,j}+\hat{f}(s)_{ij}\big(\mathbf{R}(s)\big)_{jj}\quad \text{for } s\in[0,1). \tag{4.1.7}$$

Given (4.1.7), (4.1.5) is easy. Namely,

$$\big(\mathbf{R}(s)\big)_{jj}=\frac{1-s}{1-\hat{f}(s)_{jj}}.$$

Hence, if j is transient, and therefore $\hat{f}(1)_{jj} < 1$,

$$\pi_{jj} = \lim_{s \nearrow 1} \big(\mathbf{R}(s)\big)_{jj} = 0 = \frac{1}{\mathbb{E}[\rho_j | X_0 = j]}.$$

On the other hand, if j is recurrent, then, since

$$\frac{1 - s^m}{1 - s} = \sum_{\ell=0}^{m-1} s^\ell \nearrow m,$$

the monotone convergence theorem says that

$$\frac{1 - \hat{f}(s)_{jj}}{1 - s} = \sum_{m=1}^{\infty} \frac{1 - s^m}{1 - s} f(m)_{jj} \nearrow \sum_{m=1}^{\infty} m f(m)_{jj} = \mathbb{E}[\rho_j \mid X_0 = j]$$

as $s \nearrow 1$. At the same time, when $i \neq j$,

$$\big(\mathbf{R}(s)\big)_{ij} = \hat{f}(s)_{ij} \big(\mathbf{R}(s)\big)_{jj} \nearrow \mathbb{P}(\rho_j < \infty \mid X_0 = i)\pi_{jj}.$$

4.1.3 Structure of Stationary Distributions

We will say that a probability vector $\boldsymbol{\mu}$ is $\mathbf{P}$-*stationary* and will write $\boldsymbol{\mu} \in \mathrm{Stat}(\mathbf{P})$ if $\boldsymbol{\mu} = \boldsymbol{\mu}\mathbf{P}$. Obviously, if $\boldsymbol{\mu}$ is stationary, then $\boldsymbol{\mu} = \boldsymbol{\mu}\mathbf{R}(s)$ for each $s \in [0, 1)$. Hence, by (4.1.5) and Lebesgue's dominated convergence theorem,

$$\boldsymbol{\mu} \in \mathrm{Stat}(\mathbf{P}) \quad \Longrightarrow \quad (\boldsymbol{\mu})_j = \sum_i (\boldsymbol{\mu})_i \big(\mathbf{R}(s)\big)_{ij} \longrightarrow \sum_i (\boldsymbol{\mu})_i \pi_{ij} \quad \text{as } s \nearrow 1.$$

If j is transient, then $\pi_{ij} = 0$ for all i and therefore $(\boldsymbol{\mu})_j = 0$. If j is recurrent, then either i is transient, and, by Theorem 3.1.5, $(\boldsymbol{\mu})_i = 0$, or i is recurrent, in which case, by Theorem 3.1.2, either $\pi_{ij} = 0$ or $i \leftrightarrow j$ and $\pi_{ij} = \pi_{jj}$. Hence, in either case, we have that

$$\boldsymbol{\mu} \in \mathrm{Stat}(\mathbf{P}) \quad \Longrightarrow \quad (\boldsymbol{\mu})_j = \Big(\sum_{i \leftrightarrow j} (\boldsymbol{\mu})_i\Big)\pi_{jj}. \tag{4.1.8}$$

We show next that

$$\begin{aligned} &\pi_{jj} > 0 \quad \text{and} \quad C = [j] = \{i : i \leftrightarrow j\} \\ &\Longrightarrow \quad \boldsymbol{\pi}^C \in \mathrm{Stat}(\mathbf{P}) \quad \text{when } \big(\boldsymbol{\pi}^C\big)_i \equiv \mathbf{1}_C(i)\pi_{ii}. \end{aligned} \tag{4.1.9}$$

To do this, first note that $\pi_{jj} > 0$ only if j is recurrent. Thus, all $i \in C$ are recurrent and, for each $i \in C$ and $s \in (0, 1)$, $(\mathbf{R}(s))_{ik} > 0 \iff k \in C$. In particular, $(\boldsymbol{\pi}^C)_i =$

$\lim_{s\nearrow 1}(\mathbf{R}(s))_{ji}$ for all i, and therefore, by Fatou's lemma,

$$\sum_i (\boldsymbol{\pi}^C)_i \le \underline{\lim}_{s\nearrow 1}\sum_i (\mathbf{R}(s))_{ji} = 1.$$

Similarly, for any i,

$$(\boldsymbol{\pi}^C\mathbf{P})_i = \sum_{k\in C}\pi_{kk}(\mathbf{P})_{ki} \le \underline{\lim}_{s\nearrow 1}\sum_{k\in C}(\mathbf{R}(s))_{jk}(\mathbf{P})_{ki} = (\boldsymbol{\pi}^C)_i,$$

since

$$\sum_{k\in C}(\mathbf{R}(s))_{jk}(\mathbf{P})_{ki} = \frac{(\mathbf{R}(s)-(1-s)\mathbf{I})_{ji}}{s} \longrightarrow \pi_{ji} = (\boldsymbol{\pi}^C)_i \quad \text{as } s\nearrow 1.$$

But if strict inequality were to hold for some i, then, by Fubini's theorem, Theorem 7.1.15, we would have the contradiction

$$\sum_k (\boldsymbol{\pi}^C)_k = \sum_k (\boldsymbol{\pi}^C)_k\Big(\sum_i (\mathbf{P})_{ki}\Big) = \sum_i\Big(\sum_k (\boldsymbol{\pi}^C)_k(\mathbf{P})_{ki}\Big) < \sum_i (\boldsymbol{\pi}^C)_i.$$

Hence, we now know that $\boldsymbol{\pi}^C = \boldsymbol{\pi}^C\mathbf{P}$. Finally, to prove that $\boldsymbol{\pi}^C \in \mathrm{Stat}(\mathbf{P})$, we still have to check that $\sum_i (\boldsymbol{\pi}^C)_i = 1$. However, we have already checked that $\boldsymbol{\pi}^C = \boldsymbol{\pi}^C\mathbf{P}$, and so we know that $\boldsymbol{\pi}^C = \boldsymbol{\pi}^C\mathbf{R}(s)$. Therefore, since, as we already showed, $\sum_i (\boldsymbol{\pi}^C)_i \le 1$, Lebesgue's dominated convergence theorem justifies

$$0 < \pi_{jj} = \sum_i (\boldsymbol{\pi}^C)_i(\mathbf{R}(s))_{ij} \longrightarrow \Big(\sum_i (\boldsymbol{\pi}^C)_i\Big)\pi_{jj},$$

which is possible only if $\sum_i (\boldsymbol{\pi}^C)_i = 1$.

Before summarizing the preceding as a theorem, we need to recall that a subset A of a linear space is said to be a *convex set* if $(1-\theta)a + \theta a' \in A$ for all $a, a' \in A$ and $\theta \in [0,1]$ and that $b \in A$ is an *extreme point* of the convex set A if $b = (1-\theta)a + \theta a'$ for some $\theta \in (0,1)$ and $a, a' \in A$ implies that $b = a = a'$. In addition, we need the notion of positive recurrence. Namely, we say that $j \in \mathbb{S}$ is *positive recurrent* if $\mathbb{E}[\rho_j | X_0 = j] < \infty$. Obviously, only recurrent states can be positive recurrent. On the other hand, (1.1.13) together with (1.1.15) show that there can exist *null recurrent* states, those which are recurrent but not positive recurrent.

Theorem 4.1.10 Stat(**P**) *is a convex subset of* $\mathbb{R}^{\mathbb{S}}$. *Moreover,* Stat(**P**) $\neq \emptyset$ *if and only if there is at least one positive recurrent state* $j \in \mathbb{S}$. *In fact, for any* $\boldsymbol{\mu} \in \mathrm{Stat}(\mathbf{P})$, (4.1.8) *holds, and* $\boldsymbol{\mu}$ *is an extreme point in* Stat(**P**) *if and only if there is a communicating class* C *of positive recurrent states for which (cf.* (4.1.9)) $\boldsymbol{\mu} = \boldsymbol{\pi}^C$. *In particular,* $(\boldsymbol{\mu})_j = 0$ *for any transient state* j *and, for any recurrent state* j*, either* $(\boldsymbol{\mu})_i$ *is strictly positive or it is* 0 *simultaneously for all states* i*'s which communicate with* j.

Proof The only statements not already covered are the characterization of the extreme points of Stat(**P**) and the final assertion in the case when j is recurrent.

In view of (4.1.8), the final assertion when j is recurrent comes down to showing that if j is positive recurrent and $i \leftrightarrow j$, then i is positive recurrent. To this end, suppose that j is positive recurrent, set $C = [j]$, and let $i \in C$ be given. Then $\pi^C \mathbf{P}^n = \pi^C$ for all $n \geq 0$, and therefore, by choosing n so that $(\mathbf{P}^n)_{ji} > 0$, we see that $(\pi^C)_i \geq (\pi^C)_j (\mathbf{P}^n)_{ji} > 0$.

To handle the characterization of extreme points, first suppose that $\mu \neq \pi^C$ for any communicating class C of positive recurrent states. Then, by (4.1.8), there must exist non-communicating, positive recurrent states j and j' for which $(\mu)_j > 0 < (\mu)_{j'}$. But, again by (4.1.8), this means that we can write $\mu = \theta \pi^C + (1 - \theta)\nu$, where $C = [j]$, $\theta = \sum_{i \in C} (\mu)_i \in (0, 1)$, and $(\nu)_i$ equals 0 or $(1 - \theta)^{-1}(\mu)_i$ depending on whether i is or is not in C. Clearly $\nu \in \text{Stat}(\mathbf{P})$ and, because $\nu_{j'} > 0 = (\pi^C)_{j'}$, $\nu \neq \pi^C$. Hence, μ cannot be extreme. Equivalently, every extreme $\mu = \pi^C$ for some communicating class C of positive recurrent states.

Conversely, given a communicating class C of positive recurrent states, suppose that $\pi^C = (1 - \theta)\mu + \theta\nu$ for some $\theta \in (0, 1)$ and pair $(\mu, \nu) \in \text{Stat}(\mathbf{P})^2$. Then $(\mu)_i = 0$ for all $i \notin C$, and so, by (4.1.8), we see first that $\mu = \pi^C$ and then, as a consequence, that $\nu = \pi^C$ as well. □

4.1.4 A Digression About Moments of Return Times[2]

The last part of Theorem 4.1.10 shows that *positive recurrence is a communicating class property*. In particular, if a chain is irreducible, we are justified in saying it is positive recurrent if any one of its states is. See Exercise 4.2.3 below for a criterion which guarantees positive recurrence.

I have often wondered whether the fact that positive recurrence is a communicating class property admits a direct proof, one that does not involve stationary probabilities. More generally, I wanted to know whether finiteness of other moments of return times is also a communicating class property. Affirmative answers to both these questions were given to me by Daniel Jerison, whose ideas provide the foundation on which everything in this subsection rests. To simplify the notation, I will use $\mathbb{P}_i$ and $\mathbb{E}_i$ here to denote conditional probability and expectation given $X_0 = i$.

Let i and j be a pair of distinct, recurrent, communicating states, and assume that $\mathbb{E}_i[\rho_i^p] < \infty$ for some $p \in (0, \infty)$. In order to show that $\mathbb{E}_j[\rho_j^p] < \infty$, set $A_m = \{\rho_i^{(m)} < \rho_j < \rho_i^{(m+1)}\}$ for $m \geq 1$. Then, by part (d) of Exercise 6.6.2 below,

[2]This section may be safely skipped since the material here is not used elsewhere.

$$\begin{aligned}
\mathbb{E}_j\big[\rho_j^p, A_m\big] &\le 2^{(p-1)^+}\mathbb{E}_j\big[(\rho_j - \rho_i^{(m)})^p, A_m\big] + 2^{(p-1)^+}\mathbb{E}_j\big[(\rho_i^{(m)})^p, A_m\big] \\
&\le 2^{(p-1)^+}\mathbb{E}_j\big[(\rho_j - \rho_i^{(m)})^p, A_m\big] \\
&\quad + 2^{(p-1)^+}\mathbb{E}_j\left[\left(\sum_{\ell=1}^{m}(\rho_i^{(\ell)} - \rho_i^{(\ell-1)})\right)^p, A_m\right] \\
&\le 2^{(p-1)^+}\mathbb{E}_j\big[(\rho_j - \rho_i^{(m)})^p, A_m\big] + 4^{(p-1)^+}\mathbb{E}_j\big[\rho_i^p, A_m\big] \\
&\quad + \big(4(m-1)\big)^{(p-1)^+}\sum_{2\le\ell\le m}\mathbb{E}_j\big[(\rho_i^{(\ell)} - \rho_i^{(\ell-1)})^p, A_m\big].
\end{aligned}$$

By the Markov property,

$$\mathbb{E}_j\big[(\rho_j - \rho_i^{(m)})^p, A_m\big] = \mathbb{P}_j\big(\rho_j > \rho_i^{(m)}\big)\mathbb{E}_i\big[\rho_j^p, \rho_j < \rho_i\big] \le \mathbb{P}_j\big(\rho_j > \rho_i^{(m)}\big)\mathbb{E}_i\big[\rho_i^p\big]$$

and

$$\mathbb{E}_j\big[\rho_i^p, A_m\big] = \mathbb{E}_j\big[\rho_i^p, \rho_j > \rho_i\big]\mathbb{P}_i(A_{m-1}) \le \mathbb{P}_i(A_{m-1})\mathbb{E}_j\big[\rho_i^p\big].$$

Similarly, for $2 \le \ell \le m$,

$$\begin{aligned}
&\mathbb{E}_j\big[(\rho_i^{(\ell)} - \rho_i^{(\ell-1)})^p, A_m\big] \\
&= \mathbb{E}_j\big[(\rho_i^{(\ell)} - \rho_i^{(\ell-1)})^p, \rho_j > \rho_i^{(\ell)}\big]\mathbb{P}_i(A_{m-\ell}) \\
&= \mathbb{P}_j\big(\rho_j > \rho_i^{(\ell-1)}\big)\mathbb{E}_i\big[\rho_i^p, \rho_j > \rho_i\big]\mathbb{P}_j(A_{m-\ell}) = \mathbb{E}_i\big[\rho_i^p, \rho_j > \rho_i\big]\mathbb{P}_j(A_{m-1}) \\
&\le \mathbb{P}_j(A_{m-1})\mathbb{E}_i\big[\rho_i^p\big].
\end{aligned}$$

Hence,

$$\begin{aligned}
\mathbb{E}_j\big[\rho_j^p, A_m\big] &\le 2^{(p-1)^+}\mathbb{P}_j\big(\rho_j > \rho_i^{(m)}\big)\mathbb{E}_i\big[\rho_i^p\big] + 4^{(p-1)^+}\mathbb{P}_i(A_{m-1})\mathbb{E}_j\big[\rho_i^p\big] \\
&\quad + 4^{(p-1)^+}(m-1)^{p\vee 1}\mathbb{P}_j(A_{m-1})\mathbb{E}_i\big[\rho_i^p\big].
\end{aligned}$$

After summing over $m \in \mathbb{Z}^+$, we see that $\mathbb{E}_j[\rho_j^p, \rho_j > \rho_i]$ is dominated by

$$\begin{aligned}
&\sum_{m=1}^{\infty}\big(2^{(p-1)^+}\mathbb{P}_j\big(\rho_j > \rho_i^{(m)}\big) + 4^{(p-1)^+}(m-1)^{p\vee 1}\mathbb{P}_j(A_{m-1})\big)\mathbb{E}_i\big[\rho_i^p\big] \\
&\quad + 4^{(p-1)^+}\sum_{m=1}^{\infty}\mathbb{P}_j(A_{m-1})\mathbb{E}_j\big[\rho_i^p\big].
\end{aligned}$$

Obviously $\sum_{m=1}^{\infty}\mathbb{P}_j(A_{m-1}) = 1$. In the proof of Theorem 3.1.2 we showed that

$$\alpha \equiv \mathbb{P}_i(\rho_j > \rho_i) \in (0,1) \quad \text{and} \quad \mathbb{P}_i\big(\rho_j > \rho_i^{(m)}\big) = \mathbb{P}_i(\rho_j > \rho_i)^m = \alpha^m.$$

In addition, for $m \geq 1$.

$$\mathbb{P}_j\big(\rho_j > \rho_i^{(m)}\big) = \mathbb{P}_j(\rho_j > \rho_i)\mathbb{P}_i\big(\rho_j > \rho_i^{(m-1)}\big) \leq \alpha^{m-1}.$$

Thus, since $\mathbb{P}_j(A_{m-1}) \leq \mathbb{P}_j(\rho_j > \rho_i^{(m-1)})$, we now know that there is a $C < \infty$ such that

$$\mathbb{E}_j\big[\rho_j^p, \rho_j > \rho_i\big] \leq C\big(\mathbb{E}_i\big[\rho_i^p\big] + \mathbb{E}_j\big[\rho_i^p\big]\big)$$

and therefore that

$$\mathbb{E}_j\big[\rho_j^p\big] \leq C\mathbb{E}_i\big[\rho_i^p\big] + (1+C)\mathbb{E}_j\big[\rho_i^p\big].$$

Finally, observe that

$$\mathbb{E}_i\big[\rho_i^p\big] \geq \mathbb{E}_i\big[(\rho_i - \rho_j)^p, \rho_j < \rho_i\big] = \mathbb{E}_j\big[\rho_i^p\big]\mathbb{P}_i(\rho_j < \rho_i),$$

and, as shown in the proof of Theorem 3.1.2, $\mathbb{P}_i(\rho_j < \rho_i) \in (0, 1)$. Hence, we have now shown the $\mathbb{E}_j[\rho_j^p] < \infty$.

4.1.5 A Small Improvement

Returning to our program, the next step will be the replacement of Abel convergence by Césaro convergence. That is, we will show that (4.1.5) can be replaced by

$$\lim_{n\to\infty} (\mathbf{A}_n)_{ij} = \pi_{ij}, \quad \text{where } \mathbf{A}_n = \frac{1}{n}\sum_{m=0}^{n-1} \mathbf{P}^m. \tag{4.1.11}$$

As is shown in Exercise 4.2.1, Césaro convergence does not, in general, follow from Abel convergence. In fact, general results which say when Abel convergence implies Césaro convergence can be quite delicate and, because the original one was proved by a man named Tauber, they are known as Tauberian theorems. Fortunately, the Tauberian theorem required here is quite straight-forward.

A key role in our proof will be played by the following easy estimate:

$$\begin{aligned} &\{a_m\}_0^\infty \subseteq [0, 1] \;\&\; A_n = \frac{1}{n}\sum_0^{n-1} a_\ell \\ &\implies \quad |A_n - A_{n-m}| \leq \frac{m}{n} \quad \text{for } 0 \leq m < n. \end{aligned} \tag{4.1.12}$$

The proof is:

$$A_n - A_{n-m} = \frac{1}{n}\sum_{\ell=n-m}^{n-1} a_\ell - \frac{m}{n(n-m)}\sum_{\ell=0}^{n-m-1} a_\ell \begin{cases} \geq -\frac{m}{n} \\ \leq \frac{m}{n}. \end{cases}$$

Lemma 4.1.13 *For all* (i, j), $\overline{\lim}_{n\to\infty}(\mathbf{A}_n)_{ij} \le e\pi_{jj}$. *In addition, for any* j *and any subsequence* $\{n_\ell : \ell \ge 0\} \subseteq \mathbb{N}$,

$$\lim_{\ell\to\infty}(\mathbf{A}_{n_\ell})_{jj} = \alpha \quad \Longrightarrow \quad \lim_{\ell\to\infty}(\mathbf{A}_{n_\ell})_{ij} = \mathbb{P}(\rho_j < \infty \,|\, X_0 = i)\alpha \quad \textit{for all } i.$$

Proof To prove the first part, observe that

$$(\mathbf{A}_n)_{ij} \le \frac{1}{n}\left(1-\frac{1}{n}\right)^{-n}\sum_{m=0}^{n-1}\left(1-\frac{1}{n}\right)^m (\mathbf{P}^m)_{ij} \le \left(1-\frac{1}{n}\right)^{-n}\left(\mathbf{R}\left(1-\frac{1}{n}\right)\right)_{ij},$$

which, together with (1.2.10) and (4.1.5), shows that $\overline{\lim}_{n\to\infty}(\mathbf{A}_n)_{ij} \le e\pi_{ij} \le e\pi_{jj}$.
To handle the second part, use (4.1.6) to arrive at

$$(\mathbf{A}_n)_{ij} = \sum_{m=1}^{n-1} f(m)_{ij}\left(1-\frac{m}{n}\right)(\mathbf{A}_{n-m})_{jj} \quad \text{for } i \ne j.$$

Hence,

$$\begin{aligned}\left|(\mathbf{A}_n)_{ij} - \mathbb{P}(\rho_j < n \,|\, X_0 = i)\alpha\right| &\le \sum_{m=1}^{n-1} f(m)_{ij}\left(\frac{m}{n} + \left|(\mathbf{A}_{n-m})_{jj} - \alpha\right|\right)\\ &\le 2\sum_{m=1}^{n-1}\frac{m}{n}f(m)_{ij} + \left|(\mathbf{A}_n)_{jj} - \alpha\right|,\end{aligned}$$

where, in the second inequality, I have used (4.1.12) plus $\sum_{m=1}^{\infty} f(m)_{ij} \le 1$. Finally, by Lebesgue's dominated convergence theorem, $\sum_0^{n-1}\frac{m}{n}f(m)_{ij}$ tends to 0 as $n\to\infty$, and therefore, by applying the above with $n = n_\ell$ and letting $\ell\to\infty$, we get the desired conclusion. □

We can now complete the proof of (4.1.11). If $\pi_{jj} = 0$, then the first part of Lemma 4.1.13 guarantees that $\lim_{n\to\infty}(\mathbf{A}_n)_{ij} = 0 = \pi_{ij}$ for all i. Thus, assume that $\pi_{jj} > 0$. In this case Theorem 4.1.10 says that j must be positive recurrent and $\boldsymbol{\pi}^C \in \mathrm{Stat}(\mathbf{P})$ when $C = [j]$. In particular, $\pi_{jj} = \sum_{i\in C}(\boldsymbol{\pi}^C)_i(\mathbf{A}_n)_{ij}$. At the same time, if $\alpha^+ = \overline{\lim}_{n\to\infty}(\mathbf{A}_n)_{jj}$ and the subsequence $\{n_\ell : \ell \ge 0\}$ is chosen so that $(\mathbf{A}_{n_\ell})_{jj} \longrightarrow \alpha^+$, then, by the second part of Lemma 4.1.13 and Corollary 3.1.4,

$$i \in C \quad \Longrightarrow \quad \lim_{\ell\to\infty}(\mathbf{A}_{n_\ell})_{ij} = \alpha^+.$$

Hence, after putting these two remarks together, we arrive at

$$\pi_{jj} = \lim_{\ell\to\infty}\sum_{i\in C}(\boldsymbol{\pi}^C)_i(\mathbf{A}_{n_\ell})_{ij} = \alpha^+\sum_{i\in C}(\boldsymbol{\pi}^C)_i = \alpha^+.$$

Similarly, if $\alpha^- = \underline{\lim}_{n\to\infty}(\mathbf{A}_n)_{jj}$, we can show that $\alpha^- = \pi_{jj}$, and so we now know that $\pi_{jj} > 0 \implies \lim_{n\to\infty}(\mathbf{A}_n)_{jj} = \pi_{jj}$, which, after another application of the second part of Lemma 4.1.13, means that we have proved (4.1.11).

4.1.6 The Mean Ergodic Theorem Again

Just as we were able to use Theorem 2.2.5 in Sect. 2.3.1 to prove Theorem 2.3.4, so here we can use (4.1.11) to prove the following version of the mean ergodic theorem.

Theorem 4.1.14 *Let C a communicating class of positive recurrent states. If $\mathbb{P}(X_0 \in C) = 1$, then*

$$\lim_{n\to\infty} \mathbb{E}\left[\left(\frac{1}{n}\sum_{m=0}^{n-1} \mathbf{1}_{\{j\}}(X_m) - \pi_{jj}\right)^2\right] = 0.$$

(See Exercises 4.2.10 *and* 4.2.12 *below for a more refined statement.)*

Proof Since $\mathbb{P}(X_m \in C \text{ for all } m \in \mathbb{N}) = 1$, without loss in generality we may and will assume that C is the whole state space. In keeping with this assumption, we will set $\boldsymbol{\pi} = \boldsymbol{\pi}^C$.

Next note that if $\mu_i = \mathbb{P}(X_0 = i)$, then

$$\mathbb{E}\left[\left(\frac{1}{n}\sum_{m=0}^{n-1} \mathbf{1}_{\{j\}}(X_m) - \pi_{jj}\right)^2\right]$$
$$= \sum_{i\in\mathbb{S}} \mu_i \mathbb{E}\left[\left(\frac{1}{n}\sum_{m=0}^{n-1} \mathbf{1}_{\{j\}}(X_m) - \pi_{jj}\right)^2 \,\middle|\, X_0 = i\right],$$

and so it suffices to prove that

$$\lim_{n\to\infty} \mathbb{E}\left[\left(\frac{1}{n}\sum_{m=0}^{n-1} \mathbf{1}_{\{j\}}(X_m) - \pi_{jj}\right)^2 \,\middle|\, X_0 = i\right] = 0 \quad \text{for each } i \in \mathbb{S}.$$

Further, because $\pi_{ii} > 0$ for all $i \in \mathbb{S}$ and

$$\mathbb{E}\left[\left(\frac{1}{n}\sum_{m=0}^{n-1} \mathbf{1}_{\{j\}}(X_m) - \pi_{jj}\right)^2 \,\middle|\, X_0 = i\right]$$
$$\le \frac{1}{\pi_{ii}} \sum_{k\in\mathbb{S}} \pi_{kk} \mathbb{E}\left[\left(\frac{1}{n}\sum_{m=0}^{n-1} \mathbf{1}_{\{j\}}(X_m) - \pi_{jj}\right)^2 \,\middle|\, X_0 = k\right],$$

it is enough to prove the result when $\boldsymbol{\pi}$ is the initial distribution of the Markov chain. Hence, from now on, we will be making this assumption along with $C = \mathbb{S}$.

Now let $\mathbf{f}$ the column vector whose ith component is $\mathbf{1}_{\{j\}}(i) - \pi_{jj}$. Then, just as in the proof of Theorem 2.3.4,

$$\mathbb{E}\left[\left(\frac{1}{n}\sum_{m=0}^{n-1}\mathbf{1}_{\{j\}}(X_m) - \pi_{jj}\right)^2\right] \le \frac{2}{n^2}\sum_{k=0}^{n-1}(n-k)\mathbb{E}\big[(\mathbf{f})_{X_k}(\mathbf{A}_{n-k}\mathbf{f})_{X_k}\big].$$

Since $\boldsymbol{\pi} \in \mathrm{Stat}(\mathbf{P})$,

$$\mathbb{E}\big[(\mathbf{f})_{X_k}(\mathbf{A}_{n-k}\mathbf{f})_{X_k}\big] = \boldsymbol{\pi}(f\mathbf{A}_{n-k}\mathbf{f}),$$

and therefore the preceding becomes

$$\mathbb{E}\left[\left(\frac{1}{n}\sum_{m=0}^{n-1}\mathbf{1}_{\{j\}}(X_m) - \pi_{jj}\right)^2\right] \le \frac{2}{n^2}\sum_{m=1}^{n} m\boldsymbol{\pi}(f\mathbf{A}_m\mathbf{f}),$$

where $(f\mathbf{A}_m\mathbf{f})_i \equiv (\mathbf{f})_i(\mathbf{A}_m\mathbf{f})_i$.

Finally, because $\pi_{ij} = \pi_{jj}$ for all $i \in \mathbb{S}$, (4.1.11) and Lebesgue's dominated convergence theorem say that, for each $\epsilon > 0$, there exists an $N_\epsilon \in \mathbb{Z}^+$ such that

$$\big|\boldsymbol{\pi}(f\mathbf{A}_n\mathbf{f})\big| \le \sum_i (\boldsymbol{\pi})_i\big|(\mathbf{A}_n)_{ij} - \pi_{jj}\big| < \epsilon \quad \text{for all } n \ge N_\epsilon.$$

Hence, we find that

$$\begin{aligned}
&\overline{\lim_{n\to\infty}}\,\mathbb{E}\left[\left(\frac{1}{n}\sum_{m=0}^{n-1}\mathbf{1}_{\{j\}}(X_m) - \pi_{jj}\right)^2\right] \\
&\quad \le \overline{\lim_{n\to\infty}}\,\frac{2}{n^2}\sum_{m=1}^{N_\epsilon} m\big|\boldsymbol{\pi}(f\mathbf{A}_m\mathbf{f})\big| + \overline{\lim_{n\to\infty}}\,\frac{2\epsilon}{n^2}\sum_{m=N_\epsilon+1}^{n} m \le \epsilon.
\end{aligned}$$

□

4.1.7 A Refinement in the Aperiodic Case

Our goal in this subsection is to prove that (cf. (4.1.5))

$$\text{If } j \text{ is transient or aperiodic,} \quad \text{then } \lim_{n\to\infty}\big(\mathbf{P}^n\big)_{ij} = \pi_{ij} \quad \text{for all } i \in \mathbb{S}. \tag{4.1.15}$$

Of course, the case when j is transient requires very little effort. Indeed, by (2.3.7), we have that

$$j \text{ transient} \quad \Longrightarrow \quad \sum_{n=0}^{\infty}\big(\mathbf{P}^n\big)_{ij} \le \mathbb{E}[T_j|X_0 = j] < \infty,$$

and therefore that $\lim_{n\to\infty}(\mathbf{P}^n)_{ij} = 0$. At the same time, because

$$\mathbb{P}(\rho_j = \infty | X_0 = j) > 0 \quad \text{when } j \text{ is transient,}$$

$\pi_{ij} \le \pi_{jj} = 0$. Thus, from now on, we will concentrate on the case when j is recurrent and aperiodic.

Our first step is the observation that (cf. Sect. 2.3.2)

$$\begin{aligned} &\text{if } j \text{ is aperiodic, then there exists an } N \in \mathbb{Z}^+ \text{ such that} \\ &\max_{1\le m\le n} \mathbb{P}\big(\rho_j^{(m)} = n \,\big|\, X_0 = j\big) > 0 \quad \text{for all } n \ge N. \end{aligned} \tag{4.1.16}$$

To check this, use (3.1.15) to produce an $N \in \mathbb{Z}^+$ such that $(\mathbf{P}^n)_{jj} > 0$ for all $n \ge N$. Then, since

$$\big(\mathbf{P}^n\big)_{jj} = \sum_{m=1}^{n} \mathbb{P}\big(\rho_j^{(m)} = n \,\big|\, X_0 = j\big) \quad \text{for } n \ge 1,$$

(4.1.16) is clear.

The second, and key, step is contained in the following lemma.

Lemma 4.1.17 *Assume that j is aperiodic and recurrent, and set $\alpha_j^- = \underline{\lim}_{n\to\infty}(\mathbf{P}^n)_{jj}$ and $\alpha_j^+ = \overline{\lim}_{n\to\infty}(\mathbf{P}^n)_{jj}$. Then there exist subsequences $\{n_\ell^- : \ell \ge 1\}$ and $\{n_\ell^+ : \ell \ge 1\}$ such that*

$$\alpha_j^\pm = \lim_{\ell\to\infty} \big(\mathbf{P}^{n_\ell^\pm - r}\big)_{jj} \quad \textit{for all } r \ge 0.$$

Proof Choose a subsequence $\{n_\ell : \ell \ge 1\}$ so that $(\mathbf{P}^{n_\ell})_{jj} \longrightarrow \alpha_j^+$, and, using (4.1.16), choose $N \ge 1$ so that $\max_{1\le m\le n} \mathbb{P}(\rho_j^{(m)} = n | X_0 = j) > 0$ for all $n \ge N$. Given $r \ge N$, choose $1 \le m \le r$ so that $\delta \equiv \mathbb{P}(\rho_j^{(m)} = r | X_0 = j) > 0$. Now for any $M \in \mathbb{Z}^+$, observe that, when $n_\ell \ge M + r$, $(\mathbf{P}^{n_\ell})_{jj}$ is equal to

$$\begin{aligned} &\mathbb{P}\big(X_{n_\ell} = j \;\&\; \rho_j^{(m)} = r \,\big|\, X_0 = j\big) + \mathbb{P}\big(X_{n_\ell} = j \;\&\; \rho_j^{(m)} \ne r \,\big|\, X_0 = j\big) \\ &\quad = \delta\big(\mathbf{P}^{n_\ell - r}\big)_{jj} + \mathbb{P}\big(X_{n_\ell} = j \;\&\; n_\ell - M \ge \rho_j^{(m)} \ne r \,\big|\, X_0 = j\big) \\ &\qquad + \mathbb{P}\big(X_{n_\ell} = j \;\&\; \rho_j^{(m)} > n_\ell - M \,\big|\, X_0 = j\big). \end{aligned}$$

Furthermore,

$$\begin{aligned} &\mathbb{P}\big(X_{n_\ell} = j \;\&\; n_\ell - M \ge \rho_j^{(m)} \ne r \,\big|\, X_0 = j\big) \\ &\quad = \sum_{\substack{k=1 \\ k\ne r}}^{n_\ell - M} \mathbb{P}\big(\rho_j^{(m)} = k \,\big|\, X_0 = j\big)\big(\mathbf{P}^{n_\ell - k}\big)_{jj} \le (1-\delta) \sup_{n\ge M} \big(\mathbf{P}^n\big)_{jj}, \end{aligned}$$

while

$$\mathbb{P}\big(X_{n_\ell} = j \;\&\; \rho_j^{(m)} > n_\ell - M \,\big|\, X_0 = j\big) \le \mathbb{P}\big(\rho_j^{(m)} > n_\ell - M \,\big|\, X_0 = j\big).$$

Hence, since j is recurrent and therefore $\mathbb{P}(\rho_j^{(m)} < \infty | X_0 = j) = 1$, we get

$$\alpha_j^+ \le \delta \varliminf_{\ell\to\infty} \big(\mathbf{P}^{n_\ell - r}\big)_{jj} + (1-\delta) \sup_{n\ge M} \big(\mathbf{P}^n\big)_{jj}$$

after letting $\ell \to \infty$. Since this is true for all $M \ge 1$, it leads to

$$\alpha_j^+ \le \delta \varliminf_{\ell\to\infty} \big(\mathbf{P}^{n_\ell - r}\big)_{jj} + (1-\delta)\alpha_j^+,$$

which implies $\varliminf_{\ell\to\infty}(\mathbf{P}^{n_\ell - r})_{jj} \ge \alpha_{jj}^+$. But obviously $\varlimsup_{\ell\to\infty}(\mathbf{P}^{n_\ell - r})_{jj} \le \alpha_j^+$, and so we have shown that $\lim_{\ell\to\infty}(\mathbf{P}^{n_\ell - r})_{jj} = \alpha_j^+$ for all $r \ge N$. Now choose L so that $n_L \ge N$, take $n_\ell^+ = n_{\ell+L} - N$, and conclude that $\lim_{\ell\to\infty}(\mathbf{P}^{n_\ell^+ - r})_{jj} = \alpha_j^+$ for all $r \ge 0$.

The construction of $\{n_\ell^- : \ell \ge 1\}$ is essentially the same and is left as an exercise. □

Lemma 4.1.18 *If j is aperiodic and recurrent, then $\varlimsup_{n\to\infty}(\mathbf{P}^n)_{jj} \le \pi_{jj}$. Furthermore, if the subsequences $\{n_\ell^\pm : \ell \ge 1\}$ are the ones described in Lemma* 4.1.17, *then $\lim_{\ell\to\infty}(\mathbf{P}^{n_\ell^\pm})_{ij} = \alpha_j^\pm$ for any i with $i \leftrightarrow j$.*

Proof To prove the second assertion, simply note that, by Lemma 4.1.17 and Lebesgue's dominated convergence theorem,

$$\big(\mathbf{P}^{n_\ell^\pm}\big)_{ij} = \sum_{r=1}^{n_\ell^\pm} \mathbb{P}(\rho_j = r | X_0 = i)\big(\mathbf{P}^{n_\ell^\pm - r}\big)_{jj} \longrightarrow \mathbb{P}(\rho_j < \infty | X_0 = i)\alpha_j^\pm.$$

Turning to the first assertion, we again use the result in Lemma 4.1.17 to obtain

$$\alpha_j^+ \sum_{r=1}^{N} \mathbb{P}(\rho_j \ge r | X_0 = j) = \lim_{\ell\to\infty} \sum_{r=1}^{N} \mathbb{P}(\rho_j \ge r \,|\, X_0 = j)\big(\mathbf{P}^{n_\ell^+ - r}\big)_{jj}$$

for all $N \ge 1$. Thus, if we show that

$$\sum_{r=1}^{N} \mathbb{P}(\rho_j \ge r \,|\, X_0 = j)\big(\mathbf{P}^{n-r}\big)_{jj} \le 1 \quad \text{for } n \ge N \ge 1, \tag{$*$}$$

then we will know that

$$\alpha_j^+ \mathbb{E}[\rho_j \,|\, X_0 = j] = \alpha_j^+ \sum_{r=1}^{\infty} \mathbb{P}(\rho_j \ge r \,|\, X_0 = j) \le 1,$$

which is equivalent to the first assertion.

To prove (∗), note that, for any $n \geq 1$,

$$
\begin{aligned}
\left(\mathbf{P}^n\right)_{jj} &= \sum_{r=1}^{n} \mathbb{P}(\rho_j = r \mid X_0 = j)\left(\mathbf{P}^{n-r}\right)_{jj} \\
&= \sum_{r=1}^{n} \mathbb{P}(\rho_j \geq r \mid X_0 = j)\left(\mathbf{P}^{n-r}\right)_{jj} \\
&\quad - \sum_{r=1}^{n} \mathbb{P}(\rho_j \geq r+1 \mid X_0 = j)\left(\mathbf{P}^{n-r}\right)_{jj} \\
&= \sum_{r=1}^{n} \mathbb{P}(\rho_j \geq r \mid X_0 = j)\left(\mathbf{P}^{n-r}\right)_{jj} \\
&\quad - \sum_{r=2}^{n+1} \mathbb{P}(\rho_j \geq r \mid X_0 = j)\left(\mathbf{P}^{n+1-r}\right)_{jj},
\end{aligned}
$$

and so, since $\mathbb{P}(\rho_j \geq 1 | X_0 = j) = 1$,

$$
\sum_{r=1}^{n+1} \mathbb{P}(\rho_j \geq r \mid X_0 = j)\left(\mathbf{P}^{n+1-r}\right)_{jj} = \sum_{r=1}^{n} \mathbb{P}(\rho_j \geq r \mid X_0 = j)\left(\mathbf{P}^{n-r}\right)_{jj}
$$

for all $n \geq 1$. But $\sum_{r=1}^{n} \mathbb{P}(\rho_j \geq r | X_0 = j)(\mathbf{P}^{n-r})_{jj} = 1$ when $n = 1$, and so we have now proved that

$$
\sum_{r=1}^{N} \mathbb{P}(\rho_j \geq r \mid X_0 = j)\left(\mathbf{P}^{n-r}\right)_{jj} \leq \sum_{r=1}^{n} \mathbb{P}(\rho_j \geq r \mid X_0 = j)\left(\mathbf{P}^{n-r}\right)_{jj} = 1
$$

for all $n \geq N \geq 1$. □

Remark Here is a more conceptual way to prove that

$$
\sum_{r=1}^{n} \mathbb{P}(\rho_j \geq r \mid X_0 = j)\left(\mathbf{P}^{n-r}\right)_{jj} = 1.
$$

Take $\rho_j^{(0)} = 0$ and, for $m \geq 1$, $\rho_j^{(m)}$ to be the time of the mth return to j. In addition, set $T_j^{(n-1)} = \sum_{\ell=0}^{n-1} \mathbf{1}_{\{j\}}(X_\ell)$. Then

$$
X_0 = j \quad \Longrightarrow \quad \left\{T_j^{(n-1)} = m+1\right\} = \left\{\rho_j^{(m)} < n \leq \rho_j^{(m+1)}\right\}.
$$

Hence,

$$
\begin{aligned}
&\sum_{r=1}^{n} \mathbb{P}(\rho_j \geq r \mid X_0 = j)\big(\mathbf{P}^{n-r}\big)_{jj} \\
&\quad = \sum_{r=0}^{n-1} \mathbb{P}(\rho_j \geq n-r \mid X_0 = j)\big(\mathbf{P}^{r}\big)_{jj} \\
&\quad = \sum_{r=0}^{n-1}\sum_{m=0}^{r} \mathbb{P}(\rho_j \geq n-r \mid X_0 = j)\mathbb{P}\big(\rho_j^{(m)} = r \,\big|\, X_0 = j\big) \\
&\quad = \sum_{r=0}^{n-1}\sum_{m=0}^{r} \mathbb{P}\big(\rho_j^{(m+1)} \geq n \;\&\; \rho_j^{(m)} = r \,\big|\, X_0 = j\big) \\
&\quad = \sum_{m=0}^{n-1}\sum_{r=m}^{n-1} \mathbb{P}\big(\rho_j^{(m+1)} \geq n \;\&\; \rho_j^{(m)} = r \,\big|\, X_0 = j\big) \\
&\quad = \sum_{m=0}^{n-1} \mathbb{P}\big(\rho_j^{(m)} < n \leq \rho_j^{(m+1)} \,\big|\, X_0 = j\big) \\
&\quad = \sum_{m=0}^{n-1} \mathbb{P}\big(T_j^{(n-1)} = m+1\big) = \mathbb{P}\big(T_n^{(n-1)} \leq n \,\big|\, X_0 = j\big) = 1.
\end{aligned}
$$

We can now complete the proof of (4.1.15) when j is recurrent and aperiodic. By the first part of Lemma 4.1.18, we know that $\lim_{n\to\infty}(\mathbf{P}^n)_{jj} = 0$ if $\pi_{jj} = 0$. Thus, if $\pi_{jj} = 0$, then, by Lebesgue's dominated convergence theorem, for any i,

$$
\lim_{n\to\infty}\big(\mathbf{P}^n\big)_{ij} = \lim_{n\to\infty}\sum_{r=1}^{n} \mathbb{P}(\rho_j = r | X_0 = i)\big(\mathbf{P}^{n-r}\big)_{jj} = 0 = \pi_{ij}.
$$

In order to handle the case when j is positive recurrent, set $C = [j]$, and take $\boldsymbol{\pi}^C$ accordingly. Then, $\boldsymbol{\pi}^C \in \mathrm{Stat}(\mathbf{P})$. In particular, by the last part of Lemma 4.1.18 and Lebesgue's dominated convergence theorem,

$$
\pi_{jj} = \sum_{i\in C}\big(\boldsymbol{\pi}^C\big)_i\big(\mathbf{P}^{n_\ell^\pm}\big)_{ij} \longrightarrow \alpha_j^\pm \sum_{i\in C}\big(\boldsymbol{\pi}^C\big)_i = \alpha_j^\pm,
$$

and so $(\mathbf{P}^n)_{jj} \longrightarrow \pi_{jj}$. Finally, if $i \neq j$, then, again by the Lebesgue's theorem,

$$
\big(\mathbf{P}^n\big)_{ij} = \sum_{r=1}^{n} \mathbb{P}(\rho_j = r | X_0 = i)\big(\mathbf{P}^{n-r}\big)_{jj} \longrightarrow \mathbb{P}(\rho_j < \infty | X_0 = i)\pi_{jj} = \pi_{ij}.
$$

4.1.8 Periodic Structure

The preceding result allows us to give a finer analysis even in the case when the period is not 1. Consider a Markov chain with transition probability matrix $\mathbf{P}$ which is irreducible and recurrent on $\mathbb{S}$, and assume that its period is $d \geq 2$. The basic result in this subsection is that there exists a partition of $\mathbb{S}$ into subsets $\mathbb{S}_r$, $0 \leq r < d$, with the properties that

(1) $(\mathbf{P}^{md+r})_{jk} > 0 \implies r(k) - r(j) = r \bmod d$,
(2) $r(k) - r(j) = r \bmod d \implies (\mathbf{P}^{md+r})_{jk} > 0 < (\mathbf{P}^{md-r})_{kj}$ for all sufficiently large $m \geq 1$,
(3) for each $0 \leq r < d$, the restriction of $\mathbf{P}^d$ to $\mathbb{S}_r$ is an aperiodic, recurrent, irreducible transition probability matrix,

where I have used $r(j)$ to denote the $0 \leq r < d$ for which $j \in \mathbb{S}_r$.

To prove that this decomposition exists, we begin by noting that, for any $0 \leq r < d$,

$$\exists m \geq 0 \quad \left(\mathbf{P}^{md+r}\right)_{ij} > 0 \quad \implies \quad \exists n \geq 1 \quad \left(\mathbf{P}^{nd-r}\right)_{ji} > 0. \qquad (*)$$

Indeed, by irreducibility and the Euclidean algorithm, we know that there exists an $m' \geq 0$ and $0 \leq r' < d$ such that $(\mathbf{P}^{m'd+r'})_{ji} > 0$. Furthermore, by (3.1.14), $(\mathbf{P}^{(m'+m'')d+r'})_{ji} \geq (\mathbf{P}^{m'd+r'})_{ji}(\mathbf{P}^{m''d})_{ii} > 0$ for all sufficiently large m'''s, and so we may and will assume that $m' \geq 1$. But then $(\mathbf{P}^{(m+m')d+(r+r')})_{ii} > 0$, and so $d|(r + r')$, which, because $0 \leq r, r' < d$, means that $r' = 0$ if $r = 0$ and that $r' = d - r$ if $r \geq 1$.

Starting from $(*)$ it is easy to see that, for each pair $(i, j) \in \mathbb{S}^2$, there is a unique $0 \leq r < d$ such that $(\mathbf{P}^{md+r})_{ij} > 0$ for some $m \geq 0$. Namely, suppose that $(\mathbf{P}^{md+r})_{ij} > 0 < (\mathbf{P}^{m'd+r'})_{ij}$ for some $m, m' \in \mathbb{N}$ and $0 \leq r, r' < d$. Then, by $(*)$, there exists an $n \geq 1$ such that $(\mathbf{P}^{nd-r})_{ji} > 0$ and therefore $(\mathbf{P}^{(m'+n)d+(r'-r)})_{ii} > 0$. Since this means that $d|(r' - r)$, we have proved that $r = r'$.

Now, let i_0 be a fixed reference point in $\mathbb{S}$, and, for each $0 \leq r < d$, define $\mathbb{S}_r$ to be the set of j such that there exists an $m \geq 0$ for which $(\mathbf{P}^{md+r})_{i_0 j} > 0$. By the preceding, we know that the $\mathbb{S}_r$'s are mutually disjoint. In addition, by irreducibility and the Euclidean algorithm, $\mathbb{S} = \bigcup_{r=0}^{d-1} \mathbb{S}_r$. Turning to the proof of property (1), use $(*)$ to choose $n \geq 0$ and $n' \geq 1$ so that $(\mathbf{P}^{nd+r(j)})_{i_0 j} > 0 < (\mathbf{P}^{n'd-r(k)})_{k i_0}$. Then $(\mathbf{P}^{(n+m+n')d+(r(j)+r-r(k))})_{i_0 i_0} > 0$, and so $d|(r(j) + r - r(k))$. Equivalently, $r(k) - r(j) = r \bmod d$. Conversely, if $r(k) - r(j) = r \bmod d$, choose $n \geq 0$ and $n' \geq 1$ so that $(\mathbf{P}^{nd+r(k)})_{i_0 k} > 0$ and $(\mathbf{P}^{n'd-r(j)})_{j i_0} > 0$. Then $(\mathbf{P}^{(n+m+n')d+r})_{jk} > 0$ for any $m \geq 1$ satisfying $(\mathbf{P}^{md})_{i_0 i_0} > 0$. Since, by (3.1.14), $(\mathbf{P}^{md})_{i_0 i_0} > 0$ for all sufficiently large m's, this completes the left hand inequality in (2), and the right hand inequality in (2) is proved in the same way. Finally, to check (3), note that, from (1), the restriction of $\mathbf{P}^d$ to $\mathbb{S}_r$ is a transition probability matrix, and by (2), it is both irreducible and aperiodic.

The existence of such a partition has several interesting consequences. In the first place, it says that the chain proceeds through the state space in a cyclic fashion: if it starts from i, then after n steps it is in $\mathbb{S}_{r(i)+n}$, where the addition in the subscript should be interpreted modulo d. In fact, with this convention for addition, we have that

$$\mathbf{P}^n \mathbf{1}_{\mathbb{S}_r} = \mathbf{1}_{\mathbb{S}_{r+n}}. \tag{4.1.19}$$

To see this, simply observe that, on the one hand, $\mathbf{P}^n \mathbf{1}_{\mathbb{S}_r}(i) = 0$ unless $i \in \mathbb{S}_{r+n}$, while, on the other hand, $\sum_{r'=0}^{d-1} \mathbf{P}^n \mathbf{1}_{\mathbb{S}_{r'}} = \mathbf{1}$. Hence, $i \notin \mathbb{S}_{r+n} \implies (\mathbf{P}^n \mathbf{1}_{\mathbb{S}_r})_i = 0$, whereas $i \in \mathbb{S}_{r+n} \implies 1 = \sum_{r'=0}^{d-1} (\mathbf{P}^n \mathbf{1}_{\mathbb{S}_{r'}})_i = (\mathbf{P}^n \mathbf{1}_{\mathbb{S}_{r+n}})_i$.

Secondly, because, for each $0 \le r < d$, the restriction of $\mathbf{P}^d$ to $\mathbb{S}_r$ is an irreducible, recurrent, and aperiodic transition probability matrix, we know that, for each $0 \le r < d$ and $j \in \mathbb{S}_r$, there exists a $\pi_{jj}^{(r)} \in [0, 1]$ with the property that $(\mathbf{P}^{md})_{ij} \longrightarrow \pi_{jj}^{(r)}$ for all $(i, j) \in \mathbb{S}_r^2$. More generally,

$$\lim_{m\to\infty} \left(\mathbf{P}^{md+s}\right)_{ij} = \begin{cases} \pi_{jj}^{(r(j))} & \text{if } r(j) - r(i) = s \bmod d \\ 0 & \text{otherwise.} \end{cases} \tag{4.1.20}$$

In particular, if $(i, j) \in (\mathbb{S}_r)^2$, then

$$(\mathbf{A}_{nd})_{ij} = \frac{1}{nd} \sum_{m=0}^{n-1} \sum_{s=0}^{d-1} \left(\mathbf{P}^{md+s}\right)_{ij} = \frac{1}{nd} \sum_{m=0}^{n-1} \left(\mathbf{P}^{md}\right)_{ij} \longrightarrow \frac{\pi_{jj}^{(r)}}{d}.$$

Hence, since we already know that $(\mathbf{A}_n)_{ij} \longrightarrow \pi_{jj}$, it follows that

$$\pi_{jj}^{(r)} = d\pi_{jj} \quad \text{for } 0 \le r < d \text{ and } j \in \mathbb{S}_r. \tag{4.1.21}$$

In the case when $\mathbf{P}$ is positive recurrent on $\mathbb{S}$, so is the restriction of $\mathbf{P}^d$ to each $\mathbb{S}_r$, and therefore $\sum_{j\in\mathbb{S}_r} \pi_{jj}^{(r)} = 1$. Thus, in the positive recurrent case, (4.1.21) leads to the interesting conclusion that $\boldsymbol{\pi}^{\mathbb{S}}$ assigns probability $\frac{1}{d}$ to each $\mathbb{S}_r$. See Exercises 4.2.6 and 6.6.5 below for other applications of these considerations.

4.2 Exercises

Exercise 4.2.1 Just as Césaro convergence is strictly weaker (i.e., it is implied by but does not imply) than ordinary convergence, so, in this exercise, we will show that Abel convergence is strictly weaker than Césaro convergence.

(a) Assume that the radius of convergence of $\{a_n\}_0^\infty \subseteq \mathbb{R}$ is less than or equal to 1. That is, $\overline{\lim}_{n\to\infty} |a_n|^{\frac{1}{n}} \le 1$. Set $R(s) = (1-s)\sum_0^\infty s^n a_n$ for $s \in [0, 1)$ and $A_n = \frac{1}{n}\sum_0^{n-1} a_m$ for $n \ge 1$. Show that $\overline{\lim}_{n\to\infty} |A_n| \le 1$ and that $R(s) = (1-s)^2 \sum_1^\infty n s^{n-1} A_n$ for $s \in [0, 1)$. Use this to conclude that

$$\lim_{n\to\infty} A_n = a \in \mathbb{R} \quad \implies \quad \lim_{s \nearrow 1} R(s) = a.$$

(b) Take $a_n = (-1)^{n+1}n$ for $n \geq 0$, check that the radius of convergence of $\{a_n\}_0^\infty$ is 1, and show that

$$\frac{1}{n}\sum_{m=0}^{n-1} a_m = \begin{cases} \frac{1}{2} & \text{if } n \text{ is even} \\ -\frac{1}{2} + \frac{1}{2n} & \text{if } n \text{ is odd} \end{cases} \quad \text{and} \quad (1-s)\sum_{m=0}^{\infty} s^m a_m = \frac{s(1-s)}{(1+s)^2}.$$

Hence, $\{a_n\}_0^\infty$ is Abel convergent to 0 but is Césaro divergent.

Exercise 4.2.2 Recall the queuing model in Exercise 1.3.12. Show that $\{Q_n : n \geq 0\}$ is an $\mathbb{N}$-valued Markov chain conditioned to start from 0, and write down the transition probability matrix for this chain. Further, for this chain: show that 0 is transient if $\mathbb{E}[B_1] > 0$, 0 is null recurrent if $\mathbb{E}[B_1] = 0$, and that 0 is positive recurrent if $\mathbb{E}[B_1] < 0$. In order to handle the case when $\mathbb{E}[B_1] = 0$, you might want to refer to Exercise 1.3.11.

Exercise 4.2.3 Here is a test for positive recurrence. Given a transition probability matrix $\mathbf{P}$ and an element j of the state space $\mathbb{S}$, set $C = [j]$. Assume that u is a non-negative function on C with the properties that

$$(\mathbf{P}u)_j < \infty \quad \text{and} \quad u(i) \geq (\mathbf{Pu})_i + \epsilon \quad \text{for all } i \in C \setminus \{j\} \text{ and some } \epsilon > 0.$$

(a) Begin by showing that

$$\mathbb{E}\big[u(X_{(n+1)\wedge\rho_j}) \,\big|\, X_0 = j\big] \leq \mathbb{E}\big[u(X_{n\wedge\rho_j}) \,\big|\, X_0 = j\big] - \epsilon\mathbb{P}(\rho_j > n \,|\, X_0 = j),$$

and use this to conclude that j is positive recurrent.

(b) Suppose that $\mathbb{S} = \mathbb{Z}$ and that $|i| \geq \sum_j |j|(\mathbf{P})_{ij} + \epsilon$ for all $i \in \mathbb{Z} \setminus \{0\}$. Show that 0 is positive recurrent for the chain determined by $\mathbf{P}$.

Exercise 4.2.4 Consider the nearest neighbor random walk on $\mathbb{Z}$ which moves forward with probability $p \in (\frac{1}{2}, 1)$ and backward with probability $q = 1 - p$. In other words, we are looking at the Markov chain on $\mathbb{Z}$ whose transition probability matrix $\mathbf{P}$ is given by $(\mathbf{P})_{ij} = p$ if $j = i + 1$, $(\mathbf{P})_{ij} = q$ if $j = i - 1$, and $(\mathbf{P})_{ij} = 0$ if $|j - i| \neq 1$. Obviously, this chain is irreducible, and the results in Sects. 1.2.2–1.2.1 show that 0 is transient. Thus, the considerations in Exercise 2.4.10 apply.

(a) Construct $\hat{\mathbf{P}}$ from $\mathbf{P}$ by the prescription in Exercise 2.4.10 when $j_0 = 0$, and, using (1.1.12), show that

$$(\hat{\mathbf{P}})_{ij} = \begin{cases} p & \text{if } i \leq 0 \;\&\; j = i + 1 \text{ or } i \geq 1 \;\&\; j = i - 1 \\ q & \text{if } i \leq 0 \;\&\; j = i - 1 \text{ or } i \geq 1 \;\&\; j = i + 1 \\ 0 & \text{otherwise.} \end{cases}$$

(b) On the basis of Exercise 2.4.10, we know that 0 is recurrent for the chain determined by $\hat{\mathbf{P}}$. Moreover, by part (b) of Exercise 4.2.3, one can check that it is positive recurrent. In fact, by combining part (b) of Exercise 2.4.10 with the computations in Sect. 1.1.4, show that

$$\hat{\mathbb{E}}[\rho_0|X_0 = 0] = \frac{2p}{p-q},$$

where $\hat{\mathbb{E}}$ is used to indicate that the expectation value is taken relative to the chain determined by $\hat{\mathbf{P}}$.

(c) Since $\mathbf{P}$ is irreducible, so is $\hat{\mathbf{P}}$. Hence, since 0 is positive recurrent for the chain determined by $\hat{\mathbf{P}}$, there is a unique stationary probability vector for $\hat{\mathbf{P}}$. Find this vector and use it to show that

$$\mathbb{E}^{\hat{\mathbb{P}}}[\rho_j|X_0 = j] = \begin{cases} \frac{2p}{p-q} & \text{if } j \in \{0,1\} \\ \frac{2pq^j}{p^j(p-q)} & \text{if } j \le -1 \\ \frac{2p^j}{q^{j-1}(p-q)} & \text{if } j \ge 2. \end{cases}$$

Exercise 4.2.5 As I said at the beginning of this chapter, we would be content here with statements about the convergence of either $\{(\mathbf{P}^n)_{ij} : n \ge 0\}$ or $\{(\mathbf{A}_n)_{ij} : n \ge 0\}$ for each $(i, j) \in \mathbb{S}^2$. However, as we are about to see, there are circumstances in which the pointwise results we have just obtained self-improve.

(a) Assume that j is positive recurrent, and set $C = [j]$. Given a probability vector $\boldsymbol{\mu}$ with the property that $\sum_{i \notin C} (\boldsymbol{\mu})_i = 0$, show that, in general, $(\boldsymbol{\mu}\mathbf{A}_n)_i \longrightarrow \pi_{ii}$ and, when j is aperiodic, $(\boldsymbol{\mu}\mathbf{P}^n)_i \longrightarrow \pi_{ii}$ for each $i \in C$.

(b) Here is an interesting fact about convergence of series. Namely, for each $m \in \mathbb{N}$, let $\{a_{m,n} : n \ge 0\}$ be a sequence of real numbers which converges to a real number b_m as $n \to \infty$. Further, assume that, for each $n \in \mathbb{N}$, the sequence $\{a_{m,n} : m \ge 0\}$ is absolutely summable. Finally, assume that

$$\sum_{m=0}^{\infty} |a_{m,n}| \longrightarrow \sum_{m=0}^{\infty} |b_m| < \infty \quad \text{as } n \to \infty.$$

Show that

$$\lim_{n\to\infty} \sum_{m=0}^{\infty} |a_{m,n} - b_m| = 0.$$

Hint: Using the triangle inequality, show that

$$\big||a_{m,n}| - |b_m| - |a_{m,n} - b_m|\big| \le 2|b_m|,$$

and apply Lebesgue's dominated convergence theorem to conclude that

$$\sum_{m=0}^{\infty} |a_{m,n} - b_m| \le \left|\sum_{m=0}^{\infty} \big(|a_{m,n}| - |b_m|\big)\right| + \sum_{m=0}^{\infty} \big||a_{m,n}| - |b_m| - |a_{m,n} - b_m|\big| \longrightarrow 0$$

as $n \to \infty$.

(c) Return to the setting in (a), and use (b) together with the result in (a) to show that, in general, $\|\boldsymbol{\mu}\mathbf{A}_n - \boldsymbol{\pi}^C\|_{\mathrm{v}} \longrightarrow 0$, and $\|\boldsymbol{\mu}\mathbf{P}^n - \boldsymbol{\pi}^C\|_{\mathrm{v}} \longrightarrow 0$ when j is aperiodic. In particular, for each probability vector $\boldsymbol{\mu}$ with $\sum_{i\in C}(\boldsymbol{\mu})_i = 1$,

$$\lim_{n\to\infty} \sup\big\{\big|\boldsymbol{\mu}\mathbf{P}^n\mathbf{f} - \boldsymbol{\pi}^C\mathbf{f}\big| : \|f\|_{\mathrm{u}} \le 1\big\} = 0,$$

where $\mathbf{f}$ is the column vector determined by a function f. Of course, this is still far less than what we had under Doeblin's condition since his condition provided us with a rate of convergence which was independent of $\boldsymbol{\mu}$. In general, no such uniform rate of convergence will exist.

Exercise 4.2.6 When j is aperiodic for $\mathbf{P}$, we know that $\lim_{n\to\infty}(\mathbf{P}^n)_{ij}$ exists for all $i \in \mathbb{S}$ and is 0 unless j is positive recurrent. When $d(j) > 1$ and j is positive recurrent, show that $\lim_{n\to\infty}(\mathbf{P}^n)_{jj}$ will fail to exist. On the other hand, even if $d(j) > 1$, show that $\lim_{n\to\infty}(\mathbf{P}^n)_{ij} = 0$ for any $j \in \mathbb{S}$ which is not positive recurrent.

Exercise 4.2.7 Here is an important interpretation of $\boldsymbol{\pi}^C$ when C is a positive recurrent communicating class. Let i be a recurrent state and, for $k \in \mathbb{S}$, let μ_k be the expected number of times the chain visits k before returning to i given that the chain started from i:

$$\mu_k = \mathbb{E}\left[\sum_{m=0}^{\rho_i - 1} \mathbf{1}_{\{k\}}(X_m)\,\middle|\, X_0 = i\right] \in [0,\infty].$$

Determine the row vector $\boldsymbol{\mu} \in [0,\infty]^{\mathbb{S}}$ by $(\boldsymbol{\mu})_k = \mu_k$.

(a) Show that, for all $j \in \mathbb{S}$,

$$(\boldsymbol{\mu}\mathbf{P})_j = \mathbb{E}\left[\sum_{m=1}^{\rho_i} \mathbf{1}_{\{j\}}(X_m)\,\middle|\, X_0 = i\right] = \mu_j.$$

Thus, without any further assumptions about i, $\boldsymbol{\mu}$ is $\mathbf{P}$-stationary in the sense that $\boldsymbol{\mu} = \boldsymbol{\mu}\mathbf{P}$.

(b) Clearly $\mu_i = 1$ and $\sum_j \mu_j = \infty$ unless i is positive recurrent. Nonetheless, show that $\mu_j = 0$ unless $i \leftrightarrow j$ and that $\mu_j \in (0,\infty)$ if $i \leftrightarrow j$.

Hint: Show that

$$\mathbb{P}\big(\rho_j^{(m)} < \rho_i \,\big|\, X_0 = i\big) = \mathbb{P}(\rho_j < \rho_i \,|\, X_0 = j)^{m-1}\mathbb{P}(\rho_j < \rho_i \,|\, X_0 = i).$$

(c) If i is positive recurrent, show that

$$\bar{\boldsymbol{\mu}} \equiv \frac{\boldsymbol{\mu}}{\sum_k \mu_k} = \boldsymbol{\pi}^C.$$

Equivalently, when i is positive recurrent,

$$\left(\pi^C\right)_j = \frac{\mathbb{E}[\sum_{m=0}^{\rho_i-1} \mathbf{1}_{\{j\}}(X_m) \,|\, X_0 = i]}{\mathbb{E}[\rho_i \,|\, X_0 = i]}.$$

In words, $(\pi^C)_j$ *is the relative expected amount of time that the chains spends at j before returning to i.*

Exercise 4.2.8 Let $\mathbf{P}$ be a transition probability on a finite state space $\mathbb{S}$, and assume that π is the one and only stationary probability for $\mathbf{P}$. Given a recurrent state $i \in \mathbb{S}$ and a $j \neq i$, show that (cf. Sect. 3.2.1)

$$\mathbb{P}(\rho_j < \rho_i \,|\, X_0 = i) = \frac{\det((\mathbf{I}-\mathbf{P})^{\{j\}})}{\det((\mathbf{I}-\mathbf{P})^{\{i,j\}})}.$$

You might want to take the following steps.

(a) Since $\mathbb{S}$ is finite, we know that i is positive recurrent. Now use the uniqueness of π, Exercises 4.2.7, and 3.2.7 to show that

$$\frac{\det((\mathbf{I}-\mathbf{P})^{\{j\}})}{\det((\mathbf{I}-\mathbf{P})^{\{i\}})} = \mathbb{E}\left[\sum_{m=0}^{\rho_i-1} \mathbf{1}_{\{j\}}(X_m) \,\middle|\, X_0 = i\right].$$

(b) Use the Markov property to show that

$$\mathbb{E}\left[\sum_{m=0}^{\rho_i-1} \mathbf{1}_{\{j\}}(X_m) \,\middle|\, X_0 = i\right] = \mathbb{P}(\rho_j < \rho_i \,|\, X_0 = i)\mathbb{E}\left[\sum_{m=0}^{\rho_i-1} \mathbf{1}_{\{j\}}(X_m) \,\middle|\, X_0 = j\right].$$

(c) Use (3.2.11) together with (b) to arrive at the desired conclusion.

Exercise 4.2.9 We continue with the program initiated in Exercise 4.2.7 but assume now that the reference point i is null recurrent. In this case, $\sum_j (\boldsymbol{\mu})_j = \infty$ when $\boldsymbol{\mu} \in [0,\infty)^{\mathbb{S}}$ is the $\mathbf{P}$-stationary measure introduced in Exercise 4.2.7. In this exercise we will show that, up to a multiplicative constant, $\boldsymbol{\mu}$ is the only $\mathbf{P}$-stationary $\boldsymbol{\nu} \in [0,\infty)^{\mathbb{S}}$ with the property that $(\boldsymbol{\nu})_j = 0$ unless $i \to j$ (and therefore $i \leftrightarrow j$). Equivalently, given such a $\boldsymbol{\nu}$, $\boldsymbol{\nu} = (\boldsymbol{\nu})_i \boldsymbol{\mu}$.

(a) Assume that $\boldsymbol{\nu} \in [0,\infty)^{\mathbb{S}}$ satisfies $\boldsymbol{\nu} = \boldsymbol{\nu}\mathbf{P}$. If $(\boldsymbol{\nu})_i = 1$, show that, for all $j \in \mathbb{S}$ and $n \geq 0$,

$$(\boldsymbol{\nu})_j = \sum_{k \neq i} (\boldsymbol{\nu})_k \mathbb{P}(X_n = j \;\&\; \rho_i > n \,|\, X_0 = k)$$
$$+ \mathbb{E}\left[\sum_{m=0}^{n \wedge (\rho_i - 1)} \mathbf{1}_{\{j\}}(X_m) \,\middle|\, X_0 = i\right].$$

Hint: Work by induction on $n \geq 0$. When $n = 0$ there is nothing to do. To carry out the inductive step, use (2.1.1) and Fubini's theorem to show that

$$\begin{aligned}\sum_{k \neq i} (\boldsymbol{\nu})_k \mathbb{P}(X_n = j \,\&\, \rho_i > n \,|\, X_0 = k) \\ = \sum_{k \neq i} (\boldsymbol{\nu}\mathbf{P})_k \mathbb{P}(X_n = j \,\&\, \rho_i > n \,|\, X_0 = k) \\ = \sum_{\ell} (\boldsymbol{\nu})_\ell \mathbb{P}(X_{n+1} = j \,\&\, \rho_i > n+1 \,|\, X_0 = \ell).\end{aligned}$$

(b) Assuming that $\boldsymbol{\nu} = \boldsymbol{\nu}\mathbf{P}$ and $(\boldsymbol{\nu})_j = 0$ unless $i \to j$, show that $\boldsymbol{\nu} = (\boldsymbol{\nu})_i \boldsymbol{\mu}$.

Hint: First show that $\boldsymbol{\nu} = \mathbf{0}$ if $(\boldsymbol{\nu})_i = 0$, and thereby reduce to the case when $(\boldsymbol{\nu})_i = 1$. Starting from the result in (a), apply the monotone convergence theorem to see that $(\boldsymbol{\nu})_j \geq (\boldsymbol{\mu})_j$ for all $j \in \mathbb{S}$. Now consider $\boldsymbol{\omega} \equiv \boldsymbol{\nu} - \boldsymbol{\mu}$, and conclude that $\boldsymbol{\omega} = \mathbf{0}$.

Exercise 4.2.10 Let C be a communicating class of positive recurrent states. The reason why Theorem 4.1.14 is called a "mean ergodic theorem" is that the asserted convergence is taking place in the sense of mean square convergence. Of course, mean square convergence implies convergence in probability, but, in general, it cannot be used to get convergence with probability 1. Nonetheless, as you are to show here, when $\mathbb{P}(X_0 \in C) = 1$ and $j \in C$,

$$\lim_{n \to \infty} \frac{1}{n} \sum_{m=0}^{n-1} \mathbf{1}_{\{j\}}(X_m) = \pi_{jj} \quad \text{with probability 1.} \tag{4.2.11}$$

Observe that in Sect. 2.3.3 we proved the individual ergodic theorem (2.3.10) under Doeblin's condition, and (4.2.11) says that the same sort of individual ergodic theorem holds even when Doeblin's condition is not present. In fact, there is a very general result, of which (4.2.11) is a very special case, which was proved originally by G.D. Birkhoff. However, we will not follow Birkhoff and instead, as we did in Sect. 2.3.3, we will base our proof on the strong saw of large numbers, although this time we need the full statement which holds (cf. Theorem 3.3.10 in [8])) for averages of mutually independent, identically distributed, integrable random variables.

(a) Show that it suffices to prove the result when $\mathbb{P}(X_0 = i) = 1$ for some $i \in C$.
(b) Set $\rho_i^{(0)} = 0$, and use $\rho_i^{(m)}$ to denote the time of the mth return to i. If

$$\tau_m = \rho_i^{(m)} - \rho_i^{(m-1)} \quad \text{and} \quad Y_m = \sum_{\ell = \rho_i^{(m-1)}}^{\rho_i^{(m)} - 1} \mathbf{1}_{\{j\}}(X_\ell),$$

show that, conditional on $X_0 = i$, both $\{\tau_m : m \geq 1\}$ and $\{Y_m : m \geq 1\}$ are sequences of mutually independent, identically distributed, non-negative, integrable random variables. In particular, as an application of the strong law of large numbers and (4.1.5) plus the result in Exercise 4.2.7, conclude that, conditional on $X_0 = i$,

$$\lim_{m\to\infty} \frac{\rho_i^{(m)}}{m} = \frac{1}{\pi_{ii}} \quad \text{and} \quad \lim_{m\to\infty} \frac{1}{m} \sum_{\ell=0}^{\rho_i^{(m)}-1} \mathbf{1}_{\{j\}}(X_m) = \frac{\pi_{jj}}{\pi_{ii}} \qquad (*)$$

with probability 1. Hence, $\lim_{m\to\infty} \frac{1}{\rho_i^{(m)}} \sum_{\ell=0}^{\rho_i^{(m)}-1} \mathbf{1}_{\{j\}}(X_m) = \pi_{jj}$ with probability 1.

(c) In view of the results in (a) and (b), we will be done once we check that, conditional on $X_0 = i$,

$$\lim_{n\to\infty} \left| \frac{1}{n} \sum_0^{n-1} \mathbf{1}_{\{j\}}(X_\ell) - \frac{1}{\rho_i^{(m_n)}} \sum_0^{\rho_i^{(m_n)}-1} \mathbf{1}_{\{j\}}(X_\ell) \right| = 0$$

with probability 1, where m_n is the $\mathbb{Z}^+$-valued random variable determined so that $\rho_i^{(m_n-1)} \le n < \rho_i^{(m_n)}$. To this end, first show that

$$\left| \frac{1}{n} \sum_0^{n-1} \mathbf{1}_{\{j\}}(X_\ell) - \frac{1}{\rho_i^{(m_n)}} \sum_0^{\rho_i^{(m_n)}-1} \mathbf{1}_{\{j\}}(X_\ell) \right| \le \frac{\tau_{m_n}}{m_n}.$$

Next, from the first part of $(*)$, show that $\mathbb{P}(\lim_{n\to\infty} m_n = \infty | X_0 = i) = 1$. Finally, check that, for any $0 < \epsilon < 1$,

$$\begin{aligned} &\mathbb{P}\left(\sup_{m\ge M} \frac{\tau_m}{m} \ge \epsilon \,\middle|\, X_0 = i \right) \\ &\le \sum_{m=M}^{\infty} \mathbb{P}(\rho_i \ge m\epsilon \mid X_0 = i) \le \frac{1}{\epsilon} \mathbb{E}[\rho_i, \rho_i \ge M\epsilon \mid X_0 = i], \end{aligned}$$

and use this to complete the proof of (4.2.11).

(d) Introduce the *empirical measure* $\mathbf{L}_n$, which is the random probability vector measuring the average time spent at points. That is, $(\mathbf{L}_n)_i = \frac{1}{n} \sum_0^{n-1} \mathbf{1}_{\{i\}}(X_m)$. By combining the result proved here with the one in (b) of Exercise 4.2.5, conclude that $\lim_{n\to\infty} \|\mathbf{L}_n - \boldsymbol{\pi}^C\|_{\mathrm{v}} = 0$ with probability 1 when $\mathbb{P}(X_0 \in C) = 1$.

Exercise 4.2.12 Although the statement in Exercise 4.2.10 applies only to positive recurrent states, it turns out that there is a corresponding limit theorem for states which are not positive recurrent. Namely, show that if j is not positive recurrent, then, no matter what the initial distribution,

$$\mathbb{P}\left(\lim_{n\to\infty} \frac{1}{n} \sum_{m=0}^{n-1} \mathbf{1}_{\{j\}}(X_m) = 0 \right) = 1.$$

When j is transient, $\mathbb{E}[\sum_0^\infty \mathbf{1}_{\{j\}}(X_m)] < \infty$ and therefore the result is trivial. To handle j that are null recurrent, begin by noting that it suffices to handle the case when $\mathbb{P}(X_0 = j) = 1$. Next, note that, conditional on $X_0 = j$, $\{\rho_j^{(m+1)} - \rho_j^{(m)} : m \geq 0\}$ is a sequence of independent, identically distributed, $\mathbb{Z}^+$-valued random variables, and apply the strong law of large numbers to see that, for any $R \geq 1$,

$$\mathbb{P}\left(\varliminf_{m\to\infty} \frac{\rho_j^{(m)}}{m} \geq H(R) \,\middle|\, X_0 = j\right)$$
$$\geq \mathbb{P}\left(\varliminf_{m\to\infty} \frac{\rho_j^{(m)} \wedge R}{m} \geq H(R) \,\middle|\, X_0 = j\right) = 1,$$

where $H(R) \equiv \frac{1}{2}\mathbb{E}[\rho_j \wedge R | X_0 = j] \nearrow \infty$ as $R \nearrow \infty$. Hence, given $X_0 = j$, $\frac{\rho_j^{(m)}}{m} \longrightarrow \infty$ with probability 1. Finally, check that, for any $\epsilon > 0$,

$$\mathbb{P}\left(\sup_{n\geq N} \frac{1}{n} \sum_0^{n-1} \mathbf{1}_{\{j\}}(X_m) \geq \epsilon \,\middle|\, X_0 = j\right)$$
$$\leq \mathbb{P}\left(\sup_{n\geq N} \frac{\rho_j^{(\lfloor n\epsilon \rfloor)}}{n} \leq \frac{1}{\epsilon} \,\middle|\, X_0 = j\right)$$
$$\leq \mathbb{P}\left(\sup_{m\geq n\epsilon} \frac{\rho_j^{(m)}}{m} \leq \frac{1}{\epsilon} \,\middle|\, X_0 = j\right),$$

and combine this with the preceding to reach the desired conclusion.

Chapter 5
Markov Processes in Continuous Time

Up until now we have been dealing with Markov processes for which time has a discrete parameter $n \in \mathbb{N}$. In this chapter we will introduce Markov processes for which time has a continuous parameter $t \in [0, \infty)$, even though our processes will continue to take their values in a countable state space $\mathbb{S}$.

5.1 Poisson Processes

Just as Markov chains with independent, identically distributed increments (a.k.a. random walks) on $\mathbb{Z}^d$ are the simplest discrete parameter Markov processes, so the simplest continuous time Markov processes are those whose increments are mutually independent and *homogeneous* in the sense that the distribution of an increment depends only on the length of the time interval over which the increment is taken. More precisely, we will be dealing in this section with $\mathbb{Z}^d$-valued stochastic processes $\{X(t) : t \geq 0\}$ with the property that $\mathbb{P}(X(0) = \mathbf{0}) = 1$ and

$$\begin{aligned}
&\mathbb{P}\big(X(t_1) - X(t_0) = j_1, \ldots, X(t_n) - X(t_{n-1}) = j_n\big) \\
&\quad = \prod_{m=1}^{n} \mathbb{P}\big(X(t_m - t_{m-1}) = j_m\big)
\end{aligned}$$

for all $n \geq 1$, $0 \leq t_0 < \cdots < t_n$, and $(j_1, \ldots, j_n) \in (\mathbb{Z}^d)^n$.

5.1.1 The Simple Poisson Process

The *simple Poisson process* is the $\mathbb{N}$-valued stochastic process $\{N(t) : t \geq 0\}$ that starts from 0, sits there for a unit exponential holding time E_1 (i.e., $N(t) = 0$ for $t \in [0, E_1)$ and, for $t > 0$, $\mathbb{P}(E_1 > t) = e^{-t}$), at time E_1 moves to 1 where it remains

D.W. Stroock, *An Introduction to Markov Processes*, Graduate Texts in Mathematics 230, DOI 10.1007/978-3-642-40523-5_5,

for an independent unit exponential holding time E_2, moves to 2 at time $E_1 + E_2$, etc. More precisely, if $\{E_n : n \geq 1\}$ is a sequence of mutually independent, unit exponential random variables and

$$J_n = \begin{cases} 0 & \text{when } n = 0 \\ \sum_{m=1}^{n} E_m & \text{when } n \geq 1, \end{cases}$$

then the stochastic process $\{N(t) : t \geq 0\}$ given by

$$N(t) = \max\{n \geq 0 : J_n \leq t\} \tag{5.1.1}$$

is a simple Poisson process. When thinking about the meaning of (5.1.1), keep in mind that, because,

$$\text{with probability 1,} \quad E_n > 0 \quad \text{for all } n \geq 1 \quad \text{and} \quad \sum_{m=1}^{\infty} E_m = \infty,$$

with probability 1 the path $t \in [0, \infty) \longmapsto N(t)$ is piecewise constant, right continuous, and, when it jumps, it jumps by $+1$: $N(t) - N(t-) \in \{0, 1\}$ for all $t > 0$, where $N(t-) \equiv \lim_{s \nearrow t} N(s)$ is the left limit of $N(\cdot)$ at t.

We now show that $\{N(t) : t \geq 0\}$ moves along in independent, homogeneous increments.[1] That is, for each $s, t \in [0, \infty)$, we will prove that the increment $N(s + t) - N(s)$ is independent of (cf. Sect. 7.1.3) $\sigma(\{N(\tau) : \tau \in [0, s]\})$ and has the same distribution as $N(t)$:

$$\mathbb{P}\big(N(s+t) - N(s) = n \,|\, N(\tau), \tau \in [0, s]\big) = \mathbb{P}\big(N(t) = n\big), \quad n \in \mathbb{N}. \tag{5.1.2}$$

Equivalently, what we have to check is that when $(s, t) \in [0, \infty)^2$ and $A \in \sigma(\{N(\tau) : \tau \in [0, s]\})$, $\mathbb{P}(\{N(s+t) - N(s) \geq n\} \cap A) = \mathbb{P}(N(t) \geq n)\mathbb{P}(A)$ for all $n \in \mathbb{N}$. Since this is trivial when $n = 0$, we will assume that $n \in \mathbb{Z}^+$. In addition, since we can always write A as the disjoint union of the sets $A \cap \{N(s) = m\}$, $m \in \mathbb{N}$, and each of these is again in $\sigma(\{N(\tau) : \tau \in [0, s]\})$, we may and will assume that, for some m, $N(s) = m$ on A. But in that case we can write $A = \{J_{m+1} > s\} \cap B$ where $B \in \sigma(\{E_1, \dots, E_m\})$ and $J_m \leq s$ on B. Hence, since $N(s + t) \geq m + n \iff J_{m+n} \leq s + t$, and $\sigma(\{J_\ell - J_m : \ell > m\})$ is independent of $\sigma(\{E_k : k \leq m\})$, an application of (7.4.2) shows that

$$\begin{aligned}
&\mathbb{P}\big(\{N(s+t) - N(s) \geq n\} \cap A\big) \\
&= \mathbb{P}\big(\{J_{m+n} \leq s + t\} \cap \{J_{m+1} > s\} \cap B\big) \\
&= \mathbb{P}\big(\{J_{m+n} - J_m \leq s + t - J_m\} \cap \{J_{m+1} - J_m > s - J_m\} \cap B\big) \\
&= \mathbb{E}\big[v(J_m), B\big],
\end{aligned}$$

[1] I have chosen the following proof in order to develop a line of reasoning which will serve us well later on. A more straight-forward, but less revealing, proof that the simple Poisson process has independent, homogeneous increments is given in Exercise 5.5.1 below.

where, for $\xi \in [0, s]$,

$$\begin{aligned} v(\xi) &\equiv \mathbb{P}\big(\{J_{m+n} - J_m \le s + t - \xi\} \cap \{J_{m+1} - J_m > s - \xi\}\big) \\ &= \mathbb{P}\big(\{J_n \le s + t - \xi\} \cap \{E_1 > s - \xi\}\big) = \mathbb{P}\big(\{J_n \le s + t - \xi\} \cap \{E_n > s - \xi\}\big) \\ &= \mathbb{P}\big(\{J_{n-1} + E_n \le s + t - \xi\} \cap \{E_n > s - \xi\}\big) = \mathbb{E}\big[w(\xi, E_n), E_n > s - \xi\big] \end{aligned}$$

when $w(\xi, \eta) \equiv \mathbb{P}(J_{n-1} \le s + t - \xi - \eta)$ for $\xi \in [0, s]$ and $\eta \in [s - \xi, s + t - \xi]$.

Up to this point we have not used any property of exponential random variables other than that they are positive. However, in our next step we will use their characteristic property, namely, the fact that an exponential random variable E "has no memory." That is, $\mathbb{P}(E > a + b | E > a) = \mathbb{P}(E > b)$, from which it is an easy step to

$$\mathbb{E}\big[f(E), E > a\big] = e^{-a}\mathbb{E}\big[f(a + E)\big] \tag{5.1.3}$$

for any non-negative, $\mathcal{B}_{[0,\infty)}$-measurable function f. In particular, this means that

$$\begin{aligned} v(\xi) &= \mathbb{E}\big[w(\xi, E_n), E_n > s - \xi\big] = e^{-(s-\xi)}\mathbb{E}\big[w(\xi, E_n + s - \xi)\big] \\ &= e^{-(s-\xi)}\mathbb{P}(J_{n-1} \le t - E_n) = e^{-(s-\xi)}\mathbb{P}(J_n \le t) = e^{-(s-\xi)}\mathbb{P}\big(N(t) \ge n\big). \end{aligned}$$

That is, we have shown that

$$\mathbb{P}\big(\big\{N(s + t) - N(s) \ge n\big\} \cap A\big) = \mathbb{E}\big[e^{-(s-J_m)}, B\big]\mathbb{P}\big(N(t) \ge n\big).$$

Finally, since

$$\mathbb{P}(A) = \mathbb{P}\big(\{J_{m+1} > s\} \cap B\big) = \mathbb{P}\big(\{E_{m+1} > s - J_m\} \cap B\big) = \mathbb{E}\big[e^{-(s-J_m)}, B\big],$$

the proof of (5.1.2) is complete. Hence, we now know that *the simple Poisson process* $\{N(t) : t \ge 0\}$ *has homogeneous and mutually independent increments.*

Before moving on, we must still find out what is the distribution of $N(t)$. But, the sum of n mutually independent, unit exponential random variables has a $\Gamma(n)$-distribution, and so

$$\begin{aligned} \mathbb{P}\big(N(t) = n\big) &= \mathbb{P}(J_n \le t < J_{n+1}) = \mathbb{P}(J_n \le t) - \mathbb{P}(J_{n+1} \le t) \\ &= \frac{1}{(n-1)!}\int_0^t \tau^{n-1}e^{-\tau}\,d\tau - \frac{1}{n!}\int_0^t \tau^n e^{-\tau}\,d\tau = \frac{t^n}{n!}. \end{aligned}$$

In other words, *$N(t)$ is a Poisson random variable with mean t*. More generally, when we combine this with (5.1.2), we get that, for all $s, t \in [0, \infty)$ and $n \in \mathbb{N}$,

$$\mathbb{P}\big(N(s + t) - N(s) = n \mid N(\tau), \tau \in [0, s]\big) = e^{-t}\frac{t^n}{n!}. \tag{5.1.4}$$

Alternatively, again starting from (5.1.2), we can now give the following Markovian description of $\{N(t) : t \geq 0\}$:

$$\mathbb{P}\big(N(s+t) = n \mid N(\tau), \tau \in [0,s]\big) = e^{-t} \frac{t^{n-N(s)}}{(n-N(s))!} \mathbf{1}_{[0,n]}\big(N(s)\big). \tag{5.1.5}$$

5.1.2 Compound Poisson Processes on $\mathbb{Z}^d$

Having constructed the simplest Poisson process, we can easily construct a rich class of processes which are the continuous time analogs of random walks on $\mathbb{Z}^d$. Suppose that $\boldsymbol{\mu}$ is a probability vector on $\mathbb{Z}^d$ which gives 0 mass to the origin $\mathbf{0}$. Then the *compound Poisson process with jump distribution $\boldsymbol{\mu}$ and rate $R \in (0,\infty)$* is the stochastic process $\{\mathbf{X}(t) : t \geq 0\}$ which starts at $\mathbf{0}$, sits there for an exponential holding time having mean value R^{-1}, at which time it jumps by the amount $\mathbf{k} \in \mathbb{Z}^d$ with probability $(\boldsymbol{\mu})_{\mathbf{k}}$, sits where it lands for another, independent holding time with mean R^{-1}, jumps again a random amount with distribution $\boldsymbol{\mu}$, etc. Thus, the simple Poisson process is the case when $d = 1$, $(\boldsymbol{\mu})_1 = 1$, and $R = 1$. Of course, the amount by which the simple Poisson process jumps is deterministic whereas the amount by which a compound Poisson process jumps will, in general, be random.

To construct a compound Poisson process, let $\{\mathbf{B}_n : n \geq 1\}$ be a sequence of mutually independent $\mathbb{Z}^d$-valued random variables with distribution $\boldsymbol{\mu}$, introduce the random walk $\mathbf{X}_0 = \mathbf{0}$ and $\mathbf{X}_n = \sum_{m=1}^{n} \mathbf{B}_m$ for $n \geq 1$, and define $\{\mathbf{X}(t) : t \geq 0\}$ so that $\mathbf{X}(t) = \mathbf{X}_{N(Rt)}$, where $\{N(t) : t \geq 0\}$ is a simple Poisson process that is independent of the $\mathbf{B}_m$'s. The existence of all these random variables is guaranteed (cf. the footnote 8 in Sect. 1.2.1) by Theorem 7.3.2. Obviously, $\mathbf{X}(0) = \mathbf{0}$ and $t \in [0,\infty) \longmapsto \mathbf{X}(t) \in \mathbb{Z}^d$ is a piecewise constant, right continuous $\mathbb{Z}^d$-valued path. In addition, because $(\boldsymbol{\mu})_{\mathbf{0}} = 0$, it is clear[2] that the number of jumps that $t \rightsquigarrow \mathbf{X}(t)$ makes during a time interval $(s,t]$ is precisely $N(Rt) - N(Rs)$ and that $\mathbf{X}_n - \mathbf{X}_{n-1}$ is the amount by which $t \rightsquigarrow \mathbf{X}(t)$ jumps on its nth jump. Thus, if $J_0 \equiv 0$ and, for $n \geq 1$, J_n is the time of the nth jump of $t \rightsquigarrow \mathbf{X}(t)$, then $N(Rt) = n \iff J_n \leq Rt < J_{n+1}$, and $\mathbf{X}(J_n) - \mathbf{X}(J_{n-1}) = \mathbf{X}_n - \mathbf{X}_{n-1}$. Equivalently, if $\{E_n : n \geq 1\}$ denotes the sequence of unit exponential random variables out of which $\{N(t) : t \geq 0\}$ is built, then $J_n - J_{n-1} = \frac{E_n}{R}$, $\mathbf{X}(t) - \mathbf{X}(t-) = \mathbf{0}$ for $t \in (J_{n-1}, J_n)$, and $\mathbf{X}(J_n) - \mathbf{X}(J_{n-1}) = \mathbf{B}_n$. Therefore $\{\mathbf{X}(t) : t \geq 0\}$ is indeed a compound Poisson process with jump distribution $\boldsymbol{\mu}$ and rate R.

---

[2]This is the reason for my having assumed that $(\boldsymbol{\mu})_{\mathbf{0}} = 0$. However, one should realize that this assumption causes no loss in generality. Namely, if $(\boldsymbol{\mu})_{\mathbf{0}} = 1$, then the resulting compound process would be trivial: it would never move. On the other hand, if $(\boldsymbol{\mu})_0 \in (0,1)$, then we could replace $\boldsymbol{\mu}$ by $\bar{\boldsymbol{\mu}}$, where $(\bar{\boldsymbol{\mu}})_{\mathbf{0}} = 0$ and $(\bar{\boldsymbol{\mu}})_{\mathbf{k}} = (1-(\boldsymbol{\mu})_0)^{-1}(\boldsymbol{\mu})_{\mathbf{k}}$ when $\mathbf{k} \neq \mathbf{0}$, and R by $\bar{R} = (1-(\boldsymbol{\mu})_{\mathbf{0}})R$. The compound Poisson process corresponding to $\bar{\boldsymbol{\mu}}$ and $\bar{R}$ would have exactly the same distribution of the one corresponding to $\boldsymbol{\mu}$ and R.

We show next that a compound process moves along in homogeneous, mutually independent increments:

$$\mathbb{P}\big(\mathbf{X}(s+t)-\mathbf{X}(s)=\mathbf{k}\,|\,\mathbf{X}(\tau),\tau\in[0,s]\big)=\mathbb{P}\big(\mathbf{X}(t)=\mathbf{k}\big),\quad \mathbf{k}\in\mathbb{Z}^d. \tag{5.1.6}$$

For this purpose, we use the representation $\mathbf{X}(t)=\mathbf{X}_{N(Rt)}$ introduced above. Given $A\in\sigma(\{\mathbf{X}(\tau):\tau\in[0,s]\})$, we need to show that

$$\mathbb{P}\big(\{\mathbf{X}(s+t)-\mathbf{X}(s)=\mathbf{k}\}\cap A\big)=\mathbb{P}\big(\{\mathbf{X}(s+t)-\mathbf{X}(s)=\mathbf{k}\}\big)\mathbb{P}(A),$$

and, just as in the derivation of (5.1.2), we will, without loss in generality, assume that, for some $m\in\mathbb{N}$, $N(Rs)=m$ on A. But then A is independent of $\sigma(\{\mathbf{X}_{m+n}-\mathbf{X}_m:n\geq 0\}\cup\{N(R(s+t))-N(Rs)\})$, and so

$$\begin{aligned}
&\mathbb{P}\big(\{\mathbf{X}(s+t)-\mathbf{X}(s)=\mathbf{k}\}\cap A\big)\\
&=\sum_{n=0}^{\infty}\mathbb{P}\big(\{\mathbf{X}(s+t)-\mathbf{X}(s)=\mathbf{k}\ \&\ N\big(R(s+t)\big)-N(Rs)=n\}\cap A\big)\\
&=\sum_{n=0}^{\infty}\mathbb{P}\big(\{\mathbf{X}_{m+n}-\mathbf{X}_m=\mathbf{k}\ \&\ N\big(R(s+t)\big)-N(Rs)=n\}\cap A\big)\\
&=\sum_{n=0}^{\infty}\mathbb{P}(\mathbf{X}_n=\mathbf{k})\mathbb{P}\big(N(Rt)=n\big)\mathbb{P}(A)\\
&=\sum_{n=0}^{\infty}\mathbb{P}\big(\mathbf{X}_n=\mathbf{k}\ \&\ N(Rt)=n\big)\mathbb{P}(A)=\mathbb{P}\big(\mathbf{X}(t)=\mathbf{k}\big)\mathbb{P}(A).
\end{aligned}$$

Hence, (5.1.6) is proved.

Finally, to compute the distribution of $\mathbf{X}(t)$, begin by recalling that the distribution of the sum of n independent, identically distributed random variables is the *n-fold convolution* of their distribution. Thus, $\mathbb{P}(\mathbf{X}_n=\mathbf{k})=(\boldsymbol{\mu}^{\star n})_{\mathbf{k}}$, where $(\boldsymbol{\mu}^{\star 0})_{\mathbf{k}}=\delta_{\mathbf{0},\mathbf{k}}$ is the point mass at $\mathbf{0}$ and

$$\big(\boldsymbol{\mu}^{\star n}\big)_{\mathbf{k}}=\sum_{\mathbf{j}\in\mathbb{Z}^d}\big(\boldsymbol{\mu}^{\star(n-1)}\big)_{\mathbf{k}-\mathbf{j}}(\boldsymbol{\mu})_{\mathbf{j}}\quad\text{for } n\geq 1.$$

Hence,

$$\mathbb{P}\big(\mathbf{X}(s+t)=\mathbf{k}\big)=\sum_{n=0}^{\infty}\mathbb{P}\big(\mathbf{X}_n=\mathbf{k}\ \&\ N(Rt)=n\big)=e^{-Rt}\sum_{n=0}^{\infty}\frac{(Rt)^n}{n!}\big(\boldsymbol{\mu}^{\star n}\big)_{\mathbf{k}}.$$

Putting this together with (5.1.6), we have that for $A \in \sigma(\{N(\tau) : \tau \in [0, s]\})$,

$$\begin{aligned}
&\mathbb{P}\big(\{\mathbf{X}(s+t)=\mathbf{k}\}\cap A\big)\\
&=\sum_{\mathbf{j}\in\mathbb{Z}^d}\mathbb{P}\big(\{\mathbf{X}(s+t)=\mathbf{k}\}\cap A\cap\{\mathbf{X}(s)=\mathbf{j}\}\big)\\
&=\sum_{\mathbf{j}\in\mathbb{Z}^d}\mathbb{P}\big(\{\mathbf{X}(s+t)-\mathbf{X}(s)=\mathbf{k}-\mathbf{j}\}\cap A\cap\{\mathbf{X}(s)=\mathbf{j}\}\big)\\
&=\sum_{\mathbf{j}\in\mathbb{Z}^d}\big(\mathbf{P}(t)\big)_{\mathbf{jk}}\mathbb{P}\big(A\cap\{\mathbf{X}(s)=\mathbf{j}\}\big)=\mathbb{E}\big[\big(\mathbf{P}(t)\big)_{\mathbf{X}(s)\mathbf{k}}, A\big],
\end{aligned}$$

where

$$\big(\mathbf{P}(t)\big)_{\mathbf{k}\boldsymbol{\ell}} \equiv e^{-Rt}\sum_{m=0}^{\infty}\frac{(Rt)^m}{m!}\big(\mu^{\star m}\big)_{\boldsymbol{\ell}-\mathbf{k}}. \tag{5.1.7}$$

Equivalently, we have proved that $\{\mathbf{X}(t) : t \geq 0\}$ *is a continuous time Markov process with transition probability function* $t\rightsquigarrow\mathbf{P}(t)$ in the sense that

$$\mathbb{P}\big(\mathbf{X}(s+t)=\mathbf{k}\,|\,\mathbf{X}(\tau),\tau\in[0,s]\big)=\big(\mathbf{P}(t)\big)_{\mathbf{X}(s)\mathbf{k}}. \tag{5.1.8}$$

Observe that, as a consequence of (5.1.8), we find that $\{\mathbf{P}(t) : t \geq 0\}$ is a *semigroup*. That is, $\mathbf{P}(0) = \mathbf{I}$, $\mathbf{P}(t)$ is a transition probability for each $t \geq 0$, and $t\rightsquigarrow\mathbf{P}(t)$ satisfies the *Chapman–Kolmogorov* equation

$$\mathbf{P}(s+t)=\mathbf{P}(s)\mathbf{P}(t), \quad s,t\in[0,\infty). \tag{5.1.9}$$

Indeed, it is obvious that $\mathbf{P}(0) = \mathbf{I}$ and that $\mathbf{P}(t)$ is a transition probability for each $t \geq 0$. In addition,

$$\begin{aligned}
\big(\mathbf{P}(s+t)\big)_{\mathbf{0k}} &= \sum_{\mathbf{j}\in\mathbb{Z}^d}\mathbb{P}\big(\mathbf{X}(s+t)=\mathbf{k}\ \&\ \mathbf{X}(s)=\mathbf{j}\big)\\
&=\sum_{\mathbf{j}\in\mathbb{Z}^d}\big(\mathbf{P}(t)\big)_{\mathbf{j}\boldsymbol{\ell}}\big(\mathbf{P}(s)\big)_{\mathbf{0j}}=\big(\mathbf{P}(s)\mathbf{P}(t)\big)_{\mathbf{0k}},
\end{aligned}$$

from which the asserted matrix equality follows immediately when one remembers that $(\mathbf{P}(\tau))_{\mathbf{k}\boldsymbol{\ell}} = (\mathbf{P}(\tau))_{\mathbf{0}(\boldsymbol{\ell}-\mathbf{k})}$.

5.2 Markov Processes with Bounded Rates

There are two directions in which one can generalize the preceding without destroying the Markov property. For one thing, one can make the distribution of jumps depend on where the process is at the time it makes the jump. This change comes down to replacing the random walk in the compound Poisson process by a more general Markov chain. The second way is to increase the randomness by making the rate of jumping depend on the place where the process is waiting before it jumps. That is, instead of the holding times all having the same mean value, the holding time at a particular state may depend on that state.

5.2.1 Basic Construction

Let $\mathbb{S}$ be a finite or countable state space and $\mathbf{P}$ a transition probability matrix on $\mathbb{S}$. In addition, let $\mathfrak{R} = \{R_i : i \in \mathbb{S}\}$ be a bounded set of non-negative numbers. We will show in this subsection that for each probability vector $\boldsymbol{\mu}$ there exists a stochastic process $\{X(t) : t \geq 0\}$ with the properties that

(a) $t \rightsquigarrow X(t)$ is right continuous and piecewise constant,
(b) $\mathbb{P}(X(0) = i) = (\boldsymbol{\mu})_i$ for all $i \in \mathbb{S}$ and

$$\lim_{h \searrow 0} \sup_{\substack{j \in \mathbb{S} \\ t \in [0,\infty)}} \frac{1}{h} \Big| \mathbb{P}\big(X(t+h) = j \mid X(\tau), \tau \in [0,t]\big) - (1 - hR_{X(t)})\delta_{X(t),j} + hR_{X(t)}\mathbf{P}_{X(t)j} \Big| = 0. \tag{5.2.1}$$

Furthermore, we will show that $\{X(t) : t \geq 0\}$ is a Markov process and that its distribution is uniquely determined by (a) and (b).

Set $\mathbf{Q} = \mathbf{R}(\mathbf{P} - \mathbf{I})$, where $\mathbf{R}$ is the diagonal matrix with $(\mathbf{R})_{ii} = R_i$, and define $t \rightsquigarrow \mathbf{P}(t)$ by

$$\mathbf{P}(t) = \sum_{n=0}^{\infty} \frac{t^n}{n!} \mathbf{Q}^n. \tag{5.2.2}$$

Since $\|\mathbf{Q}\|_{\mathrm{u,v}} \leq 2R$, where $R = \sup_{i \in \mathbb{S}} R_i$, it is easy to check that the series in its definition is converging in $\|\cdot\|_{\mathrm{u,v}}$-norm to $\mathbf{P}(t)$ uniformly for t's in bounded intervals. Also, it is easy to see that $t \rightsquigarrow \mathbf{P}(t)$ satisfies the Chapman–Kolmogorov (5.1.9) equation. Less obvious is the fact that $\mathbf{P}(t)$ is a transition probability for each $t \geq 0$.[3] To verify this, first note that there is nothing to do unless $R > 0$, since $\mathbf{P}(t) \equiv \mathbf{I}$ when $R = 0$. Thus, assume that $R > 0$ and define

$$(\tilde{\mathbf{P}})_{ij} = \begin{cases} 1 - \frac{R_i}{R}(1 - (\mathbf{P})_{ii}) & \text{if } j = i \\ \frac{R_i}{R}(\mathbf{P})_{ij} & \text{if } j \neq i. \end{cases} \tag{5.2.3}$$

[3]It is important to observe that $\|\mathbf{P}(t)\|_{\mathrm{u,v}}$ can be as large as e^{2tR} when $t < 0$ and therefore will not be a transition probability matrix when $R > 0$ and $t < 0$.

Clearly $\tilde{\mathbf{P}}$ is a transition probability matrix and $\mathbf{Q} = R(\tilde{\mathbf{P}} - \mathbf{I})$. Hence, another expression for $\mathbf{P}(t)$ is

$$\sum_{n=0}^{\infty} \frac{(tR)^n}{n!} \sum_{m=0}^{n} \binom{n}{m} (-1)^{n-m} \tilde{\mathbf{P}}^m = \sum_{m=0}^{\infty} \frac{(tR)^m}{m!} \tilde{\mathbf{P}}^m \sum_{n=m}^{\infty} \frac{(-tR)^{n-m}}{(n-m)!}$$
$$= e^{-tR} \sum_{m=0}^{\infty} \frac{(tR)^m}{m!} \tilde{\mathbf{P}}^m,$$

from which it is clear that $\mathbf{P}(t)$ is a transition probability matrix and therefore that $t \rightsquigarrow \mathbf{P}(t)$ is a transition probability function.

We begin by showing that if $\{X(t) : t \geq 0\}$ satisfies (5.2.1) and has initial distribution $\boldsymbol{\mu}$, then, for any $K \geq 0$, $0 = t_0 < t_1 < \cdots < t_K$, and $(j_0, \ldots, j_K) \in \mathbb{S}^{K+1}$,

$$\mathbb{P}\big(X(t_k) = j_k \text{ for } 0 \leq k \leq K\big) = \mu_{j_0} \prod_{1 \leq k \leq K} \big(\mathbf{P}(t_k - t_{k-1})\big)_{j_{k-1} j_k}, \tag{5.2.4}$$

where the product over the empty set is taken to be 1. To prove this, we will work by induction on K. When $K = 0$ there is nothing to do. Thus, suppose that $K \geq 1$ and that (5.2.4) holds for $K - 1$. Set $A = \{X(t_k) = j_k, 0 \leq k < K\}$ and define $u(t, j) = \mathbb{P}(\{X(t_{k-1} + t) = j\} \cap A)$ for $t \geq 0$ and $j \in \mathbb{S}$. By (5.2.1),

$$\lim_{h \searrow 0} \sup_{\substack{j \in \mathbb{S} \\ t \in [0,\infty)}} \frac{1}{h} \big| u(t + h, j) - u(t, j) - h\big(\mathbf{u}(t)\mathbf{Q}\big)_j \big| = 0,$$

where $\mathbf{u}(t)$ is the row vector determined by $u(t, \cdot)$. Therefore $t \rightsquigarrow u(t, j)$ is continuously differentiable and its derivative $\dot{u}(t, j)$ is equal to $(\mathbf{Q}\mathbf{u}(t))_j$. In addition, $u(0, j) = \delta_{j, j_{K-1}} \mathbb{P}(A)$. In other words, we now know that $(t, j) \rightsquigarrow u(t, j)$ is a bounded solution to

$$\dot{u}(t, j) = \big(\mathbf{Q}\mathbf{u}(t)\big)_j \quad \text{with } u(0, j) = \delta_{j, j_{K-1}} \mathbb{P}(A). \tag{$*$}$$

This information together with the following lemma will allow us to complete the inductive step.

Lemma 5.2.5 *A bounded function $w : [0, \infty) \times \mathbb{S} \longrightarrow \mathbb{R}$ that is continuously differentiable with respect to $t \geq 0$ satisfies*

$$\dot{w}(t, j) = \big(\mathbf{Q}\mathbf{w}(t)\big)_j,$$

where $\mathbf{w}(t)$ is the row vector determined by $w(t, \cdot)$, if and only if $w(t, j) = (\mathbf{w}(0)\mathbf{P}(t))_j$ for $(t, j) \in [0, \infty) \times \mathbb{S}$.

Proof Let $\mathbf{w}$ be a row vector with $\|\mathbf{w}\|_{\mathrm{u}} < \infty$, and define $w(t, j) = (\mathbf{w}\mathbf{P}(t))_j$ for $(t, j) \in [0, \infty) \times \mathbb{S}$. Then $|w(t, j)| \leq \|\mathbf{w}\|_{\mathrm{u}}$ and

$$w(t + h, j) - w(t, j) = \big(\big(\mathbf{P}(h) - \mathbf{I}\big)\mathbf{w}(t)\big)_j,$$

where $\mathbf{w}(t)$ is the row vector determined by $w(t,\cdot)$. Next note that

$$\lim_{h\searrow 0}\left\|\frac{\mathbf{P}(h)-\mathbf{I}}{h}-\mathbf{Q}\right\|_{\mathrm{u,v}}=0,$$

and from this conclude that $t\rightsquigarrow w(t,j)$ is continuously differentiable and that $\dot{w}(t,j)=(\mathbf{w}\mathbf{Q})_j$. Thus the "if" assertion is proved.

To prove the "only if" assertion, let $t>0$ and consider the function $f(s,j)=(\mathbf{w}(t-s)\mathbf{P}(s))_j$ for $s\in[0,t]$. Clearly $f(\cdot,j)$ is continuously differentiable and

$$\dot{f}(s,j)=-\big(\mathbf{w}(t-s)\mathbf{Q}\mathbf{P}(s)\big)_j+\big(\mathbf{w}(t-s)\mathbf{Q}\mathbf{P}(s)\big)_j=0$$

for $s\in(0,t)$. Hence, $w(t,j)=f(0,j)=f(t,j)=(\mathbf{w}(0)\mathbf{P}(t))_j$. □

Combining (∗) with Lemma 5.2.5, we see that

$$u(t_K-t_{K-1},j_K)=\big(\mathbf{P}(t_K-t_{K-1})\big)_{j_{K-1}j_K}\mathbb{P}(A),$$

and therefore, by the induction hypothesis, that (5.2.4) holds for K.

We next show that if (5.2.4) holds, then $\{X(t):t\ge 0\}$ is a Markov process with transition probability function $t\rightsquigarrow\mathbf{P}(t)$. That is, we will show that if $s\ge 0$, then for $t>s$ and $j\in\mathbb{S}$,

$$\mathbb{P}\big(\{X(t)=j\}\,\big|\,X(\tau),\tau\in[0,s]\big)=\big(\mathbf{P}(t-s)\big)_{X(s)j}. \tag{5.2.6}$$

Equivalently, we have to show that if $A\in\sigma(\{X(\tau):\tau\in[0,s]\})$, then

$$\mathbb{P}\big(\{X(t)=j\}\cap A\big)=\mathbb{E}\big[\big(\mathbf{P}(t-s)\big)_{X(s)j},A\big],$$

and we have to do so only when $A=\{X(t_k)=j_k,0\le k<K\}$ for some $K\ge 1$, $0=t_0<\cdots<t_{K-1}=s$, and $(j_0,\dots,j_{K-1})\in\mathbb{S}^K$. But in that case, the required equation is just (5.2.4) with $t_K=t$ and $j_K=j$.

In order to prove that there exists a stochastic process satisfying conditions (a) and (b), it will be useful to know that (5.2.4) implies (5.2.1). To see this, note that, (5.2.4) implies (5.2.6) and (5.2.6) implies

$$\begin{aligned}&\mathbb{P}\big(X(t+h)=j\mid X(\tau),\tau\in[0,t]\big)\\&=\big(\mathbf{P}(h)\big)_{X(t)j}\\&=\delta_{X(t),j}+h(\mathbf{Q})_{X(t)j}+h^2\sum_{n=2}^{\infty}\frac{h^{n-2}(\mathbf{Q}^n)_{X(t)j}}{n!}.\end{aligned}$$

Therefore, since

$$\delta_{X(t)j}+h(\mathbf{Q})_{X(t)j}=(1-hR_{X(t)})\delta_{X(t),j}+hR_{X(t)}(\mathbf{P})_{X(t)j},$$

and

$$\left\|\sum_{n=2}^{\infty}\frac{h^{n-2}\mathbf{Q}^n}{n!}\right\|_{\mathrm{u,v}} \le e^{2R} \quad \text{for } h\in[0,1],$$

(5.2.1) follows from (5.2.4).

We can now prove the existence of a stochastic process that satisfies (a) and (b). To this end, recall the transition function $\tilde{\mathbf{P}}$ in (5.2.3), and let $\{X_n : n \ge 0\}$ be a Markov chain with transition probability $\tilde{\mathbf{P}}$ and initial distribution $\boldsymbol{\mu}$. Next, let $\{N(t) : t \ge 0\}$ be a simple Poisson process that is independent of $\{X_n : n \ge 0\}$, and set $X(t) = X_{N(Rt)}$. Clearly $t\rightsquigarrow X(t)$ is right continuous and piecewise constant. In addition, if $K \ge 0$, $0 = t_0 < \cdots < t_K$, and $(j_0, \ldots, j_K) \in \mathbb{S}^{K+1}$, then

$$\begin{aligned}
&\mathbb{P}\big(X(t_k) = j_k, 0 \le k \le K\big)\\
&\quad= \sum_{0=m_0\le\cdots\le m_K} \mathbb{P}\big(X_{m_k} = j_k \;\&\; N(t_k) = m_k \text{ for } 0 \le k \le K\big)\\
&\quad= \sum_{0=m_0\le\cdots\le m_K} \mathbb{P}(X_{m_k} = j_k, 0 \le k \le K)\mathbb{P}\big(N(t_k) = m_k, 0 \le k \le K\big)\\
&\quad= (\boldsymbol{\mu})_{j_0} \sum_{0=m_0\le\cdots\le m_K} \prod_{k=1}^{K} \big(\tilde{\mathbf{P}}^{m_k-m_{k-1}}\big)_{j_{k-1}k_k}\\
&\qquad\qquad\times \frac{e^{-(t_k-t_{k-1})R}\big(R(t_k-t_{k-1})\big)^{m_k-m_{k-1}}}{(m_k-m_{k-1})!}\\
&\quad= (\boldsymbol{\mu})_{j_0} \prod_{k=1}^{K} \big(\mathbf{P}(t_k-t_{k-1})\big)_{j_{k-1}j_k}.
\end{aligned}$$

Hence, $\{X(t) : t \ge 0\}$ has initial distribution $\boldsymbol{\mu}$ and satisfies (5.2.4) and therefore (5.2.1).

The following theorem summarizes our progress thus far.

Theorem 5.2.7 *Let a bounded set $\mathfrak{R} = \{R_i : i \in \mathbb{S}\} \subseteq [0,\infty)$ and a transition probability matrix $\mathbf{P}$ be given. Then, for each probability vector $\boldsymbol{\mu}$, there is a stochastic process $\{X(t) : t \ge 0\}$ satisfying the conditions* (a) *and* (b). *Furthermore, if $\mathbf{Q} = \mathbf{R}(\mathbf{P}-\mathbf{I})$, where $(\mathbf{R})_{ij} = \delta_{i,j}R_i$, and $\mathbf{P}(t)$ is given by* (5.2.2), *then $\{X(t) : t \ge 0\}$ is a Markov process with transition probability function $t\rightsquigarrow\mathbf{P}(t)$ in the sense that it satisfies* (5.2.6). *In particular, the distribution of $\{X(t) : t \ge 0\}$ is uniquely determined by* (5.2.4).

5.2.2 An Alternative Construction

In this subsection we will give an alternative construction of processes that satisfy conditions (a) and (b), one that reveals other properties.

Given a bounded set $\mathfrak{R} = \{R_i : i \in \mathbb{S}\} \subseteq [0, \infty)$, define

$$\Phi^{\mathfrak{R}} : [0, \infty) \times (0, \infty)^{\mathbb{Z}^+} \times \mathbb{S}^{\mathbb{N}} \longrightarrow \mathbb{S}$$

by the prescription

$$\begin{aligned} &\Phi^{\mathfrak{R}}\big(t; (e_1, \ldots, e_n, \ldots), (j_0, \ldots, j_n, \ldots)\big) = j_n \quad \text{for } \xi_n \le t < \xi_{n+1} \\ &\text{where} \quad \xi_0 = 0 \quad \text{and} \quad \xi_n = \sum_{m=1}^{n} R_{j_{m-1}}^{-1} e_m \quad \text{when } n \ge 1, \end{aligned} \tag{5.2.8}$$

with $\xi_n = \infty$ if $R_{j_{m-1}} = 0$ for some $1 \le m \le n$. Next, let $\{\tilde{E}_n : n \ge 1\}$ be a sequence of mutually independent random, unit exponential random variables, and without loss in generality, assume that, for each ω, $\tilde{E}_n(\omega) > 0$ for all n. Because,

$$\mathbb{E}\left[\exp\left(-\sum_{n=1}^{\infty} \tilde{E}_n\right)\right] = \lim_{N\to\infty} \left(\int_0^{\infty} e^{-2t}\, dt\right)^N = 0,$$

we may and will also assume that $\sum_{n=1}^{\infty} \tilde{E}_n(\omega) = \infty$ for all ω. Now let $\{\tilde{X}_n : n \ge 0\}$ be a Markov chain with transition probability $\mathbf{P}$ and initial distribution $\boldsymbol{\mu}$, assume that $\sigma(\{\tilde{X}_n : n \ge 0\})$ and $\sigma(\{\tilde{E}_n : n \ge 1\})$ are independent, and set

$$\tilde{J}_0 = 0 \quad \text{and} \quad \tilde{J}_n = \sum_{m=1}^{n} R_{\tilde{X}_{m-1}}^{-1} \tilde{E}_m \quad \text{for } n \ge 1,$$

again with the understanding that $\tilde{J}_n = \infty$ if $R_{\tilde{X}_{m-1}} = 0$ for some $1 \le m \le n$. Because the R_i's are bounded and $\sum_{m=1}^{\infty} \tilde{E}_m \equiv \infty$, $\tilde{J}_n \nearrow \infty$ as $n \to \infty$. Hence, we can define a right continuous, piecewise constant process $\{\tilde{X}(t) : t \ge 0\}$ by

$$\tilde{X}(t) = \Phi^{\mathfrak{R}}\big(t; (\tilde{E}_1, \ldots, \tilde{E}_n, \ldots), (\tilde{X}_0, \ldots, \tilde{X}_n, \ldots)\big). \tag{5.2.9}$$

Theorem 5.2.10 *Refer to the preceding. The process* $\{\tilde{X}(t) : t \ge 0\}$ *given by* (5.2.9) *satisfies conditions* (a) *and* (b) *at the beginning of Sect.* 3.2.1.

Proof Since it is obvious that (a) is satisfied, what remains to be shown is that $\{\tilde{X}(t) : t \ge 0\}$ satisfies (5.2.1). We begin by observing that

$$\begin{aligned} &\Phi^{\mathfrak{R}}\big(t + h; (e_1, \ldots, e_n, \ldots), (j_0, \ldots, j_n, \ldots)\big) \\ &\quad = \Phi^{\mathfrak{R}}\big(h; \big(e_{m+1} - R_{j_m}(t - \xi_m), e_{m+2} \ldots, e_{m+n}, \ldots\big), \\ &\qquad\quad \times (j_m, \ldots, j_{m+n}, \ldots)\big) \quad \text{for } \xi_m \le t < \xi_{m+1}. \end{aligned} \tag{5.2.11}$$

Now, let $\tilde{A} \in \sigma(\{\tilde{X}(\tau) : \tau \in [0,t]\})$ be given, and assume that $\tilde{X}(t) = i$ on A. Set $\tilde{A}_m \equiv \tilde{A} \cap \{\tilde{J}_m \le s < \tilde{J}_{m+1}\}$ for $m \ge 0$. Then

$$\tilde{A}_m = \{\tilde{E}_{m+1} > R_i(t - \tilde{J}_m)\} \cap \tilde{B}_m,$$

where $\{\tilde{J}_m \le s\} \supseteq \tilde{B}_m \in \sigma(\{\tilde{E}_1, \dots, \tilde{E}_m\} \cup \{\tilde{X}_0, \dots, \tilde{X}_m\})$. In addition,

$$\begin{aligned}\mathbb{P}(\{\tilde{X}(t+j) = j\} \cap \tilde{A}) &= \sum_{m=0}^{\infty} \mathbb{P}(\{\tilde{X}(t+h) = j\} \cap \tilde{A}_m) \\ &= \sum_{m=0}^{\infty} \mathbb{P}(\{\tilde{X}(t+h) = j \;\&\; \tilde{E}_{m+1} > R_i(t - \tilde{J}_m)\} \cap \tilde{B}_m),\end{aligned}$$

and, by (5.2.9) and (5.2.11),

$$\begin{aligned}&\mathbb{P}(\{\tilde{X}(t+h) = j \;\&\; \tilde{E}_{m+1} > R_i(t - \tilde{J}_m)\} \cap \tilde{B}_m) \\ &\quad = \mathbb{P}(\{\Phi^{\mathfrak{R}}(h; (\tilde{E}_{m+1} - R_i(t - \tilde{J}_m), \tilde{E}_{m+2}, \dots, \tilde{E}_{m+n}, \dots), \\ &\qquad\qquad (i, \tilde{X}_{m+1}, \dots, \tilde{X}_{m+n}, \dots)) = j\} \cap \{\tilde{E}_{m+1} > R_i(t - \tilde{J}_m)\} \cap \tilde{B}_m).\end{aligned}$$

Since $\tilde{E}_{m+1}$ is independent of all the other quantities in the last expression, we can apply (5.1.3) to see that it is equal to

$$\mathbb{P}(\tilde{X}(h) = j \mid \tilde{X}_0 = i)\mathbb{E}[e^{-R_i(t-\tilde{J}_m)}, \tilde{B}_m] = \mathbb{P}(\tilde{X}(h) = j \mid \tilde{X}_0 = i)\mathbb{P}(\tilde{A}_m).$$

Hence, we will be done once we show that

$$\lim_{h \searrow 0} \sup_{i,j \in \mathbb{S}} \frac{1}{h} \left|\mathbb{P}(\tilde{X}(h) = j \mid \tilde{X}_0 = i) - (1 - hR_i)\delta_{i,j} - hR_i(\mathbf{P})_{ij}\right| = 0. \qquad (*)$$

Note that $\mathbb{P}(\tilde{X}(h) = j \mid \tilde{X}_0 = i)$ equals

$$\begin{aligned}&\mathbb{P}(\tilde{X}_0 = j \;\&\; \tilde{J}_1 \le h \mid \tilde{X}_0 = i) + \mathbb{P}(\tilde{X}_1 = j \;\&\; \tilde{J}_1 \le h < \tilde{J}_2 \mid \tilde{X}_0 = i) \\ &\quad + \mathbb{P}(\tilde{X}(h) = j \;\&\; \tilde{J}_2 \le h \mid \tilde{X}_0 = i).\end{aligned}$$

Obviously

$$\mathbb{P}(\tilde{X}_0 = j \;\&\; \tilde{J}_1 \le h \mid \tilde{X}_0 = i) = e^{-hR_i}\delta_{i,j}$$

and

$$\begin{aligned}&\mathbb{P}(\tilde{X}_1 = j \;\&\; \tilde{J}_1 \le h < \tilde{J}_2 \mid \tilde{X}_0 = i) = \mathbb{P}(R_i^{-1}\tilde{E}_1 \le h < R_i^{-1}\tilde{E}_1 + R_j^{-1}\tilde{E}_2)(\mathbf{P})_{ij} \\ &\quad = (1 - e^{-hR_i})(\mathbf{P})_{ij} - \mathbb{P}(R_i^{-1}\tilde{E}_1 + R_j^{-1}\tilde{E}_2 \le h)(\mathbf{P})_{ij}.\end{aligned}$$

Hence

$$\left|\mathbb{P}(\tilde{X}(h) = j \mid \tilde{X}_0 = i) - e^{-hR_i}\delta_{i,j} - (1 - e^{-hR_i})(\mathbf{P})_{ij}\right| \le 2\mathbb{P}(\tilde{E}_1 + \tilde{E}_2 \le Rh),$$

where $R = \sup_{i\in\mathbb{S}} R_i$. Finally,

$$\mathbb{P}(\tilde{E}_1 + \tilde{E}_2 \le Rh) = \int_0^{Rh} \tau e^{-\tau}\, d\tau \le (Rh)^2,$$

and so (∗) follows and the proof of theorem is complete. □

5.2.3 Distribution of Jumps and Jump Times

As a consequence of Theorem 5.2.7, we see that the distribution of $\{X(t) : t \ge 0\}$ is determined by its initial distribution μ and the matrix $\mathbf{Q} = \mathbf{R}(\mathbf{P} - \mathbf{I})$. For this reason, we say that a stochastic process satisfying conditions (a) and (b) is a Markov process *generated* $\mathbf{Q}$ with initial distribution μ.

The distinguishing characteristics of $\mathbf{Q}$ are that its off-diagonal entries are non-negative and each of its rows sums to 0. For no better reason than that Q follows P in the alphabet, probabilists call such a matrix a *Q-matrix*. Given a Q-matrix $\mathbf{Q}$, there are many diagonal matrices $\mathbf{R}$ with non-negative entries and transition matrices $\mathbf{P}$ for which $\mathbf{Q} = \mathbf{R}(\mathbf{P} - \mathbf{I})$. Indeed, we already took advantage of this fact when we wrote $\mathbf{R}(\mathbf{P} - \mathbf{I})$ as (cf. (5.2.3)) $R(\tilde{\mathbf{P}} - \mathbf{I})$ with $R = \sup_{i\in\mathbb{S}} R_i$. Nonetheless, it is possible to remove this ambiguity and make a canonical choice of $\mathbf{R}$ and $\mathbf{P}$. For reasons that will become clear shortly, we will make this canonical choice by taking $(\mathbf{R})_{ij} = \delta_{i,j} R_i$ and $\mathbf{P}$ where

$$\begin{aligned} &R_i = -(\mathbf{Q})_{ii}, \qquad (\mathbf{P})_{ii} = \mathbf{1}_{\{0\}}(R_i), \quad \text{and,} \\ &\text{for } j \ne i, \quad (\mathbf{P})_{ij} = \begin{cases} 0 & \text{if } R_i = 0 \\ R_i^{-1}(\mathbf{Q})_{ij} & \text{if } R_i > 0. \end{cases} \end{aligned} \tag{5.2.12}$$

When $\mathfrak{R} = \{R_i : i \in \mathbb{S}\}$ and $\mathbf{P}$ are those given by (5.2.12), we will call $\mathfrak{R}$ the *rates determined by* $\mathbf{Q}$ and $\mathbf{P}$ the *transition probability matrix determined by* $\mathbf{Q}$.

Our goal in this subsection is to prove the following theorem that describes how $\{X(t) : t \ge 0\}$ jumps when it jumps.

Theorem 5.2.13 *Let $\{X(t) : t \ge 0\}$ be a Markov process generated by $\mathbf{Q}$ and let $\mathfrak{R}$ and $\mathbf{P}$ be given by* (5.2.12). *Set $J_0 = 0$,*

$$J_n = \inf\{t > J_{n-1} : X(t) \ne X(J_{n-1})\} \quad \textit{for } n \ge 1,$$

$$X_n = \begin{cases} X(J_n) & \textit{if } J_n < \infty \\ X_{n-1} & \textit{if } n \ge 1 \textit{ and } J_n = \infty, \end{cases}$$

$$E_n = \begin{cases} R_{X_{n-1}}(J_n - J_{n-1}) & \textit{if } J_n < \infty \\ 0 & \textit{if } J_n = \infty, \end{cases}$$

and $\zeta = \inf\{n \geq 0 : R_{X_n} = 0\}$. *Then* $\{X_n : n \geq 0\}$ *is a Markov chain with transition matrix* $\mathbf{P}$ *and, for any* $K \geq 1$ *and* $\{t_k : 1 \leq k \leq K\} \subseteq (0, \infty)$,

$$\mathbb{P}\big(\{E_k > t_k, 1 \leq k \leq K\} \cap A\big) = e^{-\sum_{k=1}^K t_k}\mathbb{P}(A)$$

for any $A \in \sigma(\{X_n : n \geq 0\})$ *contained in* $\{\zeta > K\}$.

Proof Refer to Sect. 5.2.2 and let $\{\tilde{X}(t) : t \geq 0\}$ be the process in Theorem 5.2.10 when μ is the distribution of $X(0)$. Because it has the same distribution as $\{X(t) : t \geq 0\}$, it suffices to prove that $\{\tilde{X}(t) : t \geq 0\}$ has the asserted properties.

Obviously, $\{\tilde{J}_n : n \geq 0\}$ and $\{\tilde{X}_n : n \geq 0\}$ are is related to $\{\tilde{X}(t) : t \geq 0\}$ in the same way as $\{J_n : n \geq 0\}$ and $\{X_n : n \geq 0\}$ are related to $\{X(t) : t \geq 0\}$. Furthermore, if $\tilde{\zeta} = \inf\{n \geq 0 : R_{\tilde{X}_{n-1}} = 0\}$, then $\tilde{\zeta}$ is related to $\{\tilde{X}(t) : t \geq 0\}$ the same way that ζ is related to $\{X(t) : t \geq 0\}$. Finally, $\tilde{E}_k = R_{\tilde{X}_{k-1}}(\tilde{J}_k - \tilde{J}_{k-1})$ if $\tilde{\zeta} > k$. Now suppose that $\tilde{A} \in \sigma(\{\tilde{X}_n : n \geq 0\})$ is contained in $\{\tilde{\zeta} > K\}$. Then, because the $\tilde{E}_k$'s are independent of $\tilde{A}$,

$$\begin{aligned}\mathbb{P}\big(\{\tilde{E}_k > t_k, 1 \leq k \leq K\} \cap \tilde{A}\big) &= \mathbb{P}(\tilde{E}_k > t_k, 1 \leq k \leq K)\mathbb{P}(\tilde{A}) \\ &= e^{-\sum_{k=1}^K t_k}\mathbb{P}(\tilde{A}).\end{aligned}$$ □

Remark It is important to appreciate what Theorem 5.2.13 is saying. Namely, if $\mathfrak{R}$ and $\mathbf{P}$ are the canonical rates and transition probability matrix determined by the Q-matrix $\mathbf{Q}$ and if $\{X(t) : t \geq 0\}$ is a Markov process generated by $\mathbf{Q}$, then the distribution of $\{X(J_n \wedge \zeta) : n \geq 0\}$, where the J_n's are the successive jump times of the process, has the same distribution as that of a Markov chain with transition matrix $\mathbf{P}$ stopped at time ζ. Furthermore, conditional on $\{X(J_n \wedge \zeta) : n \geq 0\}$, the holding times $\{J_n - J_{n-1} : n \geq 1\}$ are mutually independent exponential randoms, $J_n - J_{n-1}$ being infinite if $R_{X(J_{n-1})} = 0$, and $J_n - J_{n-1}$ being a exponential random variable with mean value $R_{X(J_{n-1})}^{-1}$ if $R_{X(J_{n-1})} > 0$.

5.2.4 *Kolmogorov's Forward and Backward Equations*

As we saw at the end of the Sect. 3.2.1, apart from its initial distribution, the distribution of a Markov process is completely determined by the transition probability function $t \rightsquigarrow \mathbf{P}(t)$ in (5.2.2). Although the expression in (5.2.2) is useful from a theoretical perspective, for many applications it is not practical. For this reason it is desirable to have other ways of doing computations involving $\mathbf{P}(t)$, and we will develop some in this subsection.

It should be clear both from (5.1.9) and (5.2.2) that $\mathbf{P}(t)$ is some sort of exponential function. Indeed, (5.2.2) makes it reasonable to think that $\mathbf{P}(t)$ ought to be written as $e^{t\mathbf{Q}}$, and that is how a functional analyst would write it. Taking a hint from the usual exponential function for real numbers, one should guess that another way to describe $\mathbf{P}(t)$ is as the solution to the equation

$$\dot{\mathbf{P}}(t) \equiv \frac{d}{dt}\mathbf{P}(t) = \mathbf{P}(t)\mathbf{Q} \quad \text{with } \mathbf{P}(0) = \mathbf{I}. \tag{5.2.14}$$

In fact, we made implicit use of (5.2.14), which is called *Kolmogorov's forward equation*, in Lemma 5.2.5. The term "forward" can be understood by writing (5.2.14) in coordinates:

$$\big(\dot{\mathbf{P}}(t)\big)_{ij} = \sum_{k\in\mathbb{S}} \big(\mathbf{P}(t)\big)_{ik}(\mathbf{Q})_{kj}.$$

Thinking in terms of someone traveling along the paths of the process, in this expression i is the *backward variable* since it is the variable that specifies where the traveler was at time 0 and is therefore the one the traveler sees when he looks backward in time. Similarly, j is the *forward variable* because it is the one that says where he will be at time t. Thus (5.2.14) is an equation that gives the evolution of $t\rightsquigarrow\mathbf{P}(t)$ in terms of its forward variable when the backward variable is fixed. Of course, it is equally reasonable to expect that

$$\dot{\mathbf{P}}(t) = \mathbf{Q}\mathbf{P}(t) \quad \text{with } \mathbf{P}(0) = \mathbf{I}, \tag{5.2.15}$$

which is known as *Kolmogorov's backward equation* since its coordinate version expresses the evolution of $t\rightsquigarrow\mathbf{P}(t)$ as a function of its backward variable.

The proof that $t\rightsquigarrow\mathbf{P}(t)$ solves (5.2.14) and (5.2.15) is easy. By (5.1.9),

$$\mathbf{P}(t+h) - \mathbf{P}(t) = \mathbf{P}(t)\big(\mathbf{P}(h) - \mathbf{I}\big) = \big(\mathbf{P}(h) - \mathbf{I}\big)\mathbf{P}(t),$$

and so, for $h \in (0, 1]$,

$$\begin{aligned}
&\left\| \frac{\mathbf{P}(t+h) - \mathbf{P}(t)}{h} - \mathbf{P}(t)\mathbf{Q} \right\|_{\mathrm{u,v}} \vee \left\| \mathbf{Q}\mathbf{P}(t) - \frac{\mathbf{P}(t+h) - \mathbf{P}(t)}{h} \right\|_{\mathrm{u,v}} \\
&\quad \le \left\| \sum_{n=2}^{\infty} \frac{h^{n-1}\mathbf{Q}^n}{n!} \right\|_{\mathrm{u,v}} \le he^{2R},
\end{aligned}$$

where $R = \sup_{i\in\mathbb{S}}(\mathbf{Q})_{ii}$. In addition, one can show that $t\rightsquigarrow\mathbf{P}(t)$ is uniquely determined by either (5.2.14) or (5.2.15). In fact, suppose that $t \in [0, \infty) \longmapsto \mathbf{M}(t) \in M_{\mathrm{u,v}}(\mathbb{S})$ is a continuously differentiable and

$$\dot{\mathbf{M}}(t) = \mathbf{M}(t)\mathbf{Q} \quad \text{or} \quad \dot{\mathbf{M}}(t) = \mathbf{Q}\mathbf{M}(t).$$

Then $\mathbf{M}(t) = \mathbf{0}$ for $t \ge 0$ if $\mathbf{M}(0) = \mathbf{0}$. To see this, note that in either case,

$$\big\|\mathbf{M}(t)\big\|_{\mathrm{u,v}} \le \|\mathbf{Q}\|_{\mathrm{u,v}} \int_0^t \big\|\mathbf{M}(\tau)\big\|_{\mathrm{u,v}} \quad \text{for all } t \ge 0$$

and apply *Gronwall's inequality* (cf. Exercise 5.5.5). Applying this to the difference of two solutions of either (5.2.14) or of (5.2.15) one sees that the difference vanishes.

In applications, one often uses these ideas to compute $t\rightsquigarrow\boldsymbol{\mu}\mathbf{P}(t)$ or $t\rightsquigarrow\mathbf{P}(t)\mathbf{f}$ for probability vectors $\boldsymbol{\mu}$ or bounded column vectors $\mathbf{f}$. That is, (5.2.14) and (5.2.15) say that

$$\begin{aligned}&\frac{d}{dt}\boldsymbol{\mu}\mathbf{P}(t)=\boldsymbol{\mu}\mathbf{P}(t)\mathbf{Q}\quad\text{with }\boldsymbol{\mu}\mathbf{P}(0)=\boldsymbol{\mu}\\&\frac{d}{dt}\mathbf{P}(t)\mathbf{f}=\mathbf{Q}\mathbf{P}(t)\mathbf{f}\quad\text{with }\mathbf{P}(0)\mathbf{f}=\mathbf{f},\end{aligned}\tag{5.2.16}$$

and, just as in the preceding, one can use Gronwall's inequality to check that $t\rightsquigarrow\boldsymbol{\mu}\mathbf{P}(t)$ and $t\rightsquigarrow\mathbf{P}(t)\mathbf{f}$ are uniquely determined by these equations.

5.3 Unbounded Rates

Thus far we have been assuming that the rates $\mathfrak{R}=\{R_i : i\in\mathbb{S}\}$ are bounded. This assumption simplified our considerations in several ways. For one thing, it allowed us to use (5.2.2) to define and derive properties of $t\rightsquigarrow\mathbf{P}(t)$. Secondly, and more important, it allowed us to show that there exists a process that satisfies the conditions (a) and (b) at the beginning of Sect. 5.2.1. The easiest way to understand what might prevent one from constructing such a process is to look at the construction given in Sect. 5.2.2. No matter what the rates are, the process (5.2.9) can be used to define $\tilde{X}(t)$ as long as $t<\sum_{m=1}^{\infty}R^{-1}_{\tilde{X}_{m-1}}\tilde{E}_m$. When the R_i's are bounded, and therefore $\sum_{m=1}^{\infty}R^{-1}_{\tilde{X}_{m-1}}\tilde{E}_m=\infty$, $\tilde{X}(t)$ is defined for all $t\in[0,\infty)$. However, when the R_i's are unbounded, $\sum_{m=1}^{\infty}R^{-1}_{\tilde{X}_{m-1}}\tilde{E}_m$ may be finite with positive probability, with the result that $\tilde{X}(t)$ will not be defined for all t. In this section I will give conditions, other than boundedness, that enable one to overcome this problem.

5.3.1 Explosion

The setting here is the same as the one in Sect. 5.2.2, only now we are no longer assuming that the rates are bounded. Even so, as I just said, $\tilde{X}(t)$ is well-defined by the prescription in (5.2.9) as long as $0\le t<\tilde{J}_\infty\equiv\sum_{m=1}^{\infty}R^{-1}_{\tilde{X}_{m-1}}\tilde{E}_m$.

Our first step is to show that, with probability 1, $\tilde{J}_\infty$ coincides with the time when the process explodes out of the state space. To be precise, choose an *exhaustion* $\{F_N : N\ge 1\}$ of $\mathbb{S}$ by non-empty, finite subsets. That is, for each N, F_N is a non-empty, finite subset of $\mathbb{S}$, $F_N\subseteq F_{N+1}$, and $\mathbb{S}=\bigcup_N F_N$. Next, take $\mathfrak{R}^{(N)}$ to be the set of rates given by

$$R_i^{(N)}=\begin{cases}R_i & \text{if } i\in F_N\\ 0 & \text{if } i\notin F_N.\end{cases}$$

Then, for each $N \geq 1$, (5.2.9) with $\mathfrak{R}^{(N)}$ replacing $\mathfrak{R}$ determines a Markov process $\{\tilde{X}^{(N)}(t) : t \geq 0\}$. Moreover, if $\tilde{\zeta}_N \equiv \inf\{t \geq 0 : \tilde{X}^{(N)}(t) \notin F_N\}$, then

$$t \wedge \tilde{\zeta}_N < \tilde{J}_\infty \quad \text{and} \quad \tilde{X}(t \wedge \tilde{\zeta}_N) = \tilde{X}^{(N)}(t \wedge \tilde{\zeta}_N) \quad \text{for all } t \geq 0. \tag{5.3.1}$$

To see this, choose $n \geq 0$ so that

$$\tilde{J}_n^{(N)} \leq t \wedge \tilde{\zeta}_N < \tilde{J}_{n+1}^{(N)},$$

where $\tilde{J}_0^{(N)} = 0$ and

$$\tilde{J}_n^{(N)} = \sum_{m=1}^{n} \frac{\tilde{E}_m}{R^{(N)}_{\tilde{X}^{(N)}_{m-1}}} \quad \text{for } n \geq 1.$$

Because $\tilde{X}_{m-1} \in F_N$ for $1 \leq m \leq n+1$, $\tilde{J}_\ell^{(N)} = \tilde{J}_\ell < \tilde{J}_\infty$ for all $0 \leq \ell \leq n+1$, and so $t \wedge \tilde{\zeta}_N < J_\infty$ and $\tilde{J}_n \leq t \wedge \tilde{\zeta}_N < \tilde{J}_{n+1}$. At the same time, it also means that

$$\begin{aligned}\tilde{X}(t \wedge \tilde{\zeta}_N) &= \Phi^{\mathfrak{R}}\big(t \wedge \tilde{\zeta}_N; (\tilde{E}_1, \dots, \tilde{E}_n, \dots), (\tilde{X}_0, \dots, \tilde{X}_n, \dots)\big) \\ &= \Phi^{\mathfrak{R}^{(N)}}\big(t \wedge \tilde{\zeta}_N; (\tilde{E}_1, \dots, \tilde{E}_n, \dots), (\tilde{X}_0, \dots, \tilde{X}_n, \dots)\big) = \tilde{X}^{(N)}(t \wedge \tilde{\zeta}_N).\end{aligned}$$

As a consequence of (5.3.1), we know that $\tilde{X}^{(N+1)}(t \wedge \tilde{\zeta}_N) = X^{(N)}(t \wedge \tilde{\zeta}_N)$ for all $t \geq 0$ and therefore that $\tilde{\zeta}_N \leq \tilde{\zeta}_{N+1}$. Hence the *explosion time* $\tilde{\mathfrak{e}} \equiv \lim_{N\to\infty} \tilde{\zeta}_N$ exists (in $[0, \infty]$).

Lemma 5.3.2 $\mathbb{P}(\tilde{\mathfrak{e}} = \tilde{J}_\infty) = 1$ *and so, with probability* 1, $\tilde{X}(t)$ *is well-defined for* $t \in [0, \tilde{\mathfrak{e}})$.

Proof Because $\{\tilde{\mathfrak{e}} \neq \tilde{J}_\infty\}$ can be written as the union of the sets

$$\{\tilde{\mathfrak{e}} > T \geq \tilde{J}_\infty\} \cup \{\tilde{J}_\infty > T \geq \tilde{\mathfrak{e}}\}$$

as T runs over the positive rational numbers, in order to prove that $\tilde{\mathfrak{e}} = \tilde{J}_\infty$ with probability 1, (7.1.5) says that it suffices for us to show that

$$\mathbb{P}(\tilde{\mathfrak{e}} > T \geq \tilde{J}_\infty) = 0 = \mathbb{P}(\tilde{J}_\infty > T \geq \tilde{\mathfrak{e}}) \quad \text{for each } T > 0.$$

To this end, first suppose that $\mathbb{P}(\tilde{\mathfrak{e}} > T \geq \tilde{J}_\infty) > 0$ for some T. Then there exists an N such that $\mathbb{P}(\tilde{\zeta}_N > T \geq \tilde{J}_\infty) > 0$. On the other hand,

$$\tilde{\zeta}_N > T \geq \tilde{J}_\infty \quad \Longrightarrow \quad \tilde{J}_\infty \geq \frac{1}{R^{(N)}} \sum_1^\infty \tilde{E}_m,$$

where $R^{(N)} = \sup_{i \in \mathbb{S}} R_j^{(N)}$, and therefore we are led to the contradiction

$$0 < \mathbb{P}(\tilde{\zeta}_N > T \geq \tilde{J}_\infty) \leq \mathbb{P}\left(\sum_{m=1}^\infty E_m \leq R^{(N)} T\right) = 0.$$

Next suppose that $\mathbb{P}(\tilde{J}_\infty > T \ge \tilde{\mathfrak{e}}) > 0$. Then there exists an $n \ge 1$ such that $\mathbb{P}(\tilde{J}_n > T \ge \tilde{\mathfrak{e}}) > 0$. On the other hand, as $N \to \infty$,

$$\begin{aligned}\mathbb{P}(\tilde{J}_n > T \ge \tilde{\mathfrak{e}}) &\le \mathbb{P}(\tilde{J}_n > T \ge \zeta_N) \\ &\le \mathbb{P}(\exists 0 \le m \le n \tilde{X}_m \notin F_N) \longrightarrow 0.\end{aligned}$$

That is, we know that $\mathbb{P}(\tilde{J}_n > T \ge \tilde{\mathfrak{e}}) = 0$. □

Lemma 5.3.3 *If* $\mathbb{P}(\tilde{\mathfrak{e}} = \infty | \tilde{X}(0) = i) = 1$ *for all* $i \in \mathbb{S}$ *and*

$$\big(\mathbf{P}(t)\big)_{ij} \equiv \mathbb{P}\big(\tilde{X}(t) = j | \tilde{X}(0) = i\big),$$

then, for each initial distribution $\boldsymbol{\mu}$ *and* $T > 0$,

$$\lim_{N\to\infty} \sup_{t\in(0,T]} \big\|\boldsymbol{\mu}\mathbf{P}^{(N)}(t) - \boldsymbol{\mu}\mathbf{P}(t)\big\|_{\mathrm{v}} = 0, \tag{5.3.4}$$

where $t\rightsquigarrow\mathbf{P}^{(N)}(t)$ *is the transition probability function in* (5.2.2) *when the* Q*-matrix there is given by* $(\mathbf{Q}^{(N)})_{ij} = \mathbf{1}_{F_N}(i)R_i(\mathbf{P} - \mathbf{I})_{ij}$. *Moreover,* $t\rightsquigarrow\mathbf{P}(t)$ *satisfies* (5.1.9) *and* $\{\tilde{X}(t) : t \ge 0\}$ *is a Markov process for which it is the transition probability function. Finally,* $t\rightsquigarrow\mathbf{P}(t)$ *satisfies Kolmogorov's backward equation in the sense that, for each* $(i, j) \in \mathbb{S}^2$,

$$\big(\mathbf{P}(t)\big)_{ij} = \delta_{i,j} + \int_0^t \big(\mathbf{Q}\mathbf{P}(\tau)\big)_{ij}\, d\tau, \tag{5.3.5}$$

where $\mathbf{Q} = \mathbf{R}(\mathbf{P} - \mathbf{I})$ *with* $(\mathbf{R})_{ij} = R_i\delta_{i,j}$.

Proof First note that

$$\mathbb{P}(\mathfrak{e} = \infty) = \sum_{i\in\mathbb{S}} (\boldsymbol{\mu})_i \mathbb{P}\big(\mathfrak{e} = \infty \,|\, X(0) = i\big) = 1,$$

and therefore that $\lim_{N\to\infty} \mathbb{P}(\zeta_N \le T) = 0$ for each $T \in [0, \infty)$. At the same time, by Theorem 5.3.2,

$$\big|\big(\boldsymbol{\mu}\mathbf{P}^{(N)}(t)\big)_j - \big(\boldsymbol{\mu}\mathbf{P}(t)\big)_j\big| \le \mathbb{P}\big(\tilde{X}(t) = j \;\&\; \zeta_N \le T\big)$$

for all $0 \le t \le T$ and $j \in \mathbb{S}$. Thus,

$$\sup_{t\in(0,T]} \big\|\boldsymbol{\mu}\mathbf{P}^{(N)}(t) - \boldsymbol{\mu}\mathbf{P}(t)\big\|_{\mathrm{v}} \le \mathbb{P}(\zeta_N \le T) \longrightarrow 0 \quad \text{as } N \to \infty.$$

Given (5.3.4), we have that

$$\big(\mathbf{P}(s+t)\big)_{ij} = \lim_{N\to\infty} \big(\mathbf{P}^{(N)}(s+t)\big)_{ij} = \lim_{N\to\infty} \big(\mathbf{P}^{(N)}(s)\mathbf{P}^{(N)}(t)\big)_{ij}$$

for each $(i, j) \in \mathbb{S}^2$. In addition, by applying (5.3.4) with $\boldsymbol{\mu} = (\mathbf{P}(s))_{i\cdot}$, we know that

$$\big(\mathbf{P}(s)\mathbf{P}(t)\big)_{ij} = \lim_{N\to\infty} \big(\mathbf{P}(s)\mathbf{P}^{(N)}(t)\big)_{ij}$$

for each $(i, j) \in \mathbb{S}^2$. Finally, again by (5.3.4), for each $M \geq 1$

$$\sum_{k\in F_M} \big(\mathbf{P}(s)\big)_{ik} = \lim_{N\to\infty} \sum_{k\in F_M} \big(\mathbf{P}^{(N)}(s)\big)_{ik},$$

and therefore, since

$$\sum_{k\notin F_M} \big(\mathbf{P}(s)\big)_{ik} = 1 - \sum_{k\in F_M} \big(\mathbf{P}(s)\big)_{ik},$$

we know that

$$\sum_{k\notin F_M} \big(\mathbf{P}(s)\big)_{ik} = \lim_{N\to\infty} \sum_{k\notin F_M} \big(\mathbf{P}^{(N)}(s)\big)_{ik}.$$

Now let $\epsilon > 0$ be given, and use the preceding to choose an M such that

$$\sum_{k\notin F_M} \big(\mathbf{P}^{(N)}(s)\big)_{ik} \leq \epsilon \quad \text{for all } N \geq M.$$

Then, for $N \geq M$,

$$\big|\big(\mathbf{P}(s)\mathbf{P}(t)\big)_{ij} - \big(\mathbf{P}(s)\mathbf{P}^{(N)}(t)\big)_{ij}\big| \leq \sum_{k\in F_M} \big|\big(\mathbf{P}(s)\big)_{ik} - \big(\mathbf{P}^{(N)}(s)\big)_{ik}\big| + \epsilon.$$

Since, by (5.3.4), the first term on the right tends to 0 as $N \to \infty$, we have now shown that $\mathbf{P}(t+s) = \mathbf{P}(s)\mathbf{P}(t)$.

To prove the Markov property, let $A \in \sigma(\{\tilde{X}(\tau) : \tau \in [0, s]\})$ be given, and assume that $\tilde{X}(s) = i$ on A. Then, since, by (5.3.1),

$$A \cap \{\tilde{\zeta}_N > s\} \in \sigma\big(\big\{\tilde{X}^{(N)}(\tau) : \tau \in [0, s]\big\}\big),$$

the Markov property for $\{\tilde{X}^{(N)}(t) : t \geq 0\}$ plus (5.3.1) and the result in Lemma 5.3.2 justify

$$\begin{aligned}\mathbb{P}\big(\big\{\tilde{X}(s+t) = j\big\} \cap A\big) &= \lim_{N\to\infty} \mathbb{P}\big(\big\{\tilde{X}^{(N)}(s+t) = j\big\} \cap A \cap \{\tilde{\zeta}_N > s\}\big) \\ &= \lim_{N\to\infty} \big(\mathbf{P}^{(N)}(t)\big)_{ij} \mathbb{P}\big(A \cap \{\zeta_N > s\}\big) = \big(\mathbf{P}(t)\big)_{ij} \mathbb{P}(A).\end{aligned}$$

Finally, to check (5.3.5), note that, by (5.2.15) applied to $\{\mathbf{P}^{(N)}(t) : t \geq 0\}$,

$$\big(\mathbf{P}^{(N)}(t)\big)_{ij} = \delta_{i,j} + \int_0^t \big(\mathbf{Q}\mathbf{P}^{(N)}(\tau)\big)_{ij}\, d\tau$$

as soon as N is large enough that $i \in F_N$. Hence, since

$$1 \geq \left(\mathbf{P}^{(N)}(\tau)\right)_{kj} \longrightarrow \left(\mathbf{P}(\tau)\right)_{kj}$$

while $\sum_k |(\mathbf{Q})_{ik}| = 2R_i < \infty$, (5.3.5) follows by Lebesgue's dominated convergence theorem. □

Theorem 5.3.6 *Let* $\mathbf{Q}$ *be a Q-matrix and, for* $N \geq 1$, *determine the Q-matrix* $\mathbf{Q}^{(N)}$ *by* $(\mathbf{Q}^{(N)})_{ij} = \mathbf{1}_{F_N}(i)(\mathbf{Q})_{ij}$. *Given a point* Δ *not in* $\mathbb{S}$ *and an* $i \in \mathbb{S}$, *there is a* $\mathbb{S} \cup \{\Delta\}$-*valued, right continuous, piecewise constant stochastic process* $\{X(t) : t \geq 0\}$ *with the properties that*

(1) $\mathbb{P}(X(0) = i) = 1$,
(2) *for each* $N \geq 1$ *with* $i \in F_N$, $\{X(t \wedge \zeta^{F_N}) : t \geq 0\}$, *where* $\zeta^{F_N} = \inf\{t \geq 0 : X(t) \notin F_N\}$, *is a Markov process generated by* $\mathbf{Q}^{(N)}$,
(3) *if* $\mathfrak{e} \equiv \lim_{N\to\infty} \zeta^{F_N}$, *then* $\mathbb{P}(X(t) = \Delta \text{ for } t \in [\mathfrak{e}, \infty)) = 1$.

Moreover, the distribution of $\{X(t) : t \geq 0\}$ *is uniquely determined by these properties. Finally, if* $\mathbb{P}(\mathfrak{e} = \infty) = 1$, *then* $\{X(t) : t \geq 0\}$ *is a Markov process with the transition probability function* $t \rightsquigarrow \mathbf{P}(t)$ *described in Lemma* 5.3.3.

Proof Let $\mathfrak{R}$ and $\mathbf{P}$ be the canonical rates and transition probability determined by $\mathbf{Q}$, and define $\{\tilde{X}(t) : t \in [0, \tilde{J}_\infty)\}$ accordingly for $\mathfrak{R}$ and $\mathbf{P}$ with $\tilde{X}(0) \equiv i$. Next, set $X(t) = \tilde{X}(t)$ if $t \in [0, \tilde{J}_\infty \wedge \tilde{\mathfrak{e}})$ and $X(t) = \Delta$ if $t \in [\tilde{J}_\infty \wedge \tilde{e}, \infty)$. Obviously, $\mathbb{P}(X(0) = i) = 1$, $\zeta^{F_N} = \tilde{\zeta}_N$ for all $N \geq 1$, and, by Lemma 5.3.2, $\mathbb{P}(\mathfrak{e} = \tilde{\mathfrak{e}}) = 1$. Hence, for each $N \geq 1$, $X(t \wedge \zeta^{F_N}) = \tilde{X}^{(N)}(t)$ for all $t \geq 0$, and so, if $i \in F_N$, then $\{X(t \wedge \zeta^{F_N}) : t \geq 0\}$ is a Markov process generated by $\mathbf{Q}^{(N)}$ starting at i, and $\mathbb{P}(X(t) = \Delta, t \in [\mathfrak{e}, \infty)) = 1$.

To see that the distribution of $\{X(t) : t \geq 0\}$ is uniquely determined, let $0 = t_0 < \cdots < t_K$ and $(j_0, \ldots, j_K) \in (\mathbb{S} \cup \{\Delta\})^{K+1}$ be given, and set $A = \{X(t_k) = j_k, 0 \leq k \leq K\}$. If $j_0 \neq i$, then $\mathbb{P}(A) = 0$. Thus, assume that $j_0 = i$, in which case,

$$\mathbb{P}(A) = \sum_{\ell=1}^{K} \mathbb{P}\big(A \cap \{t_{\ell-1} < \mathfrak{e} \leq t_\ell\}\big) + \mathbb{P}\big(A \cap \{\mathfrak{e} > t_K\}\big).$$

For each $1 \leq \ell \leq K+1$, set $A_\ell = \{X(t_k) = j_k, 0 \leq k < \ell\}$. Then $A_{K+1} = A$ and

$$\mathbb{P}\big(A \cap \{t_{\ell-1} < \mathfrak{e} \leq t_\ell\}\big) = \begin{cases} 0 & \text{if } j_k = \Delta \text{ for some } 0 \leq k < \ell \\ & \text{or } j_k \neq \Delta \text{ for some } \ell \leq k \leq K \\ \mathbb{P}(A_\ell \cap \{\mathfrak{e} > t_{\ell-1}\}) & \text{otherwise.} \end{cases}$$

Assuming that $j_k \neq \Delta$ for $0 \le k < \ell$ and that $j_k = \Delta$ for $\ell \le k \le K$,

$$\begin{aligned}\mathbb{P}\big(A_\ell \cap \{\mathfrak{e} > t_{\ell-1}\}\big) &= \lim_{N\to\infty} \mathbb{P}\big(A_\ell \cap \{\zeta^{(N)} > t_{\ell-1}\}\big)\\ &= \lim_{N\to\infty} \mathbb{P}\big(X\big(t_k \wedge \zeta^{F_N}\big) = j_k, 0 \le k < \ell\big)\\ &= \lim_{N\to\infty} \mathbb{P}\big(\tilde{X}^{(N)}(t_k) = j_k, 0 \le k < \ell\big).\end{aligned}$$

Hence, in all cases $\mathbb{P}(A)$ is uniquely determined by (1), (2), and (3).

Finally, when $\mathbb{P}(\mathfrak{e} = \infty) = 1$, then the last assertion is covered by Lemma 5.3.3. □

As Theorem 5.3.6 makes clear, aside from the choice of Δ and the initial state i, the distribution of the process $\{X(t) : t \ge 0\}$ and the random variable $\mathfrak{e}$ are completely determined by $\mathbf{Q}$. For this reason, when $\mathbb{P}(\mathfrak{e} = \infty) = 1$, we say that the process *does not explode* and will call any right continuous, piecewise constant process with this distribution a *Markov process generated by* $\mathbf{Q}$ *starting from* i.

Corollary 5.3.7 *Assume that explosion does not occur for any* $i \in \mathbb{S}$. *Then, for each probability vector* $\boldsymbol{\mu}$ *on* $\mathbb{S}$ *there exists a right continuous, piecewise constant Markov process* $\{X(t) : t \ge 0\}$ *with initial distribution* $\boldsymbol{\mu}$ *such that, for each* $i \in \mathbb{S}$ *with* $\mu_i > 0$, *the conditional distribution of* $\{X(t) : t \ge 0\}$ *given that* $X(0) = i$ *is a Markov process generated by* $\mathbf{Q}$ *starting from* i. *In particular, the distribution of such a process is uniquely determined by* $\mathbf{Q}$ *and* $\boldsymbol{\mu}$, *and the function* $t \rightsquigarrow \mathbf{P}(t)$ *in Lemma* 5.3.3 *is the transition probability function for* $\{X(t) : t \ge 0\}$.

Proof For each $i \in \mathbb{S}$ let $\{X_i(t) : t \ge 0\}$ be a Markov process generated by $\mathbf{Q}$ starting from i, let X_0 be a random variable with distribution $\boldsymbol{\mu}$ which is independent of $\sigma(\{X_i(t) : i \in \mathbb{S} \;\&\; t \ge 0\})$, and set $X(t) = X_{X_0}(t)$. Then it is easy to check that $\{X(t) : t \ge 0\}$ has all the required properties. Furthermore, since the distribution of $X(0)$ is $\boldsymbol{\mu}$ and the conditional distribution of $\{X(t) : t \ge 0\}$ given that $X(0) = i$ is uniquely determined by i and $\mathbf{Q}$ when $\mu_i > 0$, the distribution of $\{X(t) : t \ge 0\}$ is uniquely determined by $\mathbf{Q}$ and $\boldsymbol{\mu}$. □

We will call a right continuous, piecewise constant process whose distribution is the one described in Corollary 5.3.7 a *Markov process generated by* $\mathbf{Q}$ *with initial distribution* $\boldsymbol{\mu}$.

Remark Unfortunately, the relationship between $\mathbf{Q}$ and the process in Corollary 5.3.7 is a quite unsatisfactory because it fails to give a direct connection. What one would like is a description like that given in the conditions (a) and (b) at the beginning of Sect. 5.2.1. Indeed, we know that our process is right continuous and piecewise constant and has the specified initial distribution. Furthermore, we know that it satisfies a weak form of (5.2.1). To be precise, because

$$\mathbb{P}\big(X(t+h) = j \mid X(\tau), \tau \in [0,t]\big) = \big(\mathbf{P}(h)\big)_{X(t)j},$$

and, by (5.3.5),

$$\big(\mathbf{P}(h)\big)_{ij} = \delta_{ij} + h\mathbf{Q} + o(h) = (1 - hR_i)\delta_{ij} + h\big(\mathbf{P}(h)\big)_{ij} + o_{ij}(h),$$

where $o_{ij}(h)$ is a function that, as $h \searrow 0$, tends to 0 faster than h, we have that, for each $(i, j) \in \mathbb{S}^2$,

$$\mathbb{P}\big(X(t+h) = j \mid X(\tau), \tau \in [0,t]\big) = (1 - hR_i)\delta_{ij} + h\big(\mathbf{P}(h)\big)_{ij} + o_{ij}(h). \quad (5.3.8)$$

What makes (5.3.8) significantly weaker than (5.2.1) is that the function $o_{ij}(h)$ depends on i and j and, when $\mathfrak{R}$ is are unbounded, the rate at which it tends to 0 will depend on them also. Using more sophisticated techniques, one can show that, under the condition in Corollary 5.3.7, (a) together with the initial distribution and (5.3.8) uniquely determine the distribution of $\{X(t) : t \geq 0\}$, but these techniques require material that we have not covered.

5.3.2 Criteria for Non-explosion or Explosion

In this subsection we will first develop two criteria that guarantee *non-explosion*: $\mathfrak{e} = \infty$ with probability 1. We will also derive a condition that guarantees that explosion occurs with probability 1.

Theorem 5.3.9 *If* $\mathbf{P}$ *is a transition probability matrix and if* $i \in \mathbb{S}$ *is* $\mathbf{P}$*-recurrent, then for every choice of rates* $\mathfrak{R}$, $\mathbb{P}(\mathfrak{e} = \infty) = 1$ *for the process in Theorem* 5.3.6 *corresponding to the* Q*-matrix determined by* $\mathfrak{R}$ *and* $\mathbf{P}$ *starting at* i.

Proof By Lemma 5.3.2, what we must show is that $\mathbb{P}(\tilde{J}_\infty = \infty) = 1$. Equivalently, if $\{\tilde{X}_n : n \geq 0\}$ is a Markov chain with transition probability $\mathbf{P}$ with $\tilde{X}_0 = i$ and if $\{\tilde{E}_n : n \geq 1\}$ is a sequence to mutually independent independent unit exponential random variables that are independent of $\sigma(\{\tilde{X}_n : n \geq 0\})$, then we must show that $\sum_{n=0}^{\infty}(R_{\tilde{X}_{n-1}})^{-1}\tilde{E}_n = \infty$ with probability 1. To this end, let $\tilde{\rho}_i^{(m)}$ be the time of the mth return of $\{\tilde{X}_n : n \geq 0\}$ to i. Then the set A on which $\tilde{\rho}_i^{(m)} < \infty$ for all $m \geq 0$ is a $\sigma(\{\tilde{X}_n : n \geq 0\})$-measurable set having probability 1, and

$$\sum_{n=0}^{\infty}(R_{\tilde{X}_{n-1}})^{-1}\tilde{E}_n \geq (R_i)^{-1}\sum_{m=1}^{\infty}\tilde{E}_{\tilde{\rho}_i^{(m)}+1} \quad \text{on } A.$$

Since, given $\sigma(\{\tilde{X}_n : n \geq 0\})$, $\{\tilde{E}_{\tilde{\rho}_i^{(m)}+1} : m \geq 1\}$ is a sequence of mutually independent, unit exponential random variables and therefore the subset of A for which $\sum_{m=1}^{\infty}\tilde{E}_{\tilde{\rho}_i^{(m)}+1} < \infty$ has probability 0, the proof is complete. □

I will group our second non-explosion criterion together with our criterion for explosion as two parts of the same theorem. In this theorem, the process is the one in Theorem 5.3.6 determined by the rates $\mathfrak{R}$ and transition probability $\mathbf{P}$.

Theorem 5.3.10 *If there exists a non-negative function u on $\mathbb{S}$ with the properties that $U_N \equiv \inf_{j \notin F_N} u(j) \longrightarrow \infty$ as $N \to \infty$ and, for some $\alpha \in [0,\infty)$,*

$$\sum_{j \in \mathbb{S}} (\mathbf{P})_{ij} u(j) \le \left(1 + \frac{\alpha}{R_i}\right) u(i) \quad \text{whenever } i \in \mathbb{S} \text{ and } R_i > 0,$$

then $\mathbb{P}(\mathfrak{e} = \infty | X(0) = i) = 1$ for all $i \in \mathbb{S}$. On the other hand, if, for some $i \in \mathbb{S}$, $R_j > 0$ whenever $i \to j$ and there exists a non-negative function u on $\mathbb{S}$ with the property that, for some $\epsilon > 0$,

$$\sum_{\{k: i \to k\}} (\mathbf{P})_{jk} u(k) \le u(j) - \frac{\epsilon}{R_j} \quad \text{whenever } i \to j,$$

then $\mathbb{P}(\mathfrak{e} = \infty | X(0) = i) = 0$.

Proof To prove the first part, for each $N \ge 1$, set $u^{(N)}(j) = u(j)$ when $j \in F_N$ and $u^{(N)}(j) = U_N$ when $j \notin F_N$. It is an easy matter to check that if $\mathbf{Q}^{(N)} = \mathbf{R}^{(N)}(\mathbf{P} - \mathbf{I})$, where $(\mathbf{R}^{(N)})_{ij} = R_i^{(N)} \delta_{i,j}$, and $\mathbf{u}^{(N)}$ is the column vector determined by $(\mathbf{u}^{(N)})_i = u^{(N)}(i)$, then, for all $i \in \mathbb{S}$, $(\mathbf{Q}^{(N)} \mathbf{u}^{(N)})_i \le \alpha u^{(N)}(i)$. Hence, by Kolmogorov's forward equation (5.2.14) applied to the transition probability function $t \rightsquigarrow \mathbf{P}^{(N)}(t)$ corresponding to $\mathbf{Q}^{(N)}$,

$$\frac{d}{dt}\big(\mathbf{P}^{(N)}(t)\mathbf{u}^{(N)}\big)_i \le \alpha \big(\mathbf{P}^{(N)}(t)\mathbf{u}^{(N)}\big)_i,$$

and so, by Gronwall's inequality, $(\mathbf{P}^{(N)}(T)\mathbf{u}^{(N)})_i \le e^{\alpha T} u^{(N)}(i)$. But, since

$$u^{(N)}\big(X^{(N)}(T)\big) = u^{(N)}\big(X^{(N)}(\zeta_N)\big) \ge U_N \quad \text{if } \zeta_N \le T,$$

this means that

$$\begin{aligned} \mathbf{P}\big(\zeta_N \le T \mid X(0) = i\big) &\le \frac{1}{U_N} \mathbb{E}\big[u^{(N)}\big(X^{(N)}(T)\big) \,\big|\, X(0) = i\big] \\ &= \frac{(\mathbf{P}^{(N)}(T)\mathbf{u}^{(N)})_i}{U_N} \le \frac{e^{\alpha T} u^{(N)}(i)}{U_N} \le \frac{e^{\alpha T} u(i)}{U_N} \quad \text{if } i \in F_N, \end{aligned}$$

and so, by the monotone convergence theorem, we have now proved that $\mathbf{P}(\mathfrak{e} \le T | X(0) = i) \le \lim_{N\to\infty} \mathbb{P}(\zeta_N \le T | X(0) = i) = 0$.

In proving the second assertion, we may and will assume that $i \to j$ for all $j \in \mathbb{S}$. Now take $u^{(N)}(j) = u(j)$ if $j \in F_N$ and $u^{(N)}(j) = 0$ if $j \notin F_N$, and use $\mathbf{u}^{(N)}$ to denote the associated column vector. Clearly $(\mathbf{Q}^{(N)}\mathbf{u}^{(N)})_j$ is either less than or equal to $-\epsilon$ or equal to 0 according to whether $j \in F_N$ or $j \notin F_N$. Hence, again by Kolmogorov's forward equation,

$$\frac{d}{dt}\big(\mathbf{P}^{(N)}(t)\mathbf{u}^{(N)}\big)_j \le -\epsilon \mathbf{1}_{F_N}(j) \sum_{k \in F_N} \big(\mathbf{P}^{(N)}(t)\big)_{jk},$$

and so

$$\mathbb{E}\big[u^{(N)}\big(X^{(N)}(T)\big)\,\big|\,X^{(N)}(0)=i\big]-u^{(N)}(i)$$
$$\le -\epsilon\mathbb{E}\left[\int_0^T \mathbf{1}_{F_N}\big(X^{(N)}(t)\big)\,dt\,\middle|\,X^{(N)}(0)=i\right]=-\epsilon\mathbb{E}\big[T\wedge\zeta_N\,\big|\,X^{(N)}(0)=i\big].$$

Since $u^{(N)}\ge 0$, this means that $\mathbb{E}[\zeta_N|X^{(N)}(0)=i]\le\frac{u(i)}{\epsilon}$ for all N, and so $\mathbb{E}[\mathfrak{e}|X(0)=i]\le\frac{u(i)}{\epsilon}<\infty$. □

5.3.3 *What to Do when Explosion Occurs*

Although I do not intend to repent, I feel compelled to admit that I have ignored what, from the mathematical standpoint, is the most interesting aspect of the theory under consideration. Namely, so far I have said nothing about the myriad options one has when explosion occurs with positive probability, and in this subsection I will discuss only the most banal of the many choices available.

In Theorem 5.3.6, when explosion occurred I banished the process to a state Δ outside of $\mathbb{S}$. This is the least imaginative option and, for reasons that I will not attempt to explain, is called the minimal extension. It always works, but it has the disadvantage that it completely ignores the manner in which the explosion took place. For example, when $\mathbb{S}=\mathbb{Z}$, explosion can occur because, given that $\mathfrak{e}<\infty$, $\lim_{t\nearrow\mathfrak{e}}X(t)=+\infty$ with probability 1 or because, although $\lim_{t\nearrow\mathfrak{e}}|X(t)|=\infty$ with probability 1, both $\lim_{t\nearrow\mathfrak{e}}X(t)=+\infty$ and $\lim_{t\nearrow\mathfrak{e}}X(t)=-\infty$ occur with positive probability. In the latter case, one might want to record which of the two possibilities occurred, and this could be done by introducing two new absorbing states, Δ_+ for those paths that escape via $+\infty$ and Δ_- for those that escape via $-\infty$.

Alternatively, rather than thinking of the explosion time as a time to banish the process from $\mathbb{S}$, one can turn it into a time of reincarnation by redistributing the process over $\mathbb{S}$ at time $\mathfrak{e}$ and running it again until it explodes, etc. Obviously, there is an infinity of possibilities. Suffice it to say that the preceding discussion barely scratches the surface.

5.4 Ergodic Properties

In this section we will examine the ergodic behavior of the Markov processes which we have been discussing in this chapter. Our running assumption will be that the process with which we are dealing is a continuous time Markov process of the sort described in Sect. 5.2 under the condition that there is no explosion.

5.4.1 Classification of States

We begin by classifying the states of $\mathbb{S}$.

Given a Q-matrix $\mathbf{Q}$, let (cf. (5.2.12)) $\mathfrak{R}$ and $\mathbf{P}$ be the canonical rates and transition probability that it determines. Then from Theorems 5.2.10 and 5.3.6, it is clear that the states visited by a Markov process generated by $\mathbf{Q}$ starting from a state i are exactly the same as those visited by the Markov chain with transition probability $\mathbf{P}$ starting from i. Hence, we will say that i is $\mathbf{Q}$-recurrent or $\mathbf{Q}$-transient if and only if it is recurrent or transient for the Markov chain with transition probability $\mathbf{P}$. In particular, i is $\mathbf{Q}$-recurrent if $R_i = 0$. Similarly, we will say that j is $\mathbf{Q}$-*accessible* from i and will write $i \overset{\mathbf{Q}}{\to} j$ if and only if $\mathbf{P}^n_{ij} > 0$ for some $n \geq 0$, and we will say that i $\mathbf{Q}$-*communicates* with j and will write $i \overset{\mathbf{Q}}{\leftrightarrow} j$ if and only if $i \overset{\mathbf{Q}}{\to} j$ and $j \overset{\mathbf{Q}}{\to} i$.

From the results in Sect. 3.1, we know that $\mathbf{Q}$-accessibility is a transitive relation and that $\mathbf{Q}$-communication is an equivalence relation. In addition, $\mathbf{Q}$-recurrence and $\mathbf{Q}$-transitivity are $\mathbf{Q}$-communicating class properties. Thus, if the state space $\mathbb{S}$ is $\mathbf{Q}$-*irreducible* in the sense that all states $\mathbf{Q}$-communicate with one another, then either all its states are $\mathbf{Q}$-recurrent or all are $\mathbf{Q}$-transient.

I next want to describe these results in terms of the paths of the associate Markov process. Thus, let $\{X(t) : t \geq 0\}$ be a Markov process generated by $\mathbf{Q}$. In order to describe the analog of the first return time to a state i, let J_n be the time when the process makes its nth jump. That is,

$$J_n = \begin{cases} \inf\{t > 0 : X(t) \neq X(t-)\} & \text{if } n = 1 \\ \inf\{t > J_{n-1} : X(t) \neq X(t-)\} & \text{if } n > 1, \end{cases}$$

where $X(t-) \equiv \lim_{s \nearrow t} X(s)$. Then the first return time to i is

$$\sigma_i \equiv \inf\{t \geq J_1 : X(t) = i\}.$$

More generally, define $\sigma_i^{(0)} = 0$, $\sigma_i^{(1)} = \sigma_i$, and, for $m > 1$,

$$\sigma_i^{(m)} = \begin{cases} \inf\{t \geq J_{\ell+1} : X(t) = i\} & \text{if } \sigma_i^{(m-1)} = J_\ell < \infty \\ \infty & \text{if } \sigma_i^{(m-1)} = \infty. \end{cases}$$

We will call $\sigma_i^{(m)}$ the mth return time to i.

Clearly, i is $\mathbf{Q}$-recurrent if and only if $\mathbb{P}(\sigma_i < \infty \,|\, X(0) = i) = 1$. Furthermore, if i is $\mathbf{Q}$-recurrent, then, with probability 1, a process started at i returns to i infinitely often. To see this, let $\{X_n : n \geq 0\}$ be a Markov chain with transition probability $\mathbf{P}$, and observe that, by Theorems 5.2.10 and 5.3.6, $\mathbb{P}(\sigma_i^{(m)} < \infty) = \mathbb{P}(\rho_i^{(m)} < \infty)$, where $\rho_i^{(m)}$ is the mth return time of $\{X_n : n \geq 0\}$ to i. Thus, by Theorem 2.3.6, for any $(i, j) \in \mathbb{S}^2$ and $m \geq 1$

$$\begin{aligned}&\mathbb{P}\big(\sigma_j^{(m)} < \infty \,\big|\, X(0) = i\big)\\&\quad = \mathbb{P}\big(\sigma_j < \infty \mid X(0) = i\big)\mathbb{P}\big(\sigma_j^{(m)} < \infty \,\big|\, X(0) = j\big)^{m-1}.\end{aligned} \tag{5.4.1}$$

Just as in the case of chains, these considerations are intimately related to the amount of time that the process spends in a state. Indeed,

$$\begin{aligned}&\mathbb{E}\bigg[\int_0^\infty \mathbf{1}_{\{j\}}\big(X(t)\big)\,dt \,\bigg|\, X(0) = i\bigg]\\&\quad = \sum_{m=0}^\infty \mathbb{E}\bigg[\int_{\sigma_j^{(m)}}^{\sigma_j^{(m+1)}} \mathbf{1}_{\{j\}}\big(X(t)\big)\,dt, \sigma_j^{(m)} < \infty \,\bigg|\, X(0) = 0\bigg].\end{aligned}$$

Furthermore, using the notation in Theorem 5.2.10 and letting $\tilde{\sigma}_j^{(m)}$ denote the mth return time of $\{\tilde{X}(t) : t \geq 0\}$ to j, we have that

$$\begin{aligned}&\mathbb{E}\bigg[\int_{\sigma_j^{(m)}}^{\sigma_j^{(m+1)}} \mathbf{1}_{\{j\}}\big(X(t)\big)\,dt, \sigma_j^{(m)} < \infty \,\bigg|\, X(0) = 0\bigg]\\&\quad = \sum_{\ell=0}^\infty \mathbb{E}\big[J_{\ell+1} - J_\ell, \sigma_j^{(m)} = J_{\ell+1} < \infty \,\big|\, X(0) = i\big]\\&\quad = \sum_{\ell=0}^\infty \mathbb{E}\big[\tilde{E}_{\ell+1}, \tilde{\sigma}_j^{(m)} = \tilde{J}_{\ell+1} < \infty \,\big|\, \tilde{X}(0) = i\big]\\&\quad = \frac{1}{R_j}\sum_{\ell=0}^\infty \mathbb{P}\big(\tilde{\sigma}_j^{(m)} = \tilde{J}_{\ell+1} < \infty \,\big|\, \tilde{X}(0) = i\big)\\&\quad = \frac{\mathbb{P}(\tilde{\sigma}_j^{(m)} < \infty \mid \tilde{X}(0) = i)}{R_j} = \frac{\mathbb{P}(\sigma_j^{(m)} < \infty \mid X(0) = i)}{R_j}\\&\quad = \begin{cases} 0 & \text{if } \mathbb{P}(\sigma_j < \infty \mid X(0) = i) = 0\\ \infty & \text{if } R_j = 0 \text{ and } \mathbb{P}(\sigma_j < \infty \mid X(0) = i) > 0\\ \frac{\mathbb{P}(\sigma_j < \infty \mid X(0) = i)\mathbb{P}(\sigma_j < \infty \mid X(0) = j)^{(m-1)^+}}{R_j} & \text{otherwise.}\end{cases}\end{aligned}$$

Hence, by exactly the argument with which we passed from Theorem 2.3.6 to (2.3.7), we now know that

$$\mathbb{E}\left[\int_0^\infty \mathbf{1}_{\{j\}}\big(X(t)\big)\,dt\,\Big|\,X(0)=i\right] = \frac{1}{R_j}\left(\delta_{i,j} + \frac{\mathbb{P}(\sigma_j<\infty|X(0)=i)}{\mathbb{P}(\sigma_j=\infty|X(0)=j)}\right)$$

$$\mathbb{E}\left[\int_0^\infty \mathbf{1}_{\{i\}}\big(X(t)\big)\,dt\,\Big|\,X(0)=i\right] = \infty$$

$$\iff \quad \mathbb{P}\left(\int_0^\infty \mathbf{1}_{\{i\}}\big(X(t)\big)\,dt = \infty\,\Big|\,X(0)=i\right) = 1 \tag{5.4.2}$$

$$\mathbb{E}\left[\int_0^\infty \mathbf{1}_{\{i\}}\big(X(t)\big)\,dt\,\Big|\,X(0)=i\right] < \infty$$

$$\iff \quad \mathbb{P}\left(\int_0^\infty \mathbf{1}_{\{i\}}\big(X(t)\big)\,dt < \infty\,\Big|\,X(0)=i\right) = 1,$$

where, in the first line when $R_j=0$, the right hand side is ∞ if either $i=j$ or $\mathbb{P}(\sigma_j \mid X(0)=i)>0$ and is 0 if $i\neq j$ and $\mathbb{P}(\sigma_j \mid X(0)=i)=0$. In particular, when i is $\mathbf{Q}$-transient, the expected amount of time that the process spends at j is uniformly bounded independent of its initial distribution.

Our final goal in this subsection is to prove the following statement.

Theorem 5.4.3 *For any given state $i\in\mathbb{S}$, the following are equivalent:*

(1) *i is $\mathbf{Q}$-recurrent.*
(2) *There is a $t\in(0,\infty)$ such that i is recurrent relative to the transition probability $\mathbf{P}(t)$.*
(3) *i is recurrent relative to $\mathbf{P}(t)$ for all $t\in(0,\infty)$.*

Proof We will prove this equivalence by checking that the same equivalence holds when "recurrent" is replaced throughout by "transient." To this end, first observe that

$$\mathbb{E}\left[\int_0^\infty \mathbf{1}_{\{i\}}\big(X(t)\big)\,dt\,\Big|\,X(0)=i\right] = \int_0^\infty \big(\mathbf{P}(t)\big)_{ii}\,dt,$$

and therefore, from the first line of (5.4.2), that

$$i \text{ is } \mathbf{Q}\text{-transient} \quad\iff\quad \int_0^\infty \big(\mathbf{P}(t)\big)_{ii}\,dt<\infty. \tag{5.4.4}$$

Next, notice that for $0\le s<t$,

$$\big(\mathbf{P}(t)\big)_{ii} \ge \big(\mathbf{P}(t-s)\big)_{ii}\big(\mathbf{P}(s)\big)_{ii} \ge e^{-(t-s)R_i}\big(\mathbf{P}(s)\big)_{ii} \tag{5.4.5}$$

since $(\mathbf{P}(h))_{ii}\ge\mathbb{P}(J_1>h|X(0)=i)=e^{-hR_i}$. Hence, for any $t>0$ and $n\in\mathbb{N}$,

$$\big(\mathbf{P}(t)^{n+1}\big)_{ii} = \big(\mathbf{P}\big((n+1)t\big)\big)_{ii} \ge e^{-tR_i}\big(\mathbf{P}(\tau)\big)_{ii} \quad \text{for all } \tau\in\big[nt,(n+1)t\big]$$

and

$$e^{-tR_i}\big(\mathbf{P}(t)^n\big)_{ii} = e^{-tR_i}\big(\mathbf{P}(nt)\big)_{ii} \le \big(\mathbf{P}(\tau)\big)_{ii} \quad \text{for all } \tau\in\big[nt,(n+1)t\big].$$

Since this means that

$$te^{-tR_i}\sum_{n=0}^{\infty}\big(\mathbf{P}(t)^n\big)_{ii} \le \int_0^{\infty}\big(\mathbf{P}(\tau)\big)_{ii}\,d\tau \le te^{tR_i}\sum_{n=0}^{\infty}\big(\mathbf{P}(t)^{n+1}\big)_{ii},$$

the asserted implications are now immediate from (5.4.4). □

5.4.2 Stationary Measures and Limit Theorems

In this subsection I will prove the analogs of the results in Sects. 4.1.6 and 4.1.7, and, because periodicity plays no role here, the statements are simpler.

Theorem 5.4.6 *For each $j \in \mathbb{S}$*

$$\hat{\pi}_{jj} \equiv \lim_{t\to\infty}\big(\mathbf{P}(t)\big)_{jj} \quad \textit{exists}$$

and

$$\lim_{t\to\infty}\big(\mathbf{P}(t)\big)_{ij} = \hat{\pi}_{ij} \equiv \mathbb{P}\big(\sigma_j < \infty \,|\, X(0) = i\big)\hat{\pi}_{jj} \quad \textit{for } i \ne j.$$

Moreover, if $\hat{\pi}_{jj} > 0$, then $\hat{\pi}_{ii} > 0$ for all $i \in C \equiv \{i : i \overset{\mathbf{Q}}{\leftrightarrow} j\}$, and when the row vector $\hat{\boldsymbol{\pi}}^C$ is determined by $(\hat{\boldsymbol{\pi}}^C)_i = \mathbf{1}_C(i)\hat{\pi}_{ii}$, $\hat{\boldsymbol{\pi}}^C$ is, for each $s > 0$, the unique probability vector $\boldsymbol{\mu} \in \mathrm{Stat}(\mathbf{P}(s))$ for which $(\boldsymbol{\mu})_k = 0$ if $k \notin C$. In fact, if $\boldsymbol{\mu} \in \mathrm{Stat}(\mathbf{P}(s))$ for some $s > 0$, then, for each $j \in C$,

$$(\boldsymbol{\mu})_j = \Bigg(\sum_{i \overset{\mathbf{Q}}{\leftrightarrow} j}(\boldsymbol{\mu})_i\Bigg)\hat{\pi}_{jj}.$$

Proof We begin with the following continuous-time version of the renewal equation (cf. (4.1.6)):

$$\big(\mathbf{P}(t)\big)_{ij} = e^{-tR_i}\delta_{i,j} + \mathbb{E}\big[\big(\mathbf{P}(t-\sigma_j)\big)_{jj}, \sigma_j \le t \,|\, X(0) = i\big]. \tag{5.4.7}$$

The proof of (5.4.7) runs as follows. First, write $(\mathbf{P}(t))_{ij}$ as

$$\mathbb{P}\big(X(t) = j \;\&\; J_1 > t \,|\, X(0) = i\big) + \mathbb{P}\big(X(t) = j \;\&\; J_1 \le t \,|\, X(0) = i\big).$$

Clearly, the first term is 0 unless $i = j$, in which case it is equal e^{-tR_i}. To handle the second term, write it as

$$\sum_{m=1}^{\infty}\mathbb{P}\big(X(t) = j \;\&\; \sigma_j = J_m \le t \,|\, X(0) = i\big),$$

and observe that (cf. (5.2.9) and (5.2.11))

$$\begin{aligned}
&\mathbb{P}\big(X(t)=j \;\&\; \sigma_j = J_m \le t \,|\, X(0)=i\big) \\
&\quad = \mathbb{P}\big(\Phi^{\mathfrak{R}}(t-\tilde{J}_m; (\tilde{E}_{m+1},\dots,\tilde{E}_{m+n},\dots),(j,\tilde{X}_{m+1},\dots,\tilde{X}_{m+n},\dots)) = j \\
&\qquad \& \; \tilde{\sigma}_j = \tilde{J}_m \le t \,|\, \tilde{X}_0 = i\big) \\
&\quad = \mathbb{E}\big[\big(\mathbf{P}(t-J_m)\big)_{jj}, \sigma_j = J_m \le t \,|\, X(0)=i\big].
\end{aligned}$$

Hence, after summing this over $m \ge 1$ and combining this result with the preceding, one arrives at (5.4.7).

Knowing (5.4.7), we see that the first part of the theorem will be proved once we treat the case when $i = j$. To this end, we begin by observing that, because, by (5.4.5), $(\mathbf{P}(s))_{ii} \ge e^{-sR_i} > 0$ for all $s > 0$ and $i \in \mathbb{S}$, each i is aperiodic relative to $\mathbf{P}(s)$, and so, by (4.1.15), we know that $\pi(s)_{ii} \equiv \lim_{n\to\infty}(\mathbf{P}(s)^n)_{ii}$ exists for all $s > 0$ and $i \in \mathbb{S}$. We need to show that $\pi(1)_{ii} = \lim_{t\to\infty}(\mathbf{P}(t))_{ii}$ for all $t > 0$, and when $\pi(1)_{ii} = 0$, this is easy. Indeed, by (5.4.5),

$$\varlimsup_{t\to\infty} \big(\mathbf{P}(t)\big)_{ii} \le e^{R_i} \varlimsup_{t\to\infty} \big(\mathbf{P}(\lfloor t \rfloor + 1)\big)_{ii} = e^{R_j}\pi(1)_{ii},$$

where $\lfloor t \rfloor$ denotes the integer part of t.

In view of the preceding, what remains to be proved in the first assertion is that $\lim_{t\to\infty}(\mathbf{P}(t))_{ii} = \pi(1)_{ii}$ when $\pi(1)_{ii} > 0$, and the key step in our proof will be the demonstration that $\pi(s)_{ii} = \pi(1)_{ii}$ for all $s > 0$, a fact which we already know when $\pi(1)_{ii} = 0$. Thus, assume that $\pi(1)_{ii} > 0$, and let C be the communicating class of i relative to $\mathbf{P}(1)$. By Theorem 5.4.3, C is also the communicating class of i relative to $\mathbf{P}(s)$ for every $s > 0$. In addition, because $\pi(1)_{ii} > 0$, i is recurrent, in fact positive recurrent, relative to $\mathbf{P}(1)$, and therefore, by Theorem 5.4.3, it is also recurrent relative to $\mathbf{P}(s)$ for all $s > 0$. Now determine the row vector $\boldsymbol{\pi}(1)^C$ so that $(\boldsymbol{\pi}(1)^C)_j = \mathbf{1}_C(j)\pi(1)_{jj}$. Then, because $\pi(1)_{ii} > 0$, we know, by Theorem 4.1.10, that $\boldsymbol{\pi}(1)^C$ is the one and only $\boldsymbol{\mu} \in \operatorname{Stat}(\mathbf{P}(1))$ which vanishes off of C. Next, given $s > 0$, consider $\boldsymbol{\mu} = \boldsymbol{\pi}(1)^C\mathbf{P}(s)$. Then $\boldsymbol{\mu}$ is a probability vector and, because $(\mathbf{P}(1))_{jk} = 0$ when $j \in C$ and $k \notin C$, $\boldsymbol{\mu}$ vanishes off of C. Also, because

$$\begin{aligned}
\boldsymbol{\mu}\mathbf{P}(1) &= \boldsymbol{\pi}(1)^C\mathbf{P}(s)\mathbf{P}(1) = \boldsymbol{\pi}(1)^C\mathbf{P}(s+1) \\
&= \boldsymbol{\pi}(1)^C\mathbf{P}(1)\mathbf{P}(s) = \boldsymbol{\pi}(1)^C\mathbf{P}(s) = \boldsymbol{\mu},
\end{aligned}$$

$\boldsymbol{\mu}$ is stationary for $\mathbf{P}(1)$. Hence, by uniqueness, we conclude that $\boldsymbol{\pi}(1)^C\mathbf{P}(s) = \boldsymbol{\pi}(1)^C$, and therefore that $\boldsymbol{\pi}(1)^C$ is a stationary probability for $\mathbf{P}(s)$ which vanishes off of C. But, by (4.1.8) and the fact that C is the communicating class of i relative to $\mathbf{P}(s)$, this means that

$$\pi(1)_{ii} = \big(\boldsymbol{\pi}(1)^C\big)_i = \Big(\sum_{j\in C}\big(\boldsymbol{\pi}(1)^C\big)_j\Big)\pi(s)_{ii} = \pi(s)_{ii}.$$

To complete the proof of the first part from here, note that, by (5.4.5),

$$e^{-sR_j}\big(\mathbf{P}(ns)\big)_{jj} \le \big(\mathbf{P}(t)\big)_{jj} \le e^{sR_j}\big(\mathbf{P}((n+1)s)\big)_{jj} \quad \text{when } ns \le t \le (n+1)s,$$

and so, since $\pi(s)_{jj} = \pi(1)_{jj}$ and $\mathbf{P}(nt) = \mathbf{P}(t)^n$,

$$e^{-sR_j}\pi(1)_{jj} \le \varliminf_{t\to\infty}\big(\mathbf{P}(t)\big)_{jj} \le \varlimsup_{t\to\infty}\big(\mathbf{P}(t)\big)_{jj} \le e^{sR_j}\pi(1)_{jj}.$$

Now let $s \searrow 0$, and simply define $\hat{\pi}_{jj} = \pi(1)_{jj}$.

Given the first part, the proof of the second part is easy. Namely, if $\hat{\pi}_{jj} > 0$ and $C = \{i : i \overset{\mathbf{Q}}{\leftrightarrow} j\}$, then, for each $s > 0$, we know that C is the communicating class of j relative to $\mathbf{P}(s)$ and that $\hat{\boldsymbol{\pi}}^C = \boldsymbol{\pi}(s)^C \in \mathrm{Stat}(\mathbf{P}(s))$. Conversely, by (4.1.8), we know that if $\boldsymbol{\mu} \in \mathrm{Stat}(\mathbf{P}(s))$ for some $s > 0$, then

$$(\boldsymbol{\mu})_j = \Bigg(\sum_{\{i:i\overset{\mathbf{Q}}{\leftrightarrow}j\}}(\boldsymbol{\mu})_i\Bigg)\pi(s)_{jj} = \Bigg(\sum_{\{i:i\overset{\mathbf{Q}}{\leftrightarrow}j\}}(\boldsymbol{\mu})_i\Bigg)\hat{\pi}_{jj}.$$

□

With the preceding result in mind, we will say that i is **Q**-*positive recurrent* if $\hat{\pi}_{ii} > 0$ and will say that i is **Q**-*null recurrent* if i is **Q**-recurrent but not **Q**-positive recurrent. From the preceding and Theorem 4.1.10, we already know that *Q-positive recurrence is a Q-communicating class property.*

The following corollary comes at very little additional cost.

Corollary 5.4.8 (Mean Ergodic Theorem) *Assume that j is* **Q**-*positive recurrent and that* $\mathbb{P}(X(0) \overset{\mathbf{Q}}{\leftrightarrow} j) = 1$. *Then,*

$$\lim_{T\to\infty}\mathbb{E}\Bigg[\bigg(\frac{1}{T}\int_0^T \mathbf{1}_{\{j\}}\big(X(t)\big)\,dt - \hat{\pi}_{jj}\bigg)^2\Bigg] = 0.$$

See Exercise 5.5.11 *for the more refined version.*

Proof The proof is really just an obvious transcription to the continuous setting of the argument used to prove Theorem 4.1.14. By precisely the argument given there, it suffices to handle the case when $\hat{\boldsymbol{\pi}}^C$ is the initial distribution, where $C = \{i : j \overset{\mathbf{Q}}{\leftrightarrow} i\}$. Thus, if $f = \mathbf{1}_{\{j\}} - \hat{\pi}_{jj}$, what we need to show is that

$$\lim_{T\to\infty}\mathbb{E}\Bigg[\bigg(\frac{1}{T}\int_0^T f\big(X(t)\big)\,dt\bigg)^2\Bigg] = 0,$$

when $\hat{\boldsymbol{\pi}}^C$ is the distribution of $X(0)$. But, because $\hat{\boldsymbol{\pi}}^C$ is $\mathbf{P}(t)$-stationary,

$$\mathbb{E}\left[\left(\frac{1}{T}\int_0^T f\big(X(t)\big)\,dt\right)^2\right] = \frac{2}{T^2}\int_0^T\left(\int_0^t \mathbb{E}\big[f\big(X(s)\big)f\big(X(t)\big)\big]\,ds\right)dt$$
$$= \frac{2}{T^2}\int_0^T\left(\int_0^t \alpha(t-s)\,ds\right)dt$$
$$= \frac{2}{T}\int_0^T\left(1-\frac{t}{T}\right)\alpha(t)\,dt,$$

where $\alpha(t) \equiv \hat{\pi}^C(f\mathbf{P}(t)\mathbf{f})$, $\mathbf{f}$ being the column vector corresponding to the function f and $f\mathbf{P}(t)\mathbf{f}$ being the column vector determined by $(f\mathbf{P}(t)\mathbf{f})_i = f(i)(\mathbf{P}(t)\mathbf{f})_i$. Finally, since, for each $i \in C$, $\lim_{t\to\infty}(\mathbf{P}(t)\mathbf{f})_i = 0$, $\lim_{t\to\infty}\alpha(t) = 0$, and therefore

$$\left|\frac{2}{T}\int_0^T\left(1-\frac{t}{T}\right)\alpha(t)\,dt\right| \le \frac{2}{T}\int_0^T \big|\alpha(t)\big|\,dt \longrightarrow 0 \quad \text{as } T\to\infty. \qquad \square$$

5.4.3 Interpreting and Computing $\hat{\pi}_{ii}$

Although we have shown that the limit $\hat{\pi}_{ij} = \lim_{t\to\infty}(\mathbf{P}(t))_{ij}$ exists, we have yet to give an expression, analogous to the one in (4.1.5), for $\hat{\pi}_{ii}$. However, as we are about to see, such an expression is readily available from (5.4.7). Namely, for each $\alpha > 0$, set

$$L(\alpha)_{ij} = \alpha\mathbb{E}\left[\int_0^\infty e^{-\alpha t}\mathbf{1}_{\{j\}}\big(X(t)\big)\,dt\,\Big|\,X(0)=i\right] = \alpha\int_0^\infty e^{-\alpha t}\big(\mathbf{P}(t)\big)_{ij}\,dt.$$

Because $\hat{\pi}_{ii} = \lim_{t\to\infty}(\mathbf{P}(t))_{ii}$, the second of these representations makes it is easy to identify $\hat{\pi}_{ii}$ as $\lim_{\alpha\searrow 0} L(\alpha)_{ii}$. At the same time, from (5.4.7), we see that

$$L(\alpha)_{ii} = \frac{\alpha}{\alpha+R_i} + \mathbb{E}\big[e^{-\alpha\sigma_i}\,\big|\,X(0)=i\big]L(\alpha)_{ii}.$$

Hence, since

$$\frac{1-\mathbb{E}[e^{-\alpha\sigma_i}\,|\,X_0=i]}{\alpha} = \frac{\mathbb{E}[1-e^{-\alpha\sigma_i}\,|\,X_0=i]}{\alpha} \longrightarrow \mathbb{E}[\sigma_i\,|\,X_0=r_i]$$

if $R_i > 0$ and $\mathbb{P}(\sigma_i = \infty\,|\,X_0 = i) = 1$ if $R_i = 0$, we have now shown that

$$\hat{\pi}_{ii} = \begin{cases} 1 & \text{if } R_i = 0 \\ \frac{1}{R_i\mathbb{E}[\sigma_i|X(0)=i]} & \text{if } R_i > 0, \end{cases}$$

which, in conjunction with the second equality in Theorem 5.4.6, proves that

$$\hat{\pi}_{ij} = \begin{cases} \delta_{i,j} + (1-\delta_{i,j})\mathbb{P}(\sigma_j < \infty | X(0)=i) & \text{if } R_j = 0 \\ \frac{\mathbb{P}(\sigma_j < \infty \,|\, X(0)=i)}{R_j \mathbb{E}[\sigma_j | X(0)=j]} & \text{if } R_j > 0. \end{cases} \tag{5.4.9}$$

Of course, as an immediate corollary of (5.4.9), we now know that i *is positive recurrent if and only if either* $R_i = 0$ *or* $\mathbb{E}[\sigma_i | X(0) = i] < \infty$.

Finally, we will apply the results in Sect. 3.2 to the setting here. To this end, first observe that as an immediate consequence of (5.2.2), for any probability vector $\boldsymbol{\mu}$,

$$\boldsymbol{\mu}\mathbf{Q} = \mathbf{0} \iff \boldsymbol{\mu}\mathbf{P}(t) \quad \text{for all } t > 0 \text{ when } \mathfrak{R} \text{ is bounded.} \tag{5.4.10}$$

Theorem 5.4.11 *Assume that* $\mathbb{S}$ *is finite and that there is only one stationary probability* $\hat{\boldsymbol{\pi}}$ *for* $t \rightsquigarrow \mathbf{P}(t)$. *Then* 0 *is a simple eigenvalue of* $\mathbf{Q}$ *and*

$$(\hat{\boldsymbol{\pi}})_i = \frac{\det((-Q)^{\{i\}})}{\Pi_{-Q}} = \frac{\det((-Q)^{\{i\}})}{\sum_{j \in \mathbb{S}} \det((-Q)^{\{j\}})}$$

for all $i \in \mathbb{S}$.

Proof The argument is essentially the same as the one for chains. To see that 0 is a simple eigenvalue, suppose not. Then there exists a column vector $\mathbf{v} \neq \mathbf{0}$ such that $\mathbf{Q}\mathbf{v} = \mathbf{0}$ and $\hat{\boldsymbol{\pi}}\mathbf{v} = 0$. By (5.4.10), $\mathbf{v} = \mathbf{P}(t)\mathbf{v}$ for all $t > 0$, and so Theorem 5.4.6 leads to the contradiction

$$\mathbf{v} = \lim_{t \to \infty} \mathbf{P}(t)\mathbf{v} = (\mathbf{v}, \hat{\boldsymbol{\pi}})_{\mathbb{C}^{\mathbb{S}}} \mathbf{1} = \mathbf{0}.$$

Knowing that 0 is a simple eigenvalue and that $\mathbf{Q}\mathbf{1} = \mathbf{0}$, the proof of the second assertion is precisely the same as the proof of the analogous result in Theorem 3.2.6. □

The reason why I wrote the equation for $\hat{\boldsymbol{\pi}}$ in terms of $-\mathbf{Q}$ instead of $\mathbf{Q}$ is that, by arguments like those in the chain case, $(-\mathbf{Q})^{\{i\}}$ is non-negative for all i and is strictly positive for $\mathbf{Q}$-recurrent i's.

5.5 Exercises

Exercise 5.5.1 The purpose of this exercise is to give another derivation of (5.1.5). Let $\{E_n : n \geq 1\}$ be a sequence of mutually independent, unit exponential random variables, and define $\{J_n : n \geq 0\}$ and $\{N(t) : t \geq 0\}$ accordingly, as in Sect. 5.1.1. Given $0 = t_0 < \cdots < t_L$ and $0 = n_0 < \cdots < n_L$, use the change of variables formula for multi-dimensional integrals to justify:

$$
\begin{aligned}
&\mathbf{P}\big(N(t_1)=n_1,\dots,N(t_L)=n_L\big)\\
&=\mathbf{P}(J_{n_1}\le t_1<J_{n_1+1},\dots,J_{n_L}\le t_\ell<J_{n_L+1})\\
&=\int_A\cdots\int\exp\left(-\sum_{\ell=1}^{n_L+1}\xi_\ell\right)d\xi_1\cdots d\xi_{n_L+1}\\
&=\int_B\cdots\int e^{-\eta_{n_L+1}}\,d\eta_1\cdots d\eta_{n_L+1}\\
&=e^{-t_L}\prod_{\ell=1}^L\operatorname{vol}(\Delta_\ell)=\prod_{\ell=1}^L e^{-(t_\ell-t_{\ell-1})}\frac{(t_\ell-t_{\ell-1})^{n_\ell-n_{\ell-1}}}{(n_\ell-n_{\ell-1})!},
\end{aligned}
$$

where

$$
\begin{aligned}
A&=\left\{(\xi_1,\dots,\xi_{n_L+1})\in(0,\infty)^{n_L+1}:\sum_1^{n_\ell}\xi_j\le t_\ell<\sum_1^{n_\ell+1}\xi_j\text{ for }1\le\ell\le L\right\},\\
B&=\big\{(\eta_1,\dots,\eta_{n_L+1})\in(0,\infty)^{n_L+1}:\\
&\qquad\eta_i<\eta_{i+1}\text{ for }1\le i\le n_L\ \&\ \eta_{n_\ell}\le t_\ell<\eta_{n_\ell+1}\text{ for }1\le\ell\le L\big\},\\
\Delta_\ell&=\big\{(u_1,\dots,u_{n_\ell-n_{\ell-1}})\in\mathbb{R}^{n_\ell-n_{\ell-1}}:t_{\ell-1}\le u_1<\cdots<u_{n_\ell-n_{\ell-1}}\le t_\ell\big\}.
\end{aligned}
$$

When $n_\ell=n_{\ell-1}$ for some ℓ's, begin by choosing $0=\ell_0<\cdots<\ell_K=L$ so that $n_{\ell_k}=\cdots=n_{\ell_{k+1}-1}<n_{\ell_{k+1}}$ for $0\le k<L$. Show that

$$
\begin{aligned}
&\mathbb{P}\big(N(t_\ell)=t_\ell,1\le\ell\le L\big)\\
&=\mathbb{P}(J_{n_{\ell_k}}\le t_{\ell_k}\ \&\ J_{n_{\ell_{k-1}}+1}<t_{\ell_k-1},1\le k\le K\text{ and }J_{n_L+1}>t_L),
\end{aligned}
$$

and, proceeding as above, conclude that

$$
\begin{aligned}
&\mathbb{P}\big(N(t_\ell)=t_\ell,1\le\ell\le L\big)\\
&=e^{-t_L}\prod_{k=1}^K\frac{(t_{\ell_k}-t_{\ell_k-1})^{n_{\ell_k}-n_{\ell_k-1}}}{(n_{\ell_k}-n_{\ell_k-1})!}=e^{-t_L}\prod_{\ell=1}^L\frac{(t_\ell-t_{\ell-1})^{n_\ell-n_{\ell-1}}}{(n_\ell-n_{\ell-1})!}.
\end{aligned}
$$

Exercise 5.5.2 Let $\mathbf{M}_1$ and $\mathbf{M}_2$ be commuting elements of $M_{\mathrm{u,v}}(\mathbb{S})$. After checking that

$$
(\mathbf{M}_1+\mathbf{M}_2)^m=\sum_{\ell=0}^m\binom{m}{\ell}\mathbf{M}_1^\ell\mathbf{M}_2^{m-\ell}
$$

for all $m\in\mathbb{N}$, verify (5.1.9).

Exercise 5.5.3 Given a $\mathbf{Q}$-recurrent state i, set $C=\{j:i\overset{\mathbf{Q}}{\leftrightarrow}j\}$, and show that $R_i>0\implies R_j>0$ for all $j\in C$.

Exercise 5.5.4 Gronwall's is an inequality which has many forms, the most elementary of which states that if $u : [0, T] \longrightarrow [0, \infty)$ is a continuous function that satisfies

$$u(t) \le A + B \int_0^t u(\tau)\,d\tau \quad \text{for } t \in [0, T],$$

then $u(t) \le Ae^{Bt}$ for $t \in [0, T]$. Prove this form of *Gronwall's inequality*.

Hint: Assume that $B \ne 0$, set $U(t) = \int_0^t u(\tau)\,d\tau$, show that $\dot{U}(t) \le A + BU(t)$, and conclude that $U(t) \le \frac{A}{B}(e^{Bt} - 1)$.

Exercise 5.5.5 In this exercise we will give the continuous-time version of the ideas in Exercise 2.4.1. For this purpose, assume that $\mathbb{S}$ is irreducible and positive recurrent with respect to $\mathbf{Q}$, and use $\hat{\boldsymbol{\pi}}$ to denote the unique probability vector which is stationary for each $\mathbf{P}(t)$, $t > 0$. Next, determine the *adjoint* semigroup $\{\mathbf{P}(t)^\top : t \ge 0\}$ so that

$$\left(\mathbf{P}(t)^\top\right)_{ij} = \frac{(\hat{\boldsymbol{\pi}})_j (\mathbf{P}(t))_{ji}}{(\hat{\boldsymbol{\pi}})_i}.$$

(a) Show that $\mathbf{P}(t)^\top$ is a transition probability matrix and that $\hat{\boldsymbol{\pi}}\mathbf{P}(t)^\top = \hat{\boldsymbol{\pi}}$ for each $t \ge 0$. In addition, check that $\{\mathbf{P}(t)^\top : t \ge 0\}$ is the semigroup determined by the Q-matrix $\mathbf{Q}^\top$, where

$$\left(\mathbf{Q}^\top\right)_{ij} = \frac{(\hat{\boldsymbol{\pi}})_j (\mathbf{Q})_{ji}}{(\hat{\boldsymbol{\pi}})_i}.$$

(b) Let $\mathbb{P}$ and $\mathbb{P}^\top$ denote the probabilities computed for the Markov processes corresponding, respectively, to $\mathbf{Q}$ and $\mathbf{Q}^\top$ with initial distribution $\hat{\boldsymbol{\pi}}$. Show that $\mathbb{P}^\top$ is the *reverse* of $\mathbb{P}$ in the sense that, for each $n \in \mathbb{N}$, $0 = t_0 < t_1 < \cdots < t_n$, and $(j_0, \ldots, j_n) \in \mathbb{S}^{n+1}$,

$$\mathbb{P}^\top\big(X(t_m) = j_m \text{ for } 0 \le m \le n\big) = \mathbb{P}\big(X(t_n - t_m) = j_m \text{ for } 0 \le m \le n\big).$$

(c) Assume that $R_i \equiv -(\mathbf{Q})_{ii} > 0$ for all $i \in \mathbb{S}$, and define $\mathbf{P}^\top$ by

$$\left(\mathbf{P}^\top\right)_{ij} = (1 - \delta_{i,j}) \frac{(\hat{\boldsymbol{\pi}})_j (\mathbf{Q})_{ji}}{R_i (\hat{\boldsymbol{\pi}})_i}.$$

Show that $\mathbf{P}^\top$ is again a transition probability matrix, and check that $\mathbf{Q}^\top = \mathbf{R}(\mathbf{P}^\top - \mathbf{I})$ where $(\mathbf{R})_{ij} = \delta_{i,j} R_i$.

Exercise 5.5.6 Take $\mathbb{S} = \mathbb{N}$ and $(\mathbf{P})_{ij}$ equal to 1 or 0 according to whether $j = i + 1$ or not. Given a set of strictly positive rates $\mathfrak{R}$, show that, no matter what its initial distribution, the Markov process determined by $\mathfrak{R}$ and $\mathbf{P}$ explodes with probability 1 if $\sum_{i \in \mathbb{N}} R_i^{-1} < \infty$ and does not explode if $\sum_{i \in \mathbb{N}} R_i^{-1} = \infty$.

Exercise 5.5.7 Here is a more interesting example of explosion. Take $\mathbb{S} = \mathbb{Z}^3$, and let (cf. Exercise 3.4.1) $\mathbf{P}$ be the transition probability matrix for the symmetric, nearest neighbor random walk on $\mathbb{Z}^3$. Given a set of positive rates $\mathfrak{R}$ with the property that $\sum_{\mathbf{k}\in\mathbb{Z}^3} R_{\mathbf{k}}^{-1} < \infty$, show that, starting at every $\mathbf{k} \in \mathbb{Z}^3$, explosion occurs with probability 1 when $(\mathbf{Q})_{\mathbf{k}\boldsymbol{\ell}} = R_{\mathbf{k}}((\mathbf{P})_{\mathbf{k}\boldsymbol{\ell}} - \delta_{\mathbf{k},\boldsymbol{\ell}})$.

Hint: Apply the criterion in the second half of Theorem 5.3.10 to a function of the form

$$\sum_{\boldsymbol{\ell}\in\mathbb{Z}^3} \frac{1}{R_{\boldsymbol{\ell}}}\left(\alpha^2 + \sum_{i=1}^{3}\big((\mathbf{k})_i - (\boldsymbol{\ell})_i\big)^2\right)^{-\frac{1}{2}},$$

and use the computation in Exercise 3.4.1.

Exercise 5.5.8 Even when $\mathbb{S}$ is finite, writing down a reasonably explicit expression for the solution to (5.2.15) is seldom easy and often impossible. Nonetheless, a little linear algebra often does quite a lot of good. Throughout this exercise, $\mathbf{Q}$ is a Q-matrix on the state space $\mathbb{S}$, and it is assumed that the associated Markov process exists (i.e., does not explode) starting from any point.

(a) If $\mathbf{u} \in \mathbb{C}^{\mathbb{S}}$ is a bounded, non-zero, right eigenvector for $\mathbf{Q}$ with eigenvalue $\alpha \in \mathbb{C}$, show that the real part of α must be less than or equal to 0.
(b) Assume that $N = \#\mathbb{S} < \infty$ and that $\mathbf{Q}$ admits a complete set of linearly independent, right eigenvectors $\mathbf{u}_1, \dots, \mathbf{u}_N \in \mathbb{C}^{\mathbb{S}}$ with associated eigenvalues $\alpha_1, \dots, \alpha_N$. Let $\mathbf{U}$ be the matrix whose mth column is $\mathbf{u}_m$, and show that $e^{t\mathbf{Q}} = \mathbf{U}\Lambda(t)\mathbf{U}^{-1}$, where $\Lambda(t)$ is the diagonal matrix whose m diagonal entry is $e^{t\alpha_m}$.

Exercise 5.5.9 Here is the continuous analog of the result in Exercise 4.2.7. Assume that i is Q-recurrent and $R_i \equiv -(\mathbf{Q})_{ii} > 0$, set

$$\hat{\mu}_j = \mathbb{E}\left[\int_0^{\sigma_i} \mathbf{1}_{\{j\}}\big(X(t)\big)\,dt \,\Big|\, X(0) = i\right] \in [0, \infty] \quad \text{for } j \in \mathbb{S},$$

and let $\hat{\boldsymbol{\mu}}$ be the row vector given by $(\hat{\boldsymbol{\mu}})_j = \hat{\mu}_j$ for each $j \in \mathbb{S}$.

(a) Show that $\hat{\mu}_i = \frac{1}{R_i}$, $\hat{\mu}_j < \infty$ for all $j \in \mathbb{S}$, and that $\hat{\mu}_j > 0$ if and only if $i \overset{\mathbf{Q}}{\leftrightarrow} j$.
(b) Show that $\hat{\boldsymbol{\mu}} = \hat{\boldsymbol{\mu}}\mathbf{P}(t)$ for all $t > 0$.

Hint: Check that

$$\big(\hat{\boldsymbol{\mu}}\mathbf{P}(s)\big)_j = \mathbb{E}\left[\int_s^{s+\sigma_i} \mathbf{1}_{\{j\}}\big(X(t)\big)\,dt \,\Big|\, X(0) = i\right]$$

and that

$$\mathbb{E}\left[\int_{\sigma_i}^{s+\sigma_i} \mathbf{1}_{\{j\}}\big(X(t)\big)\,dt \,\Big|\, X(0) = i\right] = \mathbb{E}\left[\int_0^{s} \mathbf{1}_{\{j\}}\big(X(t)\big)\,dt \,\Big|\, X(0) = i\right].$$

(c) In particular, if i is $\mathbf{Q}$-positive recurrent and $C = \{j : i \overset{\mathbf{Q}}{\leftrightarrow} j\}$, show that $\hat{\boldsymbol{\mu}} = (\sum_j \hat{\mu}_j)\hat{\boldsymbol{\pi}}^C$. Equivalently,

$$\left(\hat{\boldsymbol{\pi}}^C\right)_j = \frac{\mathbb{E}[\int_0^{\sigma_i} \mathbf{1}_{\{j\}}(X(t))\,dt \mid X(0) = i]}{\mathbb{E}[\sigma_i \mid X(0) = i]}.$$

Exercise 5.5.10 Having given the continuous-time analog of Exercise 4.2.7, we will now give the continuous-time analog of Exercise 4.2.9. For this purpose, again let i be a $\mathbf{Q}$-recurrent element of $\mathbb{S}$ with $R_i \equiv -(\mathbf{Q})_{ii} > 0$, and define the measure $\hat{\boldsymbol{\mu}}$ accordingly, as in Exercise 5.5.9. Next, assume that $\hat{\boldsymbol{\nu}} \in [0, \infty)^{\mathbb{S}}$ satisfies the conditions that:

$$(\hat{\boldsymbol{\nu}})_i > 0, \qquad (\hat{\boldsymbol{\nu}})_j = 0 \quad \text{if } \mathbb{P}\big(\sigma_j < \infty \mid X(0) = i\big) < 1, \quad \text{and} \quad \hat{\boldsymbol{\nu}}\mathbf{Q} = \mathbf{0}$$

in the sense that $R_j(\hat{\boldsymbol{\nu}})_j = \sum_{k \neq j} (\hat{\boldsymbol{\nu}})_k \mathbf{Q}_{kj}$ for all $j \in \mathbb{S}$.[4] The goal is to show that $\hat{\boldsymbol{\nu}} = R_i(\hat{\boldsymbol{\nu}})_i \hat{\boldsymbol{\mu}}$. Equivalently,

$$(\hat{\boldsymbol{\nu}})_j = R_i(\hat{\boldsymbol{\nu}})_i \mathbb{E}\left[\int_0^{\sigma_i} \mathbf{1}_{\{j\}}\big(X(t)\big)\,dt \,\Big|\, X(0) = i\right] \quad \text{for all } j \in \mathbb{S}.$$

In particular, by Exercise 5.5.9, this will mean that $\hat{\boldsymbol{\nu}} = \hat{\boldsymbol{\nu}}\mathbf{P}(t)$ for all $t > 0$.

In the following, $\mathbf{P}$ and $\mathfrak{R}$ are the canonical transition probability and rates determined by $\mathbf{Q}$.

(a) Define the row vector $\boldsymbol{\nu}$ so that $(\boldsymbol{\nu})_j = \frac{R_j(\hat{\boldsymbol{\nu}})_j}{R_i(\hat{\boldsymbol{\nu}})_i}$, and show that $\boldsymbol{\nu} = \boldsymbol{\nu}\mathbf{P}$.

(b) By combining (a) with Exercise 4.2.7, show that

$$(\boldsymbol{\nu})_j = \mathbb{E}\left[\sum_{m=0}^{\rho_i - 1} \mathbf{1}_{\{j\}}(X_m) \,\Big|\, X_0 = i\right],$$

where $\{X_n : n \geq 0\}$ is the Markov chain with transition probability matrix $\mathbf{P}$.

(c) Show that

$$R_j \mathbb{E}\left[\int_0^{\sigma_i} \mathbf{1}_{\{j\}}\big(X(t)\big)\,dt \,\Big|\, X(0) = i\right] = \mathbb{E}\left[\sum_{m=0}^{\rho_i - 1} \mathbf{1}_{\{j\}}(X_m) \,\Big|\, X_0 = i\right],$$

and, after combining this with (b), arrive at the desired conclusion.

Exercise 5.5.11 Following the strategy used in Exercise 5.5.10, show that, under the hypotheses in Corollary 5.4.8, one has the following continuous version of the individual ergodic theorem:

$$\mathbb{P}\Big(\lim_{T \to \infty} \big\|\mathbf{L}_T - \hat{\boldsymbol{\pi}}^C\big\|_{\mathrm{v}} = 0\Big) = 1,$$

---

[4] The reason why the condition $\hat{\boldsymbol{\nu}}\mathbf{Q} = \mathbf{0}$ needs amplification is that, because $\mathbf{Q}$ has negative diagonal entries, possible infinities could cause ambiguities.

where $C = \{i : j \overset{\mathbf{Q}}{\leftrightarrow} i\}$ and $\mathbf{L}_T$ is the *empirical measure* determined by

$$(\mathbf{L}_T)_i = \frac{1}{T}\int_0^T \mathbf{1}_{\{i\}}\big(X(t)\big)\,dt.$$

In addition, again following the strategy in Exercise 5.5.10, show that, for any initial distribution,

$$\mathbb{P}\left(\lim_{T\to\infty}\frac{1}{T}\int_0^T \mathbf{1}_{\{j\}}\big(X(t)\big)\,dt = 0\right) = 1$$

when j is not $\mathbf{Q}$-positive recurrent.

Exercise 5.5.12 Given a Markov process $\{X(t) : t \ge 0\}$ with Q-matrix $\mathbf{Q}$ and an $R > 0$, one can produce a Markov process with Q-matrix $R\mathbf{Q}$ by speeding up the clock of $t \rightsquigarrow X(t)$ by a factor of R. That is, $\{X(Rt) : t \ge 0\}$ is a Markov process with Q-matrix $R\mathbf{Q}$. The purpose of this exercise is to carry out the analogous procedure, known in the literature as *random time change*, for variable rates. To be precise, let $\mathbf{P}$ be a transition probability matrix on $\mathbb{S}$ with $(\mathbf{P})_{ii} = 0$ for all i. Choose and fix an $i \in \mathbb{S}$, let $\{X_n : n \ge 0\}$ be a Markov chain starting from i with transition probability matrix $\mathbf{P}$ and $\{N(t) : t \ge 0\}$ a simple Poisson process which is independent of $\sigma(\{X_n : n \ge 0\})$, and set $X^0(t) = X^0_{N(t)}$ for $t \ge 0$. That is, $\{X^0(t) : t \ge 0\}$ is a Markov process with Q-matrix $(\mathbf{P} - \mathbf{I})$ starting from i. Finally, let $\mathfrak{R}$ be a set of positive rates, and take $\mathbf{Q} = \mathbf{R}(\mathbf{P} - \mathbf{I})$ accordingly.

(a) Define

$$A(t) = \int_0^t \frac{1}{R_{X^0(\tau)}}\,d\tau \quad \text{for } t \in [0,\infty),$$

observe that $t \rightsquigarrow A(t)$ is strictly increasing, set $A(\infty) = \lim_{t\nearrow\infty} A(t) \in (0,\infty]$, and use $s \in [0, A(\infty)) \longmapsto A^{-1}(s) \in [0,\infty)$ to denote the inverse of $t \rightsquigarrow A(t)$. Show that

$$A^{-1}(s) = \int_0^s R_{X^0(A^{-1}(\sigma))}\,d\sigma, \quad s \in \big[0, A(\infty)\big).$$

(b) Set $X(s) = X^0(A^{-1}(s))$ for $s \in [0, A(\infty))$. Define $J^0_0 = 0 = J_0$ and, for $n \ge 1$, J^0_n and J_n to be the times of the nth jump of $t \rightsquigarrow X^0(t)$ and $s \rightsquigarrow X(s)$, respectively. After noting that $J_n = A(J^0_n)$, conclude that, for each $n \ge 1$, $s > 0$, and $j \in \mathbb{S}$,

$$\begin{aligned}&\mathbb{P}\big(J_n - J_{n-1} > s \;\&\; X(J_n) = j \,|\, X(\sigma), \sigma \in [0, J_n)\big)\\ &= e^{-tR_{X(J_{n-1})}}(\mathbf{P})_{X(J_{n-1})j} \quad \text{on } \{J_{n-1} < \infty\}.\end{aligned}$$

(c) Show that the explosion time for the Markov process starting from i with Q-matrix $\mathbf{Q}$ has the same distribution as $A(\infty)$. In particular, if $A(\infty) = \infty$ with probability 1, use (b) to conclude that $\{X(s) : s \ge 0\}$ is a Markov process starting from i with Q-matrix $\mathbf{Q}$.

(d) As a consequence of these considerations, show that if the Markov process corresponding to $\mathbf{Q}$ does not explode, then neither does the one corresponding to $\mathbf{Q}'$, where $\mathbf{Q}'$ is related to $\mathbf{Q}$ by $(\mathbf{Q}')_{ij} = \alpha_i(\mathbf{Q})_{ij}$ and the $\{\alpha_i : i \in \mathbb{S}\}$ is a bounded subset of $(0, \infty)$.

Chapter 6
Reversible Markov Processes

This chapter is devoted to the study of a class of Markov processes that admit an initial distribution with respect to which they are *reversible* in the sense that, on every time interval, the distribution of the process is the same when it is run backwards as when it is run forwards. That is, for any $n \geq 1$ and $(i_0, \ldots, i_n) \in E^{n+1}$, in the discrete time setting,

$$\mathbb{P}(X_m = i_m \text{ for } 0 \leq m \leq n) = \mathbb{P}(X_{n-m} = i_m \text{ for } 0 \leq m \leq n) \tag{6.0.1}$$

and in the continuous time setting,

$$\mathbb{P}\big(X(t_m) = i_m \text{ for } 0 \leq m \leq n\big) = \mathbb{P}\big(X(t_n - t_m) = i_m \text{ for } 0 \leq m \leq n\big) \tag{6.0.2}$$

whenever $0 = t_0 < \cdots < t_n$. Notice that the initial distribution of such a process is necessarily stationary. Indeed, depending on whether the setting is that of discrete or continuous time, we have

$$\mathbb{P}(X_0 = i \;\&\; X_n = j) = \mathbb{P}(X_n = i \;\&\; X_0 = j) \quad \text{or}$$
$$\mathbb{P}\big(X(0) = i \;\&\; X(t) = j\big) = \mathbb{P}\big(X(t) = i \;\&\; X(0) = j\big),$$

from which stationarity follows after one sums over j. In fact, what the preceding argument reveals is that reversibility says that the joint distribution of, depending on the setting, (X_0, X_n) or $(X(0), X(t))$ is the same as that of (X_n, X_0) or $(X(t), X(0))$. This should be contrasted with the stationarity which gives equality only for the marginal distribution of the first components of these.

In view of the preceding, one should suspect that reversible Markov processes have ergodic properties that are better than those of general stationary processes, and in this chapter we will show that this suspicion is justified.

D.W. Stroock, *An Introduction to Markov Processes*, Graduate Texts in Mathematics 230,
DOI 10.1007/978-3-642-40523-5_6,

6.1 Reversible Markov Chains

In this section we will discuss irreducible, reversible Markov chains. Because the initial distribution of such a chain is stationary, we know (cf. Theorem 4.1.10) that the chain must be positive recurrent and that the initial distribution must be the probability vector (cf. (4.1.5) and (4.1.9)) $\pi = \pi^{\mathbb{S}}$ whose ith component is $(\pi)_i \equiv \mathbb{E}[\rho_i \,|\, X_0 = i]^{-1}$. Thus, if $\mathbf{P}$ is the transition probability matrix, then, by taking $n = 1$ in (6.0.1), we see that

$$(\pi)_i(\mathbf{P})_{ij} = \mathbb{P}(X_0 = i \;\&\; X_1 = j) = \mathbb{P}(X_0 = j \;\&\; X_1 = i) = (\pi)_j(\mathbf{P})_{ji}.$$

That is, $\mathbf{P}$ satisfies[1]

$$(\pi)_i(\mathbf{P})_{ij} = (\pi)_j(\mathbf{P})_{ji}, \quad \text{the } \textit{detailed balance condition}. \tag{6.1.1}$$

Conversely, (6.1.1) implies reversibility. To see this, one works by induction on $n \geq 1$ to check that

$$\pi_{i_0}(\mathbf{P})_{i_0 i_1} \cdots (\mathbf{P})_{i_{n-1} i_n} = \pi_{i_n}(\mathbf{P})_{i_n i_{n-1}} \cdots (\mathbf{P})_{i_1 i_0},$$

which is equivalent to (6.0.1).

6.1.1 Reversibility from Invariance

As we have already seen, reversibility implies invariance, and it should be clear that the converse is false. On the other hand, there are two canonical ways in which one can pass from an irreducible transition probability $\mathbf{P}$ with stationary distribution π to a transition probability for which π is reversible. Namely, for such a $\mathbf{P}$, $(\pi)_i > 0$ for all $i \in \mathbb{S}$, and so we can define the *adjoint* $\mathbf{P}^\top$ of $\mathbf{P}$ so that

$$\left(\mathbf{P}^\top\right)_{ij} = \frac{(\pi)_j(\mathbf{P})_{ji}}{(\pi)_i}. \tag{6.1.2}$$

Obviously, π is reversible for $\mathbf{P}$ if and only if $\mathbf{P} = \mathbf{P}^\top$. Moreover, because $\pi\mathbf{P} = \pi$, $\mathbf{P}^\top$ is again a transition probability. In addition, one can easily verify that both

$$\frac{\mathbf{P} + \mathbf{P}^\top}{2} \quad \text{and} \quad \mathbf{P}^\top\mathbf{P} \tag{6.1.3}$$

are transition probabilities that are reversible with respect to π. As is explained in Exercise 6.6.7 below, each of these constructions has its own virtue.

---

[1] The reader who did Exercise 2.4.1 should recognize that the condition below is precisely the same as the statement that $\mathbf{P} = \mathbf{P}^\top$. In particular, if one knows the conclusion of that exercise, then one has no need for the discussion which follows.

6.1.2 Measurements in Quadratic Mean

For reasons which will become increasingly clear, it turns out that we will want here to measure the size of functions using a Euclidean norm rather than the uniform norm $\|f\|_u$. Namely, we will use the norm

$$\|f\|_{2,\pi} \equiv \sqrt{\langle |f|^2 \rangle_\pi} \quad \text{where } \langle g \rangle_\pi \equiv \sum_i g(i)(\pi)_i \tag{6.1.4}$$

is the expected value of g with respect to π. Because $(\pi)_i > 0$ for each $i \in \mathbb{S}$, it is clear that $\|f\|_{2,\pi} = 0 \iff f \equiv 0$. In addition, if we define the *inner product* $\langle f, g \rangle_\pi$ to be $\langle fg \rangle_\pi$, then, for any $t > 0$,

$$0 \le \left\| tf \pm t^{-1} f \right\|_{2,\pi}^2 = t^2 \|f\|_{2,\pi}^2 \pm 2\langle f, g \rangle_\pi + t^{-2} \|g\|_{2,\pi}^2,$$

and so $|\langle f, g \rangle_\pi| \le t^2 \|f\|_{2,\pi}^2 + t^{-2} \|g\|_{2,\pi}^2$ for all $t > 0$. To get the best estimate, we minimize the right hand side with respect to $t > 0$. When either f or g is identically 0, then we see that $\langle f, g \rangle_\pi = 0$ by letting $t \to \infty$ or $t \to 0$. If neither f nor g vanishes identically, we can do best by taking $t = \left(\frac{\|g\|_{2,\pi}}{\|f\|_{2,\pi}}\right)^{\frac{1}{4}}$. Hence, in any case, we arrive at *Schwarz's inequality*

$$\left| \langle f, g \rangle_\pi \right| \le \|f\|_{2,\pi} \|g\|_{2,\pi}. \tag{6.1.5}$$

Given Schwarz's inequality, we know

$$\|f + g\|_{2,\pi}^2 = \|f\|_{2,\pi}^2 + 2\langle f, g \rangle_\pi + \|g\|_{2,\pi}^2 \le \left(\|f\|_{2,\pi} + \|g\|_{2,\pi} \right)^2.$$

That is, we have the *triangle inequality*:

$$\|f + g\|_{2,\pi} \le \|f\|_{2,\pi} + \|g\|_{2,\pi}. \tag{6.1.6}$$

Thus, if $L^2(\pi)$ denotes the space of f for which $\|f\|_{2,\pi} < \infty$, then $L^2(\pi)$ is a linear space for which $(f, g) \rightsquigarrow \|f - g\|_{2,\pi}$ is a metric. In fact, this metric space is complete since, if $\lim_{m\to\infty} \sup_{n>m} \|f_n - f_m\|_{2,\pi} = 0$, then $\{f_n(i) : n \ge 0\}$ is Cauchy convergent in $\mathbb{R}$ for each $i \in \mathbb{S}$, and therefore there exists a limit function f such that $f_n(i) \longrightarrow f(i)$ for each $i \in \mathbb{S}$. Moreover, by Fatou's lemma,

$$\|f - f_m\|_{2,\pi} \le \varliminf_{n\to\infty} \|f_n - f_m\|_{2,\pi} \longrightarrow 0 \quad \text{as } m \to \infty.$$

In this chapter, we will use the notation

$$\mathbf{P}f(i) = \sum_{j \in \mathbb{S}} f(j)(\mathbf{P})_{ij} = (\mathbf{Pf})_i,$$

where $\mathbf{f}$ is the column vector determined by f. When f is bounded, it is clear that $\mathbf{P}f(i)$ is well-defined and that $\|\mathbf{P}f\|_u \le \|f\|_u$, where $\|g\|_u \equiv \sup_{i\in\mathbb{S}} |g(i)| = \|\mathbf{g}\|_u$

when $\mathbf{g}$ is the column vector determined by g. We next show that, even when $f \in L^2(\boldsymbol{\pi})$, $\mathbf{P}f(i)$ is well-defined for each $i \in \mathbb{S}$ and that $\mathbf{P}$ is a contraction in $L^2(\boldsymbol{\pi})$: $\|\mathbf{P}f\|_{2,\boldsymbol{\pi}} \le \|f\|_{2,\boldsymbol{\pi}}$. To check that $\mathbf{P}f(i)$ is well-defined, we will show that the series $\sum_{j\in\mathbb{S}} f(j)(\mathbf{P})_{ij}$ is absolutely convergent. But, by the form of Schwarz's inequality in Exercise 1.3.1,

$$\sum_{j\in\mathbb{S}} |f(j)|(\mathbf{P})_{ij} = \sum_{j\in\mathbb{S}} |f(y)|\sqrt{(\boldsymbol{\pi})_i}\frac{(\mathbf{P})_{ij}}{\sqrt{(\boldsymbol{\pi})_i}} \le \|f\|_{2,\boldsymbol{\pi}}\left(\sum_{j\in\mathbb{S}}\frac{(\mathbf{P})_{ij}^2}{(\boldsymbol{\pi})_j}\right)^{\frac{1}{2}},$$

and, by (6.1.1),

$$\sum_{j\in\mathbb{S}}\frac{(\mathbf{P})_{ij}^2}{(\boldsymbol{\pi})_j} = \frac{1}{(\boldsymbol{\pi})_i}\sum_{j\in\mathbb{S}}(\mathbf{P})_{ij}(\mathbf{P})_{ji} = \frac{(\mathbf{P}^2)_{ii}}{((\boldsymbol{\pi})_i)} < \infty.$$

As for the estimate $\|\mathbf{P}f\|_{2,\boldsymbol{\pi}} \le \|f\|_{2,\boldsymbol{\pi}}$, we use Exercise 6.6.2 below together with $\sum_{j\in\mathbb{S}}(\mathbf{P})_{ij} = 1$ to see that $(\mathbf{P}f(i))^2 \le \mathbf{P}f^2(i)$ for each i. Thus, since $\boldsymbol{\pi}$ is $\mathbf{P}$-stationary,

$$\|\mathbf{P}f\|_{2,\boldsymbol{\pi}} \le \|f\|_{2,\boldsymbol{\pi}}. \tag{6.1.7}$$

An important consequence of (6.1.7) is the fact that, in general (cf. (2.2.4)),

$$\lim_{n\to\infty}\big\|\mathbf{A}_n f - \langle f\rangle_{\boldsymbol{\pi}}\big\|_{2,\boldsymbol{\pi}} = 0 \quad \text{for all } f \in L^2(\boldsymbol{\pi}) \tag{6.1.8}$$

and

$$\mathbf{P} \text{ is aperiodic} \quad \Longrightarrow \quad \lim_{n\to\infty}\big\|\mathbf{P}^n f - \langle f\rangle_{\boldsymbol{\pi}}\big\|_{2,\boldsymbol{\pi}} = 0 \quad \text{for all } f \in L^2(\boldsymbol{\pi}). \tag{6.1.9}$$

To see these, first observe that, by (4.1.11), (4.1.15), and Lebesgue's dominated convergence theorem, there is nothing to do when f vanishes off of a finite set. Thus, if $\{F_N : N \ge 1\}$ is an exhaustion of $\mathbb{S}$ by finite sets and if, for $f \in L^2(\boldsymbol{\pi})$, $f_N \equiv \mathbf{1}_{F_N} f$, then, for each $N \in \mathbb{Z}^+$,

$$\begin{aligned}\big\|\mathbf{A}_n f - \langle f\rangle_{\boldsymbol{\pi}}\big\|_{2,\boldsymbol{\pi}} &\le \big\|\mathbf{A}_n(f - f_N)\big\|_{2,\boldsymbol{\pi}} + \big\|\mathbf{A}_n f_N - \langle f_N\rangle_{\boldsymbol{\pi}}\big\|_{2,\boldsymbol{\pi}} + \big\langle|f_N - f|\big\rangle_{\boldsymbol{\pi}}\\ &\le 2\|f - f_N\|_{2,\boldsymbol{\pi}} + \big\|\mathbf{A}_n f_N - \langle f_N\rangle_{\boldsymbol{\pi}}\big\|_{2,\boldsymbol{\pi}},\end{aligned}$$

where, in the passage to the second line, we have used $\|\mathbf{A}_n g\|_{2,\boldsymbol{\pi}} \le \|g\|_{2,\boldsymbol{\pi}}$, which follows immediately from (6.1.7). Therefore, for each N,

$$\varlimsup_{n\to\infty}\big\|\mathbf{A}_n f - \langle f\rangle_{\boldsymbol{\pi}}\big\|_{2,\boldsymbol{\pi}} \le 2\|f - f_N\|_{2,\boldsymbol{\pi}},$$

which gives (6.1.8) when $N \to \infty$. The argument for (6.1.9) is essentially the same, and, as is shown in Exercise 6.6.4 below, all these results hold even in the non-reversible setting.

Finally, (6.1.1) leads to

$$\langle g, \mathbf{P}f\rangle_{\pi} = \sum_{(i,j)} (\pi)_i g(i)(\mathbf{P})_{ij} f(j) = \sum_{(i,j)} (\pi)_j f(i)(\mathbf{P})_{ji} g(j) = \langle \mathbf{P}g, f\rangle_{\pi}.$$

In other words, $\mathbf{P}$ is *symmetric* on $L^2(\pi)$ in the sense that

$$\langle g, \mathbf{P}f\rangle_{\pi} = \langle \mathbf{P}g, f\rangle_{\pi} \quad \text{for } (f, g) \in \left(L^2(\pi)\right)^2. \tag{6.1.10}$$

6.1.3 The Spectral Gap

Equation (6.1.7) combined with (6.1.10) say that $\mathbf{P}$ is a *self-adjoint contraction* on the Hilbert space[2] $L^2(\pi)$. For the reader who is unfamiliar with these concepts at this level of abstraction, think about the case when $\mathbb{S} = \{1, \dots, N\}$. Then the space of functions $f : \mathbb{S} \longrightarrow \mathbb{R}$ can be identified with $\mathbb{R}^N$. (Indeed, we already made this identification when we gave, in Sect. 2.1.3, the relation between functions and column vectors.) After making this identification, the inner product on $L^2(\pi)$ becomes the inner product $\sum_1^N (\pi)_i (\mathbf{v})_i (\mathbf{w})_i$ for column vectors $\mathbf{v}$ and $\mathbf{w}$ in $\mathbb{R}^N$. Hence (6.1.7) says that the matrix $\mathbf{P}$ acts as a symmetric, contraction on $\mathbb{R}^N$ with respect to this inner product. Alternatively, if $\tilde{\mathbf{P}} \equiv \boldsymbol{\Pi}^{\frac{1}{2}} \mathbf{P} \boldsymbol{\Pi}^{-\frac{1}{2}}$, where $\boldsymbol{\Pi}$ is the diagonal matrix whose ith diagonal entry is $(\pi)_i$, then, by (6.1.1), $\tilde{\mathbf{P}}$ is symmetric with respect to the standard inner product $(\mathbf{v}, \mathbf{w})_{\mathbb{R}^N} \equiv \sum_1^N (\mathbf{v})_i (\mathbf{w})_i$ on $\mathbb{R}^N$. Moreover, because, by (6.1.7),

$$\|\tilde{\mathbf{P}}\mathbf{f}\|^2_{\mathbb{R}^N} \equiv (\tilde{\mathbf{P}}\mathbf{f}, \tilde{\mathbf{P}}\mathbf{f})_{\mathbb{R}^N} = \sum_i (\pi)_i \left(\mathbf{P}\boldsymbol{\Pi}^{-\frac{1}{2}}\mathbf{f}\right)_i^2 = \|\mathbf{P}g\|^2_{2,\pi} \le \|g\|^2_{2,\pi} = \|\mathbf{f}\|^2_{\mathbb{R}^N},$$

where g is the function determined by the column vector $\boldsymbol{\Pi}^{-\frac{1}{2}}\mathbf{f}$, we see that, as an operator on $\mathbb{R}^N$, $\tilde{\mathbf{P}}$ is length contracting. Now, by the standard theory of symmetric matrices on $\mathbb{R}^N$, we know that $\tilde{\mathbf{P}}$ admits eigenvalues $1 \ge \lambda_1 \ge \cdots \ge \lambda_N \ge -1$ with associated eigenvectors $(\mathbf{e}_1, \dots, \mathbf{e}_N)$ that are orthonormal for $(\cdot, \cdot)_{\mathbb{R}^N}$: $(\mathbf{e}_k, \mathbf{e}_\ell)_{\mathbb{R}^N} = \delta_{k,\ell}$. Moreover, because $\sqrt{(\pi)_i} = \sum_{j=1}^N (\tilde{\mathbf{P}})_{ij}\sqrt{(\pi)_j}$, we know that $\lambda_1 = 1$ and can take $(\mathbf{e}_1)_i = \sqrt{(\pi)_i}$. Finally, by setting $\mathbf{g}_\ell = (\boldsymbol{\Pi})^{-\frac{1}{2}}\mathbf{e}_\ell$ and letting g_ℓ be the associated function on $\mathbb{S}$, we see that $\mathbf{P}g_\ell = \lambda_\ell g_\ell$, $g_1 \equiv 1$, and $\langle g_k, g_\ell\rangle_\pi = \delta_{k,\ell}$. To summarize, when $\mathbb{S}$ has N elements, we have shown that $\mathbf{P}$ on $L^2(\pi)$ has eigenvalues $1 = \lambda_1 \ge \cdots \ge \lambda_N \ge -1$ with corresponding eigenfunctions $g_1, \dots, g_N$ which are orthonormal with respect to $\langle \cdot, \cdot\rangle_\pi$. Of course, since $L^2(\pi)$ has dimension N and, by ortho-normality, the g_ℓ's are linearly independent,

[2]A Hilbert space is a vector space equipped with an inner product which determines a norm for which the associated metric is complete.

$(g_1, \ldots, g_N)$ is an orthonormal basis in $L^2(\pi)$. In particular,

$$\mathbf{P}^n f - \langle f \rangle_\pi = \sum_{\ell=2}^{N} \lambda_\ell^n \langle f, g_\ell \rangle_\pi g_\ell \quad \text{for all } n \geq 0 \text{ and } f \in L^2(\pi), \tag{6.1.11}$$

and so

$$\left\| \mathbf{P}^n f - \langle f \rangle_\pi \right\|_{2,\pi}^2 = \sum_{\ell=2}^{N} \lambda_\ell^{2n} \langle f, g_\ell \rangle_\pi^2 \leq (1-\beta)^{2n} \left\| f - \langle f \rangle_\pi \right\|_{2,\pi}^2,$$

where $\beta = (1-\lambda_2) \wedge (1+\lambda_N)$ is the *spectral gap* between $\{-1, 1\}$ and $\{\lambda_2, \ldots, \lambda_N\}$. In other words,

$$\begin{aligned} &\left\| \mathbf{P}^n f - \langle f \rangle_\pi \right\|_{2,\pi} \leq (1-\beta)^n \left\| f - \langle f \rangle_\pi \right\|_{L^2(\pi)} \\ &\text{for all } n \geq 0 \text{ and } f \in L^2(\pi). \end{aligned} \tag{6.1.12}$$

When $\mathbb{S}$ is not finite, it will not be true in general that one can find an orthonormal basis of eigenfunctions for $\mathbf{P}$. Instead, the closest approximation to the preceding development requires a famous result, known as the Spectral Theorem (cf. Sect. 107 in [7]), about bounded, symmetric operators on a Hilbert space. Nonetheless, seeing as it is the estimate (6.1.12) in which we are most interested, we can get away without having to invoke the Spectral Theorem. To be more precise, observe that (6.1.12) holds when

$$\beta \equiv 1 - \sup\left\{ \left\| \mathbf{P}f - \langle f \rangle_\pi \right\|_{2,\pi} : f \in L^2(\pi) \text{ with } \|f\|_{2,\pi} = 1 \right\}. \tag{6.1.13}$$

To see this, first note that, when β is given by (6.1.13),

$$\begin{aligned} \left\| \mathbf{P}f - \langle f \rangle_\pi \right\|_{2,\pi} &= \|f\|_{2,\pi} \left\| \mathbf{P}\left(\frac{f}{\|f\|_{2,\pi}} \right) - \left\langle \frac{f}{\|f\|_{2,\pi}} \right\rangle_\pi \right\|_{2,\pi} \\ &\leq (1-\beta)\|f\|_{2,\pi} \quad \text{when } f \in L^2(\pi) \setminus \{0\}, \end{aligned}$$

and that $\|\mathbf{P}f - \langle f \rangle_\pi\|_{2,\pi} \leq (1-\beta)\|f\|_{2,\pi}$ trivially when $f = 0$. Next, because $\langle \mathbf{P}^n f \rangle_\pi = \langle f \rangle_\pi$ for all $n \geq 0$,

$$\begin{aligned} \left\| \mathbf{P}^{n+1} f - \langle f \rangle_\pi \right\|_{2,\pi} &= \left\| \mathbf{P}\left(\mathbf{P}^n f - \langle \mathbf{P}^n f \rangle_\pi\right) \right\|_{2,\pi} \\ &\leq (1-\beta) \left\| \mathbf{P}^n f - \langle \mathbf{P}^n f \rangle_\pi \right\|_{2,\pi} = (1-\beta) \left\| \mathbf{P}^n f - \langle f \rangle_\pi \right\|_{2,\pi}. \end{aligned}$$

By induction on n, we get $\|\mathbf{P}^n f - \langle f \rangle_\pi\|_{2,\pi} \leq (1-\beta)^n \|f\|_{2,\pi}$ for all $n \geq 0$ and $f \in L^2(\pi)$. Hence, if $f \in L^2(\pi)$ and $\bar{f} = f - \langle f \rangle_\pi$, then

$$\left\| \mathbf{P}^n f - \langle f \rangle_\pi \right\|_{2,\pi} = \left\| \mathbf{P}^n \bar{f} \right\|_{2,\pi} \leq (1-\beta)^n \|\bar{f}\|_{2,\pi} = (1-\beta)^n \left\| f - \langle f \rangle_\pi \right\|_{2,\pi}.$$

That is, (6.1.12) holds with the β in (6.1.13). Observe that when $\mathbb{S}$ is finite the β in (6.1.13) coincides with the one in the preceding paragraph. Hence, we have made a first step toward generalizing the contents of that paragraph to situations in which the analog of (6.1.11) does not exist.

6.1.4 Reversibility and Periodicity

Clearly, the constant β in (6.1.13) can be as small as 0, in which case (6.1.12) tells us nothing. There are three ways in which this might happen. One way is that there exist an $f \in L^2(\boldsymbol{\pi})$ with the property that $\|f\|_{2,\boldsymbol{\pi}} = 1$, $\langle f \rangle_{\boldsymbol{\pi}} = 0$, and $\mathbf{P}f = f$. However, irreducibility rules out the existence of such an f. Indeed, because of irreducibility, we would have (cf. (6.1.8)) the contradiction that $0 = \langle f \rangle_{\boldsymbol{\pi}} = \lim_{n\to\infty}(\mathbf{A}_n\mathbf{f})_i = f(i)$ for all $i \in \mathbb{S}$ and that $f(i) \neq 0$ for some $i \in \mathbb{S}$. Thus, we can ignore this possibility because it never occurs. A second possibility is that there exists an $f \in L^2(\boldsymbol{\pi})$ with $\|f\|_{2,\boldsymbol{\pi}} = 1$ such that $\mathbf{P}f = -f$. In fact, if f is such a function, then $\langle f \rangle_{\boldsymbol{\pi}} = \langle \mathbf{P}f \rangle_{\boldsymbol{\pi}} = -\langle f \rangle_{\boldsymbol{\pi}}$, and so $\langle f \rangle_{\boldsymbol{\pi}} = 0$. Hence, we would have that $\|\mathbf{P}f - \langle f \rangle_{\boldsymbol{\pi}}\|_{2,\boldsymbol{\pi}} = 1$ and therefore that $\beta = 0$. The third possibility is that there is no non-zero solution to $\mathbf{P}f = -f$ but that, nonetheless, there exists a sequence $\{f_n\}_1^\infty \subseteq L^2(\boldsymbol{\pi})$ with $\|f_n\|_{2,\boldsymbol{\pi}} = 1$ and $\langle f_n \rangle_{\boldsymbol{\pi}} = 0$ such that $\|\mathbf{P}f_n\|_{2,\boldsymbol{\pi}}$ tends to 1.

Because the analysis of this last possibility requires the Spectral Theorem, we will not deal with it. However, as the next theorem shows, the second possibility has a pleasing and simple probabilistic interpretation. See Exercise 6.6.5 below for an extension of these considerations to non-reversible $\mathbf{P}$'s.

Theorem 6.1.14 *If* $\mathbf{P}$ *is an irreducible transition probability for which there is a reversible initial distribution, which is necessarily* $\boldsymbol{\pi}$, *then the period of* $\mathbf{P}$ *is either* 1 *or* 2. *Moreover, the period is* 2 *if and only if there exists an* $f \in L^2(\boldsymbol{\pi}) \setminus \{\mathbf{0}\}$ *for which* $f = -\mathbf{P}f$.

Proof We begin by showing that the period d must be less than or equal to 2. To this end, remember that, because of irreducibility, $(\boldsymbol{\pi})_i > 0$ for all i's. Hence, the detailed balance condition, (6.1.1), implies that $(\mathbf{P})_{ij} > 0 \iff (\mathbf{P})_{ji} > 0$. In particular, since, for each i, $(\mathbf{P})_{ij} > 0$ for some j and therefore $(\mathbf{P}^2)_{ii} = \sum_j (\mathbf{P})_{ij}(\mathbf{P})_{ji} > 0$, we see that the period must divide 2.

To complete the proof at this point, first suppose that $d = 1$. If $f \in L^2(\boldsymbol{\pi})$ satisfies $f = -\mathbf{P}f$, then, as noted before, $\langle f \rangle_{\boldsymbol{\pi}} = 0$, and yet, because of aperiodicity and (6.1.9), $\lim_{n\to\infty} \mathbf{P}^n f(i) = \langle f \rangle_{\boldsymbol{\pi}} = 0$ for each $i \in \mathbb{S}$. Since $f = \mathbf{P}^{2n} f$ for all $n \geq 0$, this means that $f \equiv 0$. Conversely, if $d = 2$, take $\mathbb{S}_0$ and $\mathbb{S}_1$ accordingly, as in Sect. 4.1.7, and consider $f = \mathbf{1}_{\mathbb{S}_0} - \mathbf{1}_{\mathbb{S}_1}$. Because of (4.1.19), $\mathbf{P}f = -f$, and clearly $\|f\|_{2,\boldsymbol{\pi}} = 1$. □

As an immediate corollary of the preceding, we can give the following graph theoretic picture of aperiodicity for irreducible, reversible Markov chains. Namely, if we use $\mathbf{P}$ to define a graph structure in which the elements of $\mathbb{S}$ are the "vertices" and an "edge" between i and j exists if and only if $(\mathbf{P})_{ij} > 0$, then the first part of Theorem 6.1.14, in combination with the considerations in Sect. 4.1.7, says that the resulting graph is *bipartite* (i.e., splits into two parts in such a way that all edges run from one part to the other) if and only if the chain fails to be aperiodic, and the second part says that this is possible if and only if there exists an $f \in L^2(\pi) \setminus \{\mathbf{0}\}$ satisfying $\mathbf{P}f = -f$.

6.1.5 Relation to Convergence in Variation

Before discussing methods for finding or estimating the β in (6.1.13), it might be helpful to compare the sort of convergence result contained in (6.1.12) to the sort of results we have been getting heretofore. To this end, first observe that

$$\left\|\mathbf{P}^n f - \langle f\rangle_\pi\right\|_{2,\pi} \le \left\|\mathbf{P}^n f - \langle f\rangle_\pi\right\|_{\mathrm{u}} \le \sup_i \left\|\delta_i\mathbf{P}^n - \pi\right\|_{\mathrm{v}}\|f\|_{\mathrm{u}}.$$

In particular, if one knows, as one does when Theorem 2.2.1 applies, that

$$\sup_i \left\|\delta_i\mathbf{P}^n - \pi\right\|_{\mathrm{v}} \le C(1-\epsilon)^n \tag{$*$}$$

for some $C < \infty$ and $\epsilon \in (0,1]$, then one has that

$$\left\|\mathbf{P}^n f - \langle f\rangle_\pi\right\|_{2,\pi} \le C(1-\epsilon)^n\|f\|_{\mathrm{u}},$$

which looks a lot like (6.1.12). Indeed, the only difference is that on the right hand side of (6.1.12), $C = 1$ and the norm is $\|f\|_{2,\pi}$ instead of $\|f\|_{\mathrm{u}}$. Thus, one should suspect that $(*)$ implies that the β in (6.1.12) is at least as large as the ϵ in $(*)$. If, as will always be the case when $\mathbb{S}$ is finite, there exists a $g \in L^2(\pi)$ with the properties that $\|g\|_{2,\pi} = 1$, $\langle g\rangle_\pi = 0$, and either $\mathbf{P}g = (1-\beta)g$ or $\mathbf{P}g = -(1-\beta)g$, this suspicion is easy to verify. Namely, set $f = g\mathbf{1}_{[-R,R]}(g)$ where $R > 0$ is chosen so that $a \equiv \langle f, g\rangle_\pi \ge \frac{1}{2}$, and set $\bar{f} = f - \langle f\rangle_\pi$. Then, after writing $\bar{f} = ag + (\bar{f} - ag)$ and noting that $\langle g, \bar{f} - ag\rangle_\pi = 0$, we see that, for any $n \ge 0$,

$$\begin{aligned}\left\|\mathbf{P}^n\bar{f}\right\|_{2,\pi}^2 &= a^2(1-\beta)^{2n} \pm 2a(1-\beta)^n\big\langle g, \mathbf{P}^n(\bar{f} - ag)\big\rangle_\pi + \left\|\mathbf{P}^n(\bar{f} - ag)\right\|_{2,\pi}^2 \\ &\ge \frac{1}{4}(1-\beta)^{2n},\end{aligned}$$

since

$$\pm\big\langle g, \mathbf{P}^n(\bar{f} - ag)\big\rangle_\pi = \big\langle \mathbf{P}^n g, \bar{f} - ag\big\rangle_\pi = (1-\beta)^n\langle g, \bar{f} - ag\rangle_\pi = 0.$$

On the other hand, $\|\mathbf{P}^n\bar{f}\|_{2,\pi}^2 \le C^2(1-\epsilon)^{2n}\|\bar{f}\|_{\mathrm{u}}^2 \le (CR)^2(1-\epsilon)^{2n}$. Thus, $\frac{1}{4}(1-\beta)^{2n} \le (CR)^2(1-\epsilon)^{2n}$ for all $n \ge 0$, which is possible only if $\beta \ge \epsilon$. When no such g exists, the same conclusion holds, only one has to invoke the Spectral Theorem in order to arrive at it.

As the preceding shows, uniform estimates on the variation distance between $\mu\mathbf{P}^n$ and π imply estimates like the one in (6.1.12). However, going in the opposite direction is not always possible. To examine what can be done, let a probability vector μ be given, and define f so that $f(i) = \frac{(\mu)_i}{(\pi)_i}$. Then, since $\langle f\rangle_\pi = 1$ and $\langle g\rangle_\pi^2 \le \langle g^2\rangle_\pi$ for all $g \in L^2(\pi)$,

$$\begin{aligned}\|\mu\mathbf{P}^n - \pi\|_{\mathrm{v}} &= \sum_j \left|(\mu\mathbf{P}^n)_j - (\pi)_j\right| = \sum_j \left|\langle f, \mathbf{P}^n\mathbf{1}_{\{j\}}\rangle_\pi - (\pi)_j\right| \\ &= \sum_j \left|\langle \mathbf{P}^n f - \mathbf{1}, \mathbf{1}_{\{j\}}\rangle_\pi\right| = \sum_j \left|\langle \mathbf{P}^n f - \langle f\rangle_\pi, \mathbf{1}_{\{j\}}\rangle_\pi\right| \\ &= \sum_j (\pi)_j \left|\mathbf{P}^n f(j) - \langle f\rangle_\pi\right| \le \left(\sum_j (\pi)_j \left|\mathbf{P}^n f(j) - \langle f\rangle_\pi\right|^2\right)^{\frac{1}{2}} \\ &= \|\mathbf{P}^n f - \langle f\rangle_\pi\|_{2,\pi},\end{aligned}$$

where, in the second to last line, I used Exercise 6.6.2. Hence, (6.1.12) implies that

$$\|\mu\mathbf{P}^n - \pi\|_{\mathrm{v}} \le \left(\sum_i \frac{(\mu)_i^2}{(\pi)_i} - 1\right)^{\frac{1}{2}} (1-\beta)^n. \tag{6.1.15}$$

In the case when $\mathbb{S}$ is finite, and therefore there exists an $\lambda \in (0,1)$ for which $(\pi)_i \ge \lambda$, (6.1.15) yields the Doeblin type estimate

$$\|\mu\mathbf{P}^n - \pi\|_{\mathrm{v}} \le \left(\frac{1-\lambda}{\lambda}\right)^{\frac{1}{2}} (1-\beta)^n.$$

However, when $\mathbb{S}$ is infinite, (6.1.15), as distinguished from Doeblin, does not give a rate of convergence which is independent of μ. In fact, unless $\sum_i \frac{(\mu)_i^2}{(\pi)_i} < \infty$, it gives no information at all. Thus one might ask why we are considering estimates like (6.1.15) when Doeblin does as well and sometimes does better. The answer is that, although Doeblin may do well when it works, it seldom works when $\mathbb{S}$ is infinite and, even when $\mathbb{S}$ is finite, it usually gives a far less than optimal rate of convergence. See, for example, Exercise 6.6.15 below where an example is given of a situation to which Doeblin's condition does not apply but the considerations here do.

6.2 Dirichlet Forms and Estimation of β

Our purpose in this section will be to find methods for estimating the optimal (i.e., largest) value of β for which (6.1.12) holds. Again, we will assume that the chain is reversible and irreducible. Later, we will add the assumption that it is aperiodic.

6.2.1 The Dirichlet Form and Poincaré's Inequality

Our first step requires us to find other expressions the right hand side of (6.1.13). To this end, begin with the observation that

$$\begin{aligned} 1-\beta = \sup\big\{\|\mathbf{P}f\|_{2,\pi} : f \in L^2_0(\pi) \text{ with } \|f\|_{2,\pi}=1\big\} \\ \text{where } L^2_0(\pi) \equiv \big\{f \in L^2(\pi) : \langle f\rangle_\pi = 0\big\}. \end{aligned} \tag{6.2.1}$$

It is obvious that the supremum on the right hand side of (6.1.13) dominates the supremum on the right above. On the other hand, if $f \in L^2(\pi)$ with $\|f\|_{2,\pi}=1$, then either $\|f-\langle f\rangle_\pi\|_{2,\pi}=0$ and therefore $\|\mathbf{P}f-\langle f\rangle_\pi\|_{2,\pi}=0$, or $1 \ge \|f-\langle f\rangle_\pi\|_{2,\pi}>0$, in which case

$$\big\|\mathbf{P}f-\langle f\rangle_\pi\big\|_{2,\pi} \le \left\|\mathbf{P}\left(\frac{f-\langle f\rangle_\pi}{\|f-\langle f\rangle_\pi\|_{2,\pi}}\right)\right\|_{2,\pi}$$

is also dominated by the right hand side of (6.2.1).

To go further, we need to borrow the following simple result from the theory of symmetric operators on a Hilbert space. Namely,

$$\begin{aligned} &\sup\big\{\|\mathbf{P}f\|_{2,\pi} : f \in L^2_0(\pi) \;\&\; \|f\|_{2,\pi}=1\big\} \\ &\quad = \sup\big\{\big|\langle f,\mathbf{P}f\rangle_\pi\big| : f \in L^2_0(\pi) \;\&\; \|f\|_{2,\pi}=1\big\}. \end{aligned} \tag{6.2.2}$$

That the right hand side of (6.2.2) is dominated by the left is Schwarz's inequality: $|\langle f,\mathbf{P}f\rangle_\pi| \le \|f\|_{2,\pi}\|\mathbf{P}f\|_{2,\pi}$. To prove the opposite inequality, let $f \in L^2_0(\pi)$ with $\|f\|_{2,\pi}=1$ be given, assume that $\|\mathbf{P}f\|_{2,\pi}>0$, and set $g=\frac{\mathbf{P}f}{\|\mathbf{P}f\|_{2,\pi}}$. Then $g \in L^2_0(\pi)$ and $\|g\|_{2,\pi}=1$. Hence, if γ denotes the supremum on the right hand side of (6.2.2), then, from the symmetry of $\mathbf{P}$,

$$\begin{aligned} 4\|\mathbf{P}f\|_{2,\pi} &= 4\langle g,\mathbf{P}f\rangle_\pi = \big\langle (f+g),\mathbf{P}(f+g)\big\rangle_\pi - \big\langle (f-g),\mathbf{P}(f-g)\big\rangle_\pi \\ &\le \gamma\big(\|f+g\|^2_{2,\pi}+\|f-g\|^2_{2,\pi}\big) = 2\gamma\big(\|f\|^2_{2,\pi}+\|g\|^2_{2,\pi}\big) = 4\gamma. \end{aligned}$$

The advantage won for us by (6.2.2) is that, in conjunction with (6.2.1), it shows that

$$\begin{aligned} &\beta = \beta_+ \wedge \beta_- \\ &\quad \text{where } \beta_\pm \equiv \inf\big\{\big\langle f,(\mathbf{I}\mp\mathbf{P})f\big\rangle_\pi : f \in L^2_0(\pi) \;\&\; \|f\|_{2,\pi}=1\big\}. \end{aligned} \tag{6.2.3}$$

Notice that, because $(\mathbf{I}-\mathbf{P})f = (\mathbf{I}-\mathbf{P})(f-\langle f\rangle_\pi)$,

$$\beta_+ = \inf\big\{\big\langle f,(\mathbf{I}-\mathbf{P})f\big\rangle_\pi : f \in L^2(\pi) \;\&\; \mathrm{Var}_\pi(f) \equiv \big\|f-\langle f\rangle_\pi\big\|^2_\pi = 1\big\}. \tag{6.2.4}$$

At the same time, because

$$\big\langle (f+c),(\mathbf{I}+\mathbf{P})(f+c)\big\rangle_\pi = \big\langle f,(\mathbf{I}+\mathbf{P})f\big\rangle_\pi + c^2 \quad \text{for } f \in L^2_0(\pi) \text{ and } c \in \mathbb{R},$$

it is clear that the infimum in the definition of β_- is the same whether we consider all $f \in L^2(\pi)$ or just those in $L^2_0(\pi)$. Hence, another expression for β_- is

$$\beta_- = \inf\{\langle f, (\mathbf{I}+\mathbf{P})f\rangle_\pi : f \in L^2_0(\pi)\ \&\ \|f\|_{2,\pi} = 1\}. \tag{6.2.5}$$

Two comments are in order here. First, when $\mathbf{P}$ is *non-negative definite*, abbreviated by $\mathbf{P} \geq 0$, in the sense that $\langle f, \mathbf{P}f\rangle_\pi \geq 0$ for all $f \in L^2(\pi)$, then $\beta_+ \leq 1 \leq \beta_-$, and so $\beta = \beta_+$. Hence,

$$\mathbf{P} \geq 0 \quad \Longrightarrow \quad \beta = \inf\{\langle f, (\mathbf{I}-\mathbf{P})f\rangle_\pi : f \in L^2(\pi)\ \&\ \mathrm{Var}_\pi(f) = 1\}. \tag{6.2.6}$$

Second, by Theorem 6.1.14, we know that, in general, $\beta_- = 0$ unless the chain is aperiodic, and (cf. (6.1.11) and the discussion at the beginning of Sect. 6.1.4), when $\mathbb{S}$ is finite, that $\beta > 0$ if and only if $\mathbb{S}$ is aperiodic.

The expressions for β_+ and β_- in (6.2.4) and (6.2.5) lead to important calculational tools. Observe that, by (6.1.1),

$$\begin{aligned}
\langle f, (\mathbf{I}-\mathbf{P})f\rangle_\pi &= \sum_{(i,j)} f(i)(\pi)_i(\mathbf{P})_{ij}\big(f(i) - f(j)\big) \\
&= \sum_{(i,j)} f(i)(\pi)_j(\mathbf{P})_{ji}\big(f(i) - f(j)\big) \\
&= \sum_{(i,j)} f(j)(\pi)_i(\mathbf{P})_{ij}\big(f(j) - f(i)\big).
\end{aligned}$$

Hence, when we add the second expression to the last, we find that

$$\langle f, (\mathbf{I}-\mathbf{P})f\rangle_\pi = \mathcal{E}(f, f) \equiv \frac{1}{2}\sum_{i \neq j}(\pi)_i(\mathbf{P})_{ij}\big(f(j) - f(i)\big)^2. \tag{6.2.7}$$

Because the quadratic form $\mathcal{E}(f, f)$ is a discrete analog of the famous quadratic form $\frac{1}{2}\int |\nabla f|^2(x)\,dx$ introduced by Dirichlet, it is called a *Dirichlet form*. Extending this metaphor, one interprets β_+ as the *Poincaré constant*

$$\beta_+ = \inf\{\mathcal{E}(f, f) : f \in L^2(\pi)\ \&\ \mathrm{Var}_\pi(f) = 1\} \tag{6.2.8}$$

in the *Poincaré inequality*

$$\beta_+ \mathrm{Var}_\pi(f) \leq \mathcal{E}(f, f), \quad f \in L^2(\pi). \tag{6.2.9}$$

To make an analogous application of (6.2.5), observe that

$$\begin{aligned}
\sum_{(i,j)}\big(f(j) + f(i)\big)^2(\pi)_i(\mathbf{P})_{ij} &= \langle \mathbf{P}f^2\rangle_\pi + 2\langle f, \mathbf{P}f\rangle_\pi + \|f\|^2_{2,\pi} \\
&= 2\|f\|^2_{2,\pi} + 2\langle f, \mathbf{P}f\rangle_\pi,
\end{aligned}$$

and therefore

$$\langle f, (\mathbf{I}+\mathbf{P})f\rangle_\pi = \tilde{\mathcal{E}}(f,f) \equiv \frac{1}{2}\sum_{(i,j)}(\boldsymbol{\pi})_i(\mathbf{P})_{ij}\big(f(j)+f(i)\big)^2. \tag{6.2.10}$$

Hence

$$\beta_- = \inf\big\{\tilde{\mathcal{E}}(f,f) : f \in L^2_0(\boldsymbol{\pi}) \ \&\ \|f\|_{2,\pi} = 1\big\}. \tag{6.2.11}$$

In order to give an immediate application of (6.2.9) and (6.2.11), suppose (cf. Exercise 2.4.3) that $(\mathbf{P})_{ij} \ge \epsilon_j$ for all (i,j), set $\epsilon = \sum_j \epsilon_j$, assume that $\epsilon > 0$, and define the probability vector $\boldsymbol{\mu}$ so that $(\boldsymbol{\mu})_i = \frac{\epsilon_i}{\epsilon}$. Then, by Schwarz's inequality for expectations with respect to $\boldsymbol{\mu}$ and the variational characterization of $\mathrm{Var}_\pi(f)$ as the minimum value of $a \rightsquigarrow \langle (f-a)^2\rangle_\pi$,

$$\begin{aligned} 2\mathcal{E}(f,f) &= \sum_{(i,j)}(\boldsymbol{\pi})_i(\mathbf{P})_{ij}\big(f(j)-f(i)\big)^2 \\ &\ge \epsilon\sum_{(i,j)}\big(f(j)-f(i)\big)^2(\boldsymbol{\pi})_i(\boldsymbol{\mu})_j \ge \epsilon\sum_i\big(\langle f\rangle_\mu - f(i)\big)^2(\boldsymbol{\pi})_i \ge \epsilon\mathrm{Var}_\pi(f), \end{aligned}$$

and, similarly, $\tilde{\mathcal{E}}(f,f) \ge \frac{\epsilon}{2}\mathrm{Var}_\pi(f)$. Hence, by (6.2.9), (6.2.11), and (6.2.3), $\beta \ge \frac{\epsilon}{2}$. Of course, this result is a significantly weaker than the one we get by combining Exercise 2.4.3 with the reasoning in Sect. 6.1.5. Namely, by that exercise we know that $\|\boldsymbol{\delta}_i\mathbf{P}^n - \boldsymbol{\pi}\|_{\mathrm{v}} \le 2(1-\epsilon)^n$, and so the reasoning in Sect. 6.1.5 tells us that $\beta \ge \epsilon$, which is twice as good as the estimate we are getting here.

6.2.2 Estimating β_+

The origin of many applications of (6.2.9) and (6.2.11) to the estimation of β_+ and β_- is the simple observation that

$$\mathrm{Var}_\pi(f) = \frac{1}{2}\sum_{ij}\big(f(i)-f(j)\big)^2(\boldsymbol{\pi})_i(\boldsymbol{\pi})_j, \tag{6.2.12}$$

which is easily checked by expanding $(f(i)-f(j))^2$ and seeing that the sum on the right equals $2\langle f^2\rangle_\pi - 2\langle f\rangle_\pi^2$.

The importance of (6.2.12) is that it expresses $\mathrm{Var}_\pi(f)$ in terms of difference between the values of f at different points in $\mathbb{S}$, and clearly $\mathcal{E}(f,f)$ is also given in terms of such differences. However, the differences that appear in $\mathcal{E}(f,f)$ are only between the values of f at pairs of points (i,j) for which $(\mathbf{P})_{ij} > 0$, whereas the right hand of side of (6.2.12) entails sampling all pairs (i,j). Thus, in order to estimate $\mathrm{Var}_\pi(f)$ in terms of $\mathcal{E}(f,f)$, it is necessary to choose, for each (i,j) with

$i \neq j$, a directed path $p(i,j) = (k_0, \dots, k_n) \in \mathbb{S}^{n+1}$ with $k_0 = i$ and $k_n = j$, which is *allowable* in the sense $(\mathbf{P})_{k_{m-1}k_m} > 0$ for each $1 \le m \le n$, and to write

$$\big(f(i) - f(j)\big)^2 = \left(\sum_{e\in p(i,j)} \Delta_e f\right)^2,$$

where the summation in e is taken over the oriented segments (k_{m-1}, k_m) in the path $p(i,j)$, and, for $e = (k,\ell)$, $\Delta_e f \equiv f(\ell) - f(k)$. At this point there are various ways in which one can proceed. For example, given any $\{\alpha(e) : e \in p(i,j)\} \subseteq (0,\infty)$, Schwarz's inequality (cf. Exercise 1.3.1) shows that the quantity on the right hand side of (6.2.12) is dominated by $\frac{1}{2}$ times

$$\begin{aligned}
&\left(\sum_{e\in p(i,j)} \frac{\alpha(e)}{\rho(e)}\right)\left(\sum_{e\in p(i,j)} \frac{\rho(e)}{\alpha(e)}(\Delta_e f)^2\right)\\
&\quad\le \max_{e\in p(i,j)} \frac{1}{\alpha(e)}\left(\sum_{e\in p(i,j)} \frac{\alpha(e)}{\rho(e)}\right) \sum_{e\in p(i,j)} (\Delta_e f)^2 \rho(e),
\end{aligned}$$

where $\rho(e) \equiv (\boldsymbol{\pi})_k (\mathbf{P})_{k\ell}$ when $e = (k,\ell)$. Thus, for any selection of paths $\mathcal{P} = \{p(i,j) : (i,j) \in \mathbb{S}^2 \setminus D\}$ (D here denotes the diagonal $\{(i,j) \in \mathbb{S}^2 : i = j\}$) and coefficients $\mathcal{A} = \{\alpha(e,p) : e \in p \in \mathcal{P}\} \subseteq (0,\infty)$,

$$\begin{aligned}
\operatorname{Var}_{\pi}(f) &\le \frac{1}{2} \sum_{p\in\mathcal{P}} w_{\mathcal{A}}(p) \sum_{e\in p} (\Delta_e f)^2 \rho(e)\\
&= \frac{1}{2} \sum_{e} (\Delta_e f)^2 \rho(e) \left(\sum_{p\ni e} w_{\mathcal{A}}(p)\right) \le W(\mathcal{P},\mathcal{A}) \mathcal{E}(f,f),
\end{aligned}$$

where

$$w_{\mathcal{A}}(p) \equiv (\boldsymbol{\pi})_i (\boldsymbol{\pi})_j \left(\max_{e\in p} \frac{1}{\alpha(e,p)}\right) \sum_{e\in p} \frac{\alpha(e,p)}{\rho(e)} \quad \text{if } p = p(i,j),$$

and

$$W(\mathcal{P},\mathcal{A}) = \sup_{e} \sum_{p\ni e} w_{\mathcal{A}}(p).$$

Hence, we have now shown that

$$\beta_+ \ge \frac{1}{W(\mathcal{P},\mathcal{A})} \tag{6.2.13}$$

for every choice of allowable paths $\mathcal{P}$ and coefficients $\mathcal{A}$.

The most effective applications of (6.2.13) depend on making a choice of $\mathcal{P}$ and $\mathcal{A}$ takes advantage of the particular situation under consideration. Given a selection

$\mathcal{P}$ of paths, one of the most frequently made choices of $\mathcal{A}$ is to take $\alpha(e, p) \equiv 1$. In this case, (6.2.13) gives

$$\beta_+ \geq \frac{1}{W(\mathcal{P})} \quad \text{where } W(\mathcal{P}) \equiv \sup_e \sum_{p \ni e} \sum_{e' \in p} \frac{(\pi(p))_-(\pi(p))_+}{\rho(e')}, \tag{6.2.14}$$

where $(\pi(p))_- = (\pi)_i$ and $(\pi(p))_+ = (\pi)_j$ when p begins at i and ends at j.

Finally, it should be recognized that, in general, (6.2.13) gives no information. Indeed, although irreducibility guarantees that there always is at least one path connecting every pair of points, when $\mathbb{S}$ is infinite there is no guarantee that $\mathcal{P}$ and $\mathcal{A}$ can be chosen so that $W(\mathcal{P}, \mathcal{A}) < \infty$. Moreover, even when $\mathbb{S}$ is finite, and therefore $W(\mathcal{P}, \mathcal{A}) < \infty$ for every choice, only a judicious choice will make (6.2.13) yield a good estimate.

6.2.3 Estimating β_-

The estimation of β_- starting from (6.2.11) is a bit more contrived than the one of β_+ starting from (6.2.9). For one thing, we already know (cf. Theorem 6.1.14) that $\beta_- = 0$ unless the chain is aperiodic. Thus, we will now require that the chain be aperiodic. As a consequence of aperiodicity and irreducibility, we know (cf. (3.1.14)) that, for each $i \in \mathbb{S}$, there always is a path $p(i) = (k_0, \dots, k_{2n+1})$ which is allowable (i.e. $(\mathbf{P})_{k_{m-1}k_m} > 0$ for each $1 \leq m \leq 2n+1$), is closed (i.e., $k_0 = k_{2n+1}$), and starts at i (i.e., $k_0 = i$). Note our insistence that this path have an odd number of steps. The reason for our doing so is that when the number of steps is odd, an elementary exercise in telescoping sums shows that

$$2f(i) = \sum_{m=0}^{2n} (-1)^m \big(f(k_m) + f(k_{m+1})\big).$$

Thus, if $\tilde{\mathcal{P}}$ be a selection of such paths, one path $p(i)$ for each $i \in \mathbb{S}$, and we make an associated choice of coefficients $\mathcal{A} = \{\alpha(e, p) : e \in p \in \tilde{\mathcal{P}}\} \subseteq (0, \infty)$, then, just as in the preceding section,

$$\begin{aligned} 4\|f\|_{2,\pi}^2 &= \sum_i \Big(\sum_{e \in p(i)} (-1)^{m(e)} \tilde{\Delta}_e f \Big)^2 (\pi)_i \\ &\leq \sum_{p \in \tilde{\mathcal{P}}} \pi(p) \Big(\sum_{e \in p} \frac{\alpha(e, p)}{\rho(e)} \Big) \Big(\sum_{e \in p} (\tilde{\Delta}_e f)^2 \frac{\rho(e)}{\alpha(e, p)} \Big), \end{aligned}$$

where, when $p = (k_0, \dots, k_{2n+1})$, $\boldsymbol{\pi}(p) = (\boldsymbol{\pi})_{k_0}$, and, for $0 \le m \le 2n$, $m(e) = m$ and $\tilde{\Delta}_e f \equiv f(k_m) + f(k_{m+1})$ if $e = (k_m, k_{m+1})$. Hence, if

$$\tilde{w}(p) = \boldsymbol{\pi}(p)\left(\max_{e\in p} \frac{1}{\alpha(e,p)}\right)\sum_{e\in p} \frac{\alpha(e,p)}{\rho(e)},$$

then

$$2\|f\|_{2,\boldsymbol{\pi}}^2 \le \tilde{W}(\tilde{\mathcal{P}}, \mathcal{A})\tilde{\mathcal{E}}(f,f) \quad \text{where } \tilde{W}(\tilde{\mathcal{P}}, \mathcal{A}) \equiv \sup_e \sum_{p\ni e} \tilde{w}(p),$$

and, by (6.2.11), this proves that

$$\beta_- \ge \frac{2}{\tilde{W}(\tilde{\mathcal{P}}, \mathcal{A})}$$

for any choice of allowable paths $\tilde{\mathcal{P}}$ and coefficients $\mathcal{A}$ satisfying the stated requirements. When we take $\alpha(e,p) \equiv 1$, this specializes to

$$\beta_- \ge \frac{2}{\tilde{W}(\tilde{\mathcal{P}})} \quad \text{where } \tilde{W}_1(\tilde{\mathcal{P}}) \equiv \sup_e \sum_{p\ni e} \sum_{e'\in p} \frac{\boldsymbol{\pi}(p)}{\rho(e')}, \tag{6.2.15}$$

It should be emphasized that the preceding method for getting estimates on β_- is inherently flawed in that it appears incapable of recognizing spectral properties of $\mathbf{P}$ like non-negative definiteness. In particular, when $\mathbf{P}$ is non-negative definite, then $\beta_- \ge 1$, but it seems unlikely that one could get that conclusion out of the arguments being used here. Thus, (6.2.15) should used only as a last resort, when nothing else seems to work.

6.3 Reversible Markov Processes in Continuous Time

Here we will see what the preceding theory looks like in the continuous time context and will learn that it is both easier and more æsthetically pleasing there.

We will be working with the notation and theory developed in Chap. 5. In particular, $\mathbf{Q}$ will be a Q-matrix, and we will assume that $\mathbb{S}$ is irreducible with respect to $\mathbf{Q}$.

6.3.1 Criterion for Reversibility

Let $\mathbf{Q}$ be given, assume that the associated Markov process never explodes (cf. Sect. 5.3.1), and use $t \rightsquigarrow \mathbf{P}(t)$ to denote the transition probability function determined by $\mathbf{Q}$ (cf. Theorem 5.3.6). Our purpose in this subsection is to show that if $\hat{\mu}$ is a probability vector for which the *detailed balance condition*

$$(\hat{\mu})_i(\mathbf{Q})_{ij} = (\hat{\mu})_j(\mathbf{Q})_{ji} \quad \text{for all } (i,j) \in \mathbb{S}^2, \tag{6.3.1}$$

holds relative to $\mathbf{Q}$, then the detailed balance condition

$$(\hat{\mu})_i\big(\mathbf{P}(t)\big)_{ij} = (\hat{\mu})_j\big(\mathbf{P}(t)\big)_{ji} \quad \text{for all } t > 0 \text{ and } (i,j) \in \mathbb{S}^2 \tag{6.3.2}$$

also holds.

The proof that (6.3.1) implies (6.3.2) is trivial in the case when the canonical rates for $\mathbf{Q}$ are bounded. Indeed, all that we need to do in that case is first use a simple inductive argument to check that (6.3.1) implies $(\hat{\mu})_i(\mathbf{Q}^n)_{ij} = (\hat{\mu})_j(\mathbf{Q}^n)_{ji}$ for all $n \geq 0$ and then use the expression for $\mathbf{P}(t)$ given in (5.1.9). When the rates are unbounded, we will use the approximation procedure introduced in Sect. 5.3.1. Namely, refer to Sect. 5.3.1, and take $\mathbf{Q}^{(N)}$ corresponding to the choice rates $\mathfrak{R}^{(N)}$ described there. Equivalently, take $(\mathbf{Q}^{(N)})_{ij}$ to be $(\mathbf{Q})_{ij}$ if $i \in F_N$ and 0 when $i \notin F_N$. Using induction, one finds first that $((\mathbf{Q}^{(N)})^n)_{ij} = 0$ for all $n \geq 1$ and $i \notin F_N$ and second that

$$(\hat{\mu})_i\big((\mathbf{Q}^{(N)})^n\big)_{ij} = (\hat{\mu})_j\big((\mathbf{Q}^{(N)})^n\big)_{ji} \quad \text{for all } n \geq 0 \text{ and } (i,j) \in F_N^2.$$

Hence, if $\{\mathbf{P}^{(N)}(t) : t > 0\}$ is the semigroup determined by $\mathbf{Q}^{(N)}$, then, since the rates for $\mathbf{Q}^{(N)}$ are bounded, (5.1.9) shows that

$$\begin{aligned} \big(\mathbf{P}^{(N)}(t)\big)_{ij} &= \delta_{i,j} && \text{if } i \notin F_N \\ (\hat{\mu})_i\big(\mathbf{P}^{(N)}(t)\big)_{ij} &= (\hat{\mu})_j\big(\mathbf{P}^{(N)}(t)\big)_{ji} && \text{if } (i,j) \in F_N^2. \end{aligned} \tag{6.3.3}$$

Therefore, because, by (5.3.4), $(\mathbf{P}^{(N)}(t))_{ij} \longrightarrow (\mathbf{P}(t))_{ij}$, we are done.

As a consequence of the preceding, we now know that (6.3.1) implies that $\hat{\mu}$ is stationary for $\mathbf{P}(t)$. Hence, because we are assuming that $\mathbf{Q}$ is irreducible, the results in Sect. 5.4.2 and Sect. 5.4.3 allow us to identify $\hat{\mu}$ as the probability vector $\hat{\pi} \equiv \hat{\pi}^{\mathbb{S}}$ introduced in Theorem 5.4.6 and discussed in Sect. 5.4.3. To summarize, *if* $\hat{\mu}$ *is a probability vector for which* (6.3.1) *holds, then* $\hat{\mu} = \hat{\pi}$.

6.3.2 Convergence in $L^2(\hat{\pi})$ for Bounded Rates

In view of the results just obtained, from now on we will be assuming that $\hat{\pi}$ is a probability vector for which (6.3.1) holds when $\hat{\mu} = \hat{\pi}$. In particular, this means that

$$(\hat{\pi})_i\big(\mathbf{P}(t)\big)_{ij} = (\hat{\pi})_j\big(\mathbf{P}(t)\big)_{ji} \quad \text{for all } t > 0 \text{ and } (i,j) \in \mathbb{S}^2. \tag{6.3.4}$$

Knowing (6.3.4), one is tempted to use the ideas in Sect. 6.2 to get an estimate on the rate, as measured by convergence in $L^2(\hat{\pi})$, at which $\mathbf{P}(t)f$ tends to $\langle f\rangle_{\hat{\pi}}$. To be precise, first note that, for each $h>0$,

$$\langle f,\mathbf{P}(h)f\rangle_{\hat{\pi}} = \left\|\mathbf{P}\left(\frac{h}{2}\right)f\right\|_{2,\hat{\pi}}^2 \ge 0,$$

and therefore (cf. (6.1.13), (6.2.6), and (6.2.7)) that

$$\begin{aligned}\beta(h) &\equiv 1-\sup\bigl\{\bigl|\langle f,\mathbf{P}(h)f\rangle_{\hat{\pi}}\bigr| : f\in L_0^2(\hat{\pi}) \text{ with } \|f\|_{2,\hat{\pi}}=1\bigr\}\\ &= \inf\bigl\{\langle f,\mathbf{I}-\mathbf{P}(h)f\rangle_{\hat{\pi}} : \mathrm{Var}_{\hat{\pi}}(f)=1\bigr\} = \inf\bigl\{\mathcal{E}_h(f,f) : \mathrm{Var}_{\hat{\pi}}(f)=1\bigr\},\end{aligned}$$

where

$$\mathcal{E}_h(f,f)\equiv\frac{1}{2}\sum_{i\neq j}(\hat{\pi})_i\bigl(\mathbf{P}(h)\bigr)_{ij}\bigl(f(j)-f(i)\bigr)^2$$

is the Dirichlet form for $\mathbf{P}(h)$ on $L^2(\hat{\pi})$. Hence (cf. (6.1.12)), for any $t>0$ and $n\in\mathbb{Z}^+$,

$$\bigl\|\mathbf{P}(t)f-\langle f\rangle_{\hat{\pi}}\bigr\|_{L^2(\hat{\pi})} \le \left(1-\beta\left(\frac{t}{n}\right)\right)^n \bigl\|f-\langle f\rangle_{\hat{\pi}}\bigr\|_{2,\hat{\pi}}.$$

To take the next step, we add the assumption that the canonical rates are bounded. Then, because, by (5.2.14),

$$\begin{gathered}\mathbf{P}(t)=\mathbf{I}+\int_0^t \mathbf{Q}\mathbf{P}(\tau)\,d\tau=\mathbf{I}+t\mathbf{Q}+\int_0^t(t-\tau)\mathbf{Q}^2\mathbf{P}(\tau)\,d\tau,\\ \bigl\|\mathbf{P}(t)-\mathbf{I}-t\mathbf{Q}\bigr\|_{\mathrm{u,v}}\le\frac{\|\mathbf{Q}\|_{\mathrm{u,v}}^2t^2}{2}.\end{gathered}$$

From this it follows that, uniformly in f satisfying $\mathrm{Var}_{\hat{\pi}}(f)=1$,

$$\lim_{h\searrow 0}\frac{\mathcal{E}_h(f,f)}{h}=\mathcal{E}^{\mathbf{Q}}(f,f)$$

where

$$\mathcal{E}^{\mathbf{Q}}(f,f)\equiv\frac{1}{2}\sum_{i\neq j}(\hat{\pi})_i(\mathbf{Q})_{ij}\bigl(f(j)-f(i)\bigr)^2; \tag{6.3.5}$$

and from this it follows that the limit $\lim_{h\searrow 0}h^{-1}\beta(h)$ exists and is equal to

$$\lambda\equiv\inf\bigl\{\mathcal{E}^{\mathbf{Q}}(f,f) : f\in L^2(\pi)\ \&\ \mathrm{Var}_{\hat{\pi}}(f)=1\bigr\}. \tag{6.3.6}$$

Thus, at least when the rates are bounded, we know that

$$\bigl\|\mathbf{P}(t)f-\langle f\rangle_{\hat{\pi}}\bigr\|_{2,\hat{\pi}}\le e^{-\lambda t}\bigl\|f-\langle f\rangle_{\hat{\pi}}\bigr\|_{2,\hat{\pi}}. \tag{6.3.7}$$

6.3.3 $L^2(\hat{\pi})$-Convergence Rate in General

When the rates are unbounded, the preceding line of reasoning is too naïve. In order to treat the unbounded case, one needs to make some additional observations, all of which have their origins in the following lemma.[3]

Lemma 6.3.8 *Given $f \in L^2(\hat{\pi})$, the function $t \in [0,\infty) \longmapsto \|\mathbf{P}(t)f\|_{2,\hat{\pi}}^2$ is continuous, non-increasing, non-negative, and convex. In particular, $t \in (0,\infty) \longmapsto \frac{\langle f,(\mathbf{I}-\mathbf{P}(t))f\rangle_{\hat{\pi}}}{t}$ is non-increasing, and therefore*

$$\lim_{h\searrow 0} \frac{\|f\|_{2,\hat{\pi}}^2 - \|\mathbf{P}(h)f\|_{2,\hat{\pi}}^2}{h} \quad \textit{exists in } [0,\infty].$$

Moreover (*cf.* (6.3.5)),

$$\lim_{h\searrow 0} \frac{\|f\|_{2,\hat{\pi}}^2 - \|\mathbf{P}(h)f\|_{2,\hat{\pi}}^2}{h} \geq 2\mathcal{E}^{\mathbf{Q}}(f,f).$$

Proof Let $f \in L^2(\hat{\pi})$ be given, and (cf. the notation in Sect. 5.3.1) define $f_N = \mathbf{1}_{F_N} f$. Then, by Lebesgue's dominated convergence theorem,

$$\|f - f_N\|_{2,\hat{\pi}} \longrightarrow 0 \quad \text{as } N \to \infty.$$

Because $\|\mathbf{P}(t)g\|_{2,\hat{\pi}} \leq \|g\|_{2,\hat{\pi}}$ for all $t > 0$ and $g \in L^2(\hat{\pi})$, we know that

$$\big|\|\mathbf{P}(t)f\|_{2,\hat{\pi}} - \|\mathbf{P}(t)f_N\|_{2,\hat{\pi}}\big| \leq \|\mathbf{P}(t)(f - f_N)\|_{2,\hat{\pi}} \leq \|f - f_N\|_{2,\hat{\pi}} \longrightarrow 0$$

uniformly in $t \in (0,\infty)$ as $N \to \infty$. Hence, by part (a) of Exercise 6.6.1 below, in order to prove the initial statement, it suffices to do so when f vanishes off of F_M for some M. Now let f be a function which vanishes off of F_M, and set $\psi(t) = \|\mathbf{P}(t)f\|_{2,\hat{\pi}}^2$. At the same time, set $\psi_N(t) = \|\mathbf{P}^{(N)}(t)f\|_{2,\hat{\pi}}^2$ for $N \geq M$. Then, because by (5.3.4), $\psi_N \longrightarrow \psi$ uniformly on finite intervals, another application of part (a) in Exercise 6.6.1 allows us to restrict our attention to the ψ_N's. That is, we will have proved that ψ is a continuous, non-increasing, non-negative, convex function as soon as we show that each ψ_N is. The non-negativity requires no comment. To prove the other properties, we apply (5.2.14) to see that

$$\dot{\psi}_N(t) = \big\langle \mathbf{Q}^{(N)}\mathbf{P}^{(N)}(t)f, \mathbf{P}^{(N)}(t)f\big\rangle_{\hat{\pi}} + \big\langle \mathbf{P}^{(N)}(t)f, \mathbf{Q}^{(N)}\mathbf{P}^{(N)}(t)f\big\rangle_{\hat{\pi}}.$$

Next, by the first line of (6.3.3), we know that, because $N \geq M$, $\mathbf{P}^{(N)}(t)f$ vanishes off of F_N, and so, because $(\hat{\pi})_i \mathbf{Q}_{ij}^{(N)} = (\hat{\pi})_j \mathbf{Q}_{ji}^{(N)}$ for $(i,j) \in F_N^2$, the preceding becomes

$$\dot{\psi}_N(t) = 2\big\langle \mathbf{P}^{(N)}(t)f, \mathbf{Q}^{(N)}\mathbf{P}^{(N)}(t)f\big\rangle_{\hat{\pi}}.$$

[3] If one knows spectral theory, especially Stone's theorem, the rather cumbersome argument that follows can be avoided.

Similarly, we see that

$$\begin{aligned}\ddot{\psi}_N(t) &= 2\langle \mathbf{Q}^{(N)}\mathbf{P}^{(N)}(t)f, \mathbf{Q}^{(N)}\mathbf{P}^{(N)}(t)f\rangle_{\hat{\pi}} \\ &\quad + 2\langle \mathbf{P}^{(N)}(t)f, (\mathbf{Q}^{(N)})^2\mathbf{P}^{(N)}(t)f\rangle_{\hat{\pi}} \\ &= 4\langle \mathbf{Q}^{(N)}\mathbf{P}^{(N)}(t)f, \mathbf{Q}^{(N)}\mathbf{P}^{(N)}(t)f\rangle_{\hat{\pi}} = 4\|\mathbf{Q}^{(N)}\mathbf{P}^{(N)}(t)f\|_{2,\hat{\pi}}^2 \ge 0.\end{aligned}$$

Clearly, the second of these proves the convexity of ψ_N. In order to see that the first implies that ψ_N is non-increasing, we will show that

$$\langle g, -\mathbf{Q}^{(N)}g\rangle_{\hat{\pi}} = \frac{1}{2}\sum_{\substack{(i,j)\in F_N^2\\ i\ne j}} (\hat{\pi})_i(\mathbf{Q})_{ij}\big(g(j)-g(i)\big)^2 + \sum_{i\in F_N}(\hat{\pi})_i V^{(N)}(i)g(i)^2$$

$$\text{if } g=0 \text{ off } F_N \quad\text{and}\quad V^{(N)}(i)\equiv\sum_{j\notin F_N} Q_{ij} \quad \text{for } i\in F_N. \tag{$*$}$$

To check $(*)$, first observe that

$$\begin{aligned}\langle g, -\mathbf{Q}^{(N)}g\rangle_{\hat{\pi}} &= -\sum_{(i,j)\in F_N^2}(\hat{\pi})_i(\mathbf{Q})_{ij}g(i)g(j) \\ &= -\sum_{\substack{(i,j)\in F_N^2\\ i\ne j}}(\hat{\pi})_i(\mathbf{Q})_{ij}g(i)\big(g(j)-g(i)\big) + \sum_{i\in F_N}(\hat{\pi})_i V^{(N)}(i)g(i)^2.\end{aligned}$$

Next, use $(\hat{\pi})_i(\mathbf{Q})_{ij} = (\hat{\pi})_j(\mathbf{Q})_{ji}$ for $(i,j)\in F_N^2$ to see that

$$-\sum_{\substack{(i,j)\in F_N\\ i\ne j}}(\hat{\pi})_i(\mathbf{Q})_{ij}g(i)\big(g(j)-g(i)\big) = \sum_{\substack{(i,j)\in F_N\\ i\ne j}}(\hat{\pi})_i(\mathbf{Q})_{ij}g(j)\big(g(j)-g(i)\big),$$

and thereby arrive at $(*)$. Finally, apply $(*)$ with $g=\mathbf{P}^{(N)}(t)f$ to conclude that $\dot{\psi}_N\le 0$.

Turning to the second and third assertions, let f be any element of $L^2(\hat{\pi})$. Now that we know that the corresponding ψ is a continuous, non-increasing, non-negative, convex function, it is easy (cf. part (d) in Exercise 6.6.1) to check that $t\rightsquigarrow\frac{\psi(0)-\psi(t)}{t}$ is non-increasing and therefore that $\lim_{h\searrow 0}\frac{\psi(0)-\psi(h)}{h}$ exists in $[0,\infty]$. Next, remember (6.2.7), apply it when $\mathbf{P}=\mathbf{P}(2h)$, and conclude that

$$\psi(0)-\psi(h) = \langle f, (\mathbf{I}-\mathbf{P}(2h))f\rangle_{\hat{\pi}} = \frac{1}{2}\sum_{\substack{(i,j)\\ i\ne j}}(\hat{\pi})_i\big(\mathbf{P}(2h)\big)_{ij}\big(f(j)-f(i)\big)^2.$$

Hence, since, by (5.3.5),

$$\lim_{h\searrow 0}\frac{(\mathbf{P}(2h))_{ij}}{h}=2(\mathbf{Q})_{ij}\quad\text{for } i\neq j,$$

the required inequality follows after an application of Fatou's lemma. □

Lemma 6.3.9 *If* $0<s<t$, *then for any* $f\in L^2(\hat{\pi})$

$$\frac{\|f\|_{2,\hat{\pi}}^2}{s}\geq\frac{\|\mathbf{P}(s)f\|_{2,\hat{\pi}}^2-\|\mathbf{P}(t)f\|_{2,\hat{\pi}}^2}{t-s}\geq 2\mathcal{E}^{\mathbf{Q}}\big(\mathbf{P}(t)f,\mathbf{P}(t)f\big).$$

Proof Set $\psi(t)=\|\mathbf{P}(t)f\|_{2,\hat{\pi}}^2$. We know that ψ is a continuous, non-increasing, non-negative, convex function. Hence, by part (a) of Exercise 5.5.2,

$$\frac{\psi(0)}{s}\geq\frac{\psi(0)-\psi(s)}{s}\geq\frac{\psi(s)-\psi(t)}{t-s}\geq\frac{\psi(t)-\psi(t+h)}{h}$$

for any $h>0$. Moreover, because,

$$\frac{\psi(t)-\psi(t+h)}{h}=\frac{\|\mathbf{P}(t)f\|_{2,\hat{\pi}}^2-\|\mathbf{P}(h)\mathbf{P}(t)f\|_{2,\hat{\pi}}^2}{h}$$

the last part of Lemma 6.3.8 applied with $\mathbf{P}(t)f$ replacing f yields the second asserted estimate. □

With the preceding at hand, we can now complete our program. Namely, by writing

$$\|f\|_{2,\hat{\pi}}^2-\|\mathbf{P}(t)f\|_{2,\hat{\pi}}^2=\sum_{m=0}^{n-1}\left(\left\|\mathbf{P}\left(\frac{mt}{n}\right)\right\|_{2,\hat{\pi}}^2-\left\|\mathbf{P}\left(\frac{(m+1)t}{n}\right)\right\|_{2,\hat{\pi}}^2\right),$$

we can use the result in Lemma 6.3.9 to obtain the estimate

$$\|f\|_{2,\hat{\pi}}^2-\|\mathbf{P}(t)f\|_{2,\hat{\pi}}^2\geq\frac{2t}{n}\sum_{m=1}^{n}\mathcal{E}^{\mathbf{Q}}\left(\mathbf{P}\left(\frac{mt}{n}\right)f,\mathbf{P}\left(\frac{mt}{n}\right)f\right).$$

Hence, if λ is defined as in (6.3.6), then, for any $f\in L_0^2(\hat{\pi})$,

$$\|f\|_{2,\hat{\pi}}^2-\|\mathbf{P}(t)f\|_{2,\hat{\pi}}^2\geq\frac{2\lambda t}{n}\sum_{m=1}^{n}\left\|\mathbf{P}\left(\frac{mt}{n}\right)f\right\|_{2,\hat{\pi}}^2,$$

which, when $n\to\infty$, leads to

$$\|f\|_{2,\hat{\pi}}^2-\|\mathbf{P}(t)f\|_{2,\hat{\pi}}^2\geq 2\lambda\int_0^t\|\mathbf{P}(\tau)f\|_{2,\hat{\pi}}^2\,d\tau.$$

Finally, by Gronwall's inequality (cf. Exercise 4.2.2), the preceding yields the estimate $\|\mathbf{P}(t)f\|_{2,\hat{\pi}}^2 \le e^{-2\lambda t}\|f\|_{2,\hat{\pi}}^2$. After replacing a general $f \in L^2(\hat{\pi})$ by $f - \langle f\rangle_{\hat{\pi}}$, we have now proved that (6.3.7) holds even when the canonical rates are unbounded.

6.3.4 *Estimating* λ

Proceeding in the exactly the same way that we did in Sect. 6.2.2, we can estimate the λ in (6.3.6) in the same way as we estimated β_+ there. Namely, we make a selection $\mathcal{P}$ consisting of paths $p(i,j)$, one for each from pair $(i,j) \in \mathbb{S}\setminus D$, with the properties that if $p(i,j) = (k_0,\dots,k_n)$, then $k_0 = i$, $k_n = j$, and $p(i,j)$ is *allowable* in the sense that $(\mathbf{Q})_{k_{m-1}k_m} > 0$ for each $1 \le m \le n$. Then, just as in Sect. 6.2.2, we can say that

$$\lambda \ge \frac{1}{W(\mathcal{P})} \quad \text{where } W(\mathcal{P}) \equiv \sup_e \sum_{p\ni e}\sum_{e'\in p} \frac{(\hat{\pi}(p))_-(\hat{\pi}(p))_+}{\rho(e')}, \tag{6.3.10}$$

where the supremum is over oriented edges $e = (k,\ell)$ with $(\mathbf{Q})_{k\ell} > 0$, the first sum is over $p \in \mathcal{P}$ in which the edge e appears, the second sum is over edges e' which appear in the path p, $(\hat{\pi}(p))_- = (\hat{\pi})_i$ if the path p starts at i, $(\hat{\pi}(p))_+ = (\hat{\pi})_j$ if the path p ends at j, and $\rho(e') = (\hat{\pi})_k(\mathbf{Q})_{k\ell}$ if $e' = (k,\ell)$. See [1] for further applications of these results.

6.4 Gibbs States and Glauber Dynamics

Loosely speaking, the physical principle underlying statistical mechanics can be summarized in the statement that, when a system is in equilibrium, *states with lower energy are more likely than those with higher energy*. In fact, J.W. Gibbs sharpened this statement by saying that the probability of a state i will be proportional to $e^{-\frac{H(i)}{kT}}$, where k is the Boltzmann constant, T is temperature, and $H(i)$ is the energy of the system when it is in state i. For this reason, a distribution which assigns probabilities in this Gibbsian manner is called a *Gibbs state*.

Since a Gibbs state is to be a model of equilibrium, it is only reasonable to ask what is the dynamics for which it is the equilibrium. From our point of view, this means that we should seek a Markov process for which the Gibbs state is the stationary distribution. Further, because dynamics in physics should be reversible, we should be looking for Markov processes which are reversible with respect to the Gibbs state, and, because such processes were introduced in this context by R. Glauber, we will call a Markov process which is reversible with respect to a Gibbs state a *Glauber dynamics* for that Gibbs state.

In this section, we will give a rather simplistic treatment of Gibbs states and their associated Glauber dynamics.

6.4.1 Formulation

Throughout this section, we will be working in the following setting. As usual, $\mathbb{S}$ is either a finite or countably infinite space. On $\mathbb{S}$ there is given some "natural" background assignment $\boldsymbol{\nu} \in (0,\infty)^{\mathbb{S}}$ of (not necessarily summable) weights, which should be thought of as a row vector. In many applications, $\boldsymbol{\nu}$ is uniform: it assigns each i weight 1, but in other situations it is convenient to not have to assume that it is uniform. Next, there is a function $H : \mathbb{S} \longrightarrow [0,\infty)$ (alias, the energy function) with the property that

$$Z(\beta) \equiv \sum_{i \in \mathbb{S}} e^{-\beta H(i)} (\boldsymbol{\nu})_i < \infty \quad \text{for each } \beta \in (0,\infty). \tag{6.4.1}$$

In the physics metaphor, $\beta = \frac{1}{kT}$ is, apart from Boltzmann's constant k, the reciprocal temperature, and physicists would call $\beta \rightsquigarrow Z(\beta)$ the *partition function*. Finally, for each $\beta \in (0,\infty)$, the *Gibbs state* $\boldsymbol{\gamma}(\beta)$ is the probability vector given by

$$\big(\boldsymbol{\gamma}(\beta)\big)_i = \frac{1}{Z(\beta)} e^{-\beta H(i)} (\boldsymbol{\nu})_i \quad \text{for } i \in \mathbb{S}. \tag{6.4.2}$$

From a physical standpoint, everything of interest is encoded in the partition function. For example, it is elementary to compute both the average and variance of the energy by taking logarithmic derivatives:

$$\langle H \rangle_{\boldsymbol{\gamma}(\beta)} = -\frac{d}{d\beta} \log Z(\beta) \quad \text{and} \quad \operatorname{Var}_{\boldsymbol{\gamma}(\beta)}(H) = \frac{d^2}{d\beta^2} \log Z(\beta). \tag{6.4.3}$$

The final ingredient is the description of the Glauber dynamics. For this purpose, we start with a matrix $\mathbf{A}$ all of whose entries are non-negative and whose diagonal entries are 0. Further, we assume that $\mathbf{A}$ is irreducible in the sense that

$$\sup_{n \ge 0} \big(\mathbf{A}^n\big)_{ij} > 0 \quad \text{for all } (i,j) \in \mathbb{S}^2 \tag{6.4.4}$$

and that it is reversible in the sense that

$$(\boldsymbol{\nu})_i (\mathbf{A})_{ij} = (\boldsymbol{\nu})_j (\mathbf{A})_{ji} \quad \text{for all } (i,j) \in \mathbb{S}^2. \tag{6.4.5}$$

Finally, we insist that

$$\sum_{j \in \mathbb{S}} e^{-\beta H(j)} (\mathbf{A})_{ij} < \infty \quad \text{for each } i \in \mathbb{S} \text{ and } \beta > 0. \tag{6.4.6}$$

At this point there are many ways in which to construct a Glauber dynamics. However, for our purposes, the one which will serve us best is the one whose

Q-matrix is given by

$$\begin{aligned}\big(\mathbf{Q}(\beta)\big)_{ij} &= e^{-\beta(H(j)-H(i))^+}(\mathbf{A})_{ij} \quad \text{when } j \neq i\\ \big(\mathbf{Q}(\beta)\big)_{ii} &= -\sum_{j\neq i}\big(\mathbf{Q}(\beta)\big)_{ij},\end{aligned} \tag{6.4.7}$$

where $a^+ \equiv a \vee 0$ is the non-negative part of the number $a \in \mathbb{R}$. Because,

$$\big(\boldsymbol{\gamma}(\beta)\big)_i\big(\mathbf{Q}(\beta)\big)_{ij} = Z(\beta)^{-1}e^{-\beta H(i)\vee H(j)}(\boldsymbol{\nu})_i(\mathbf{A})_{ij} \quad \text{for } i \neq j, \tag{6.4.8}$$

$\boldsymbol{\gamma}(\beta)$ clearly is reversible for $\mathbf{Q}(\beta)$. There are many other possibilities, and the optimal choice is often dictated by special features of the situation under consideration. However, whatever choice is made, it should be made in such a way that, for each $\beta > 0$, $\mathbf{Q}(\beta)$ *determines a Markov process which never explodes.*

6.4.2 The Dirichlet Form

In this subsection we will modify the ideas developed in Sect. 6.2.3 to get a lower bound on

$$\begin{aligned}&\lambda_\beta \equiv \inf\{\mathcal{E}_\beta(f,f) : \operatorname{Var}_\beta(f) = 1\}\\ &\quad\text{where } \mathcal{E}_\beta(f,f) \equiv \frac{1}{2}\sum_{j\neq i}\big(\boldsymbol{\gamma}(\beta)\big)_i\big(\mathbf{Q}(\beta)\big)_{ij}\big(f(j)-f(i)\big)^2\end{aligned} \tag{6.4.9}$$

and $\operatorname{Var}_\beta(f)$ is shorthand for $\operatorname{Var}_{\boldsymbol{\gamma}(\beta)}(f)$, the variance of f with respect to $\boldsymbol{\gamma}(\beta)$. For this purpose, we introduce the notation

$$\operatorname{Elev}(p) = \max_{0\le m\le n} H(i_m) \quad \text{and} \quad e(p) = \operatorname{Elev}(p) - H(i_0) - H(i_n)$$

for an allowable path $p = (i_0, \dots, i_n)$. Assuming that $\mathbf{Q}(\beta)$ is given by (6.4.7), one sees that, when $p = (i_0, \dots, i_n)$

$$\begin{aligned}&w_\beta(p) \equiv \sum_{m=1}^{n}\frac{(\boldsymbol{\gamma}(\beta))_{i_0}(\boldsymbol{\gamma}(\beta))_{i_n}}{(\boldsymbol{\gamma}(\beta))_{i_{m-1}}(\mathbf{Q}(\beta))_{i_{m-1}i_m}} \le Z(\beta)^{-1}e^{\beta e(p)}w(p),\\ &\quad\text{where } w(p) \equiv \sum_{m=1}^{n}\frac{(\boldsymbol{\nu})_{i_0}(\boldsymbol{\nu})_{i_n}}{(\boldsymbol{\nu})_{i_{m-1}}\mathbf{A}_{i_{m-1}i_m}}.\end{aligned}$$

Hence, for any choice of paths $\mathcal{P}$, we know that (cf. (6.3.10))

$$W_\beta(\mathcal{P}) \equiv \sup_e \sum_{p \ni e} w_\beta(p) \le Z(\beta)^{-1} e^{\beta E(\mathcal{P})} W(\mathcal{P}),$$

$$\text{where } W(\mathcal{P}) \equiv \sup_e \sum_{p \ni e} w(p) \text{ and } E(\mathcal{P}) \equiv \sup_{p \in \mathcal{P}} e(p),$$

and therefore that

$$\lambda_\beta \ge \frac{Z(\beta) e^{-\beta E(\mathcal{P})}}{W(\mathcal{P})}. \tag{6.4.10}$$

On the one hand, it is clear that (6.4.10) gives information only when $W(\mathcal{P}) < \infty$. At the same time, it shows that, at least if ones interest is in large β's, then it is important to choose $\mathcal{P}$ so that $E(\mathcal{P})$ is as small as possible. When $\mathbb{S}$ is finite, reconciling these two creates no problem. Indeed, the finiteness of $\mathbb{S}$ guarantees that $W(\mathcal{P})$ will be finite for every choice of allowable paths. In addition, finiteness allows one to find for each (i, j) a path $p(i, j)$ which minimizes $\mathrm{Elev}(p)$ among allowable paths from i to j, and clearly any $\mathcal{P}$ consisting of such paths will minimize $E(\mathcal{P})$. Of course, it is sensible to choose such a $\mathcal{P}$ so as to minimize $W(\mathcal{P})$ as well. In any case, whenever $\mathbb{S}$ is finite and $\mathcal{P}$ consists of paths $p(i, j)$ which minimize $\mathrm{Elev}(p)$ among paths p between i and j, $E(\mathcal{P})$ has a nice interpretation. Namely, think of $\mathbb{S}$ as being sites on a map and of H as giving the altitude of the sites. That is, in this metaphor, $H(i)$ is the distance of i "above sea level." Without loss in generality, we will assume that at least one site k_0 is at sea level: $H(k_0) = 0$.[4] When such an k_0 exists, the metaphorical interpretation of $E(\mathcal{P})$ is as the least upper bound on the altitude a hiker must gain, no matter where he starts or what allowable path he chooses to follow, in order to reach sea level. To see this, first observe that if p and p' are a pair of allowable paths and if the end point of p is the initial point of p', then the path q is allowable and $\mathrm{Elev}(q) = \mathrm{Elev}(p) \vee \mathrm{Elev}(p')$ when q is obtained by concatenating p and p': if $p = (i_0, \dots, i_n)$ and $p' = (i'_0, \dots, i'_{n'})$, then $q = (i_0, \dots, i_n, i'_1, \dots, i'_{n'})$. Hence, for any (i, j), $e(p(i, j)) \le e(p(i, k_0)) \vee e(p(j, k_0))$, from which it should be clear that $E(\mathcal{P}) = \max_i e(p(i, k_0))$. Finally, since, for each i, $e(p(i, k_0)) = H(\ell) - H(i)$, where ℓ is a highest point along the path $p(i, k_0)$, the explanation is complete. When $\mathbb{S}$ is infinite, the same interpretation is valid in various circumstances. For example, it applies when H "tends to infinity at infinity" in the sense that $\{i : H(i) \le M\}$ is finite for each $M < \infty$.

When $\mathbb{S}$ is finite, we can show that, at least for large β, (6.4.10) is quite good. To be precise, we have the following result.

Theorem 6.4.11 *Assume that $\mathbb{S}$ is finite and that $\mathbf{Q}(\beta)$ is given by (6.4.7). Set $\mathfrak{m} = \min_{i \in \mathbb{S}} H(i)$ and $\mathbb{S}_0 = \{i : H(i) = \mathfrak{m}\}$, and let $\mathfrak{e}$ be the minimum value $E(\mathcal{P})$*

[4] If that is not already so, we can make it so by choosing k_0 to be a point at which H takes its minimum value and replacing H by $H - H(k_0)$. Such a replacement leaves both $\gamma(\beta)$ and $\mathbf{Q}(\beta)$ as well as the quantity on the right hand side of (6.4.10) unchanged.

takes as $\mathcal{P}$ *runs over all selections of allowable paths. Then* $\mathfrak{e} \geq -\mathfrak{m}$*, and* $\mathfrak{e} = -\mathfrak{m}$ *if and only if for each* $(i, j) \in \mathbb{S} \times \mathbb{S}_0$ *there is an allowable path* p *from* i *to* j *with* $\mathrm{Elev}(p) = H(i)$. *(See also Exercise* 6.6.13 *below.) More generally, whatever the value of* $\mathfrak{e}$*, there exist constants* $0 < c_- \leq c_+ < \infty$*, which are independent of* H*, such that*

$$c_- e^{-\beta(\mathfrak{e}+\mathfrak{m})} \leq \lambda_\beta \leq c_+ e^{-\beta(\mathfrak{e}+\mathfrak{m})} \quad \textit{for all } \beta \geq 0.$$

Proof Because neither $\boldsymbol{\gamma}(\beta)$ nor $\mathbf{Q}(\beta)$ is changed if H is replaced by $H - \mathfrak{m}$ whereas $\mathfrak{e}$ changes to $\mathfrak{e} + \mathfrak{m}$, we may and will assume that $\mathfrak{m} = 0$.

Choose a collection $\mathcal{P} = \{p(i, j) : (i, j) \in \mathbb{S}^2\}$ of allowable paths so that, for each (i, j), $e(p(i, j))$ minimizes $e(p)$ over allowable paths from i to j. Next choose and fix a $k_0 \in \mathbb{S}_0$. By the reasoning given above,

$$\mathfrak{e} = \max_{i \in \mathbb{S}} e\big(p(i, k_0)\big). \tag{$*$}$$

In particular, since $e(p(i, k_0)) = \mathrm{Elev}(p(i, k_0)) - H(i) \geq 0$ for all $i \in \mathbb{S}$, this proves that $\mathfrak{e} \geq 0$ and makes it clear that $\mathfrak{e} = 0$ if and only if $H(i) = \mathrm{Elev}(p(i, k_0))$ for all $i \in \mathbb{S}$.

Turning to the lower bound for λ_β, observe that, because $\mathfrak{m} = 0$, $Z(\beta) \geq (\boldsymbol{\nu})_{k_0} > 0$ and therefore, by (6.4.10), that we can take $c_- = \frac{(\boldsymbol{\nu})_{k_0}}{W(\mathcal{P})}$.

Finally, to prove the upper bound, choose $\ell_0 \in \mathbb{S} \setminus \{k_0\}$ so that $e(p_0) = \mathfrak{e}$ when $p_0 \equiv p(\ell_0, k_0)$, let Γ be the set of $i \in \mathbb{S}$ with the property that either $i = k_0$ or $\mathrm{Elev}(p(i)) < \mathrm{Elev}(p_0)$ for the path $p(i) \equiv p(i, k_0) \in \mathcal{P}$ from i to k_0, and set $f = \mathbf{1}_\Gamma$. Then, because $k_0 \in \Gamma$ and $\ell_0 \notin \Gamma$, (cf. (6.2.12)),

$$\begin{aligned}\mathrm{Var}_\beta(f) &= \Big(\sum_{i \in \Gamma} \big(\boldsymbol{\gamma}(\beta)\big)_i\Big)\Big(\sum_{j \notin \Gamma} \big(\boldsymbol{\gamma}(\beta)\big)_j\Big) \\ &\geq \big(\boldsymbol{\gamma}(\beta)\big)_{k_0} \big(\boldsymbol{\gamma}(\beta)\big)_{\ell_0} = \frac{(\boldsymbol{\nu})_{k_0} (\boldsymbol{\nu})_{\ell_0}}{Z(\beta)^2} e^{-\beta H(\ell_0)}.\end{aligned}$$

At the same time

$$\mathcal{E}_\beta(f, f) = \sum_{(i,j) \in \Gamma \times \Gamma \complement} \big(\boldsymbol{\gamma}(\beta)\big)_i \big(\mathbf{Q}(\beta)\big)_{ij} = \frac{1}{Z(\beta)} \sum_{(i,j) \in \Gamma \times \Gamma \complement} (\boldsymbol{\nu})_i (\mathbf{A})_{ij} e^{-\beta H(i) \vee H(j)}.$$

If $i \in \Gamma \setminus \{k_0\}$, $j \notin \Gamma$, and $(\mathbf{A})_{ij} > 0$, then $H(j) \geq \mathrm{Elev}(p_0)$. To see this, consider the path q obtained by going in one step from j to i and then following $p(i)$ from i to k_0. Clearly, q is an allowable path from j to k_0, and therefore $\mathrm{Elev}(q) \geq \mathrm{Elev}(p(j)) \geq \mathrm{Elev}(p_0)$. But this means that

$$\mathrm{Elev}(p_0) \leq \mathrm{Elev}\big(p(j)\big) \leq \mathrm{Elev}(q) = \mathrm{Elev}\big(p(i)\big) \vee H(j),$$

which, together with $\mathrm{Elev}(p(i)) < \mathrm{Elev}(p_0)$, forces the conclusion that $H(j) \geq \mathrm{Elev}(p_0)$. Even easier is the observation that $H(j) \geq \mathrm{Elev}(p_0)$ if $j \notin \Gamma$ and

$(\mathbf{A})_{k_0 j} > 0$, since in that case the path (j, k_0) is allowable and

$$H(j) = \text{Elev}\big((j, k_0)\big) \geq \text{Elev}\big(p(j)\big) \geq \text{Elev}(p_0).$$

Hence, after plugging this into the preceding expression for $\mathcal{E}_\beta(f, f)$, we get

$$\mathcal{E}_\beta(f, f) \leq \frac{e^{-\beta \text{Elev}(p_0)}}{Z(\beta)} \sum_{(i,j) \in \Gamma \times \Gamma \complement} (\boldsymbol{\nu})_i (\mathbf{A})_{ij},$$

which, because $\mathfrak{e} = \text{Elev}(p_0) - H(\ell_0)$, means that

$$\lambda_\beta \leq \frac{\mathcal{E}_\beta(f, f)}{\text{Var}_\beta(f)} \leq \frac{Z(\beta)}{(\boldsymbol{\nu})_{k_0} (\boldsymbol{\nu})_{\ell_0}} \left(\sum_{(i,j) \in \Gamma \times \Gamma \complement} (\boldsymbol{\nu})_i (\mathbf{A})_{ij} \right) e^{-\beta \mathfrak{e}}.$$

Finally, since $Z(\beta) \leq \|\boldsymbol{\nu}\|_{\text{v}}$, the upper bound follows. □

6.5 Simulated Annealing

This concluding section deals with an application of the ideas in the preceding section. Namely, given a function $H : \mathbb{S} \longrightarrow [0, \infty)$, we want to describe a procedure, variously known as the *simulated annealing* or the *Metropolis algorithm*, for locating a place where H achieves its minimum value.

In order to understand the intuition that underlies this procedure, let $\mathbf{A}$ be a matrix of the sort discussed in Sect. 6.4, assume that 0 is the minimum value of H, set $\mathbb{S}_0 = \{i : H(i) = 0\}$, and think about dynamic procedures which would lead you from any initial point to $\mathbb{S}_0$ via paths which are allowable according to $\mathbf{A}$ (i.e., $\mathbf{A}_{k\ell} > 0$ if k and ℓ are successive points along the path). One procedure is based on the steepest decent strategy. That is, if one is at k, one moves to any one of the points ℓ for which $\mathbf{A}_{k\ell} > 0$ and $H(\ell)$ is minimal if $H(\ell) \leq H(k)$ for at least one such point, and one stays put if $H(\ell) > H(k)$ for every ℓ with $(\mathbf{A})_{k\ell} > 0$. This procedure works beautifully as long as you avoid, in the topographic metaphor suggested earlier, getting trapped in some "mountain valley." The point is that the steepest decent procedure is the most efficient strategy for getting to some local minimum of H. However, if that minimum is not global, then, in general, you will get stuck! Thus, if you are going to avoid this fate, occasionally you will have to go "up hill" even when you may have the option to go "down hill." However, unless you have a detailed *a priori* knowledge of the whole terrain, there is no way to know when you should decide to do so. For this reason, it may be best to abandon rationality and let the decision be made randomly. Of course, after a while, you should hope that you will have worked your way out of the mountain valleys and that a steepest decent strategy should become increasingly reasonable.

6.5.1 The Algorithm

In order to eliminate as many technicalities as possible, we will assume throughout that $\mathbb{S}$ is finite and has at least 2 elements. Next, let $H : \mathbb{S} \longrightarrow [0, \infty)$ be the function for which we want to locate a place where it achieves its minimum, and, without loss in generality, we will assume that 0 is its minimum. Now take $\boldsymbol{\nu}$ so that $(\boldsymbol{\nu})_i = 1$ for all $i \in \mathbb{S}$, and choose a matrix $\mathbf{A}$ so that $(\mathbf{A})_{ii} = 0$ for all $i \in \mathbb{S}$, $(\mathbf{A})_{ij} = (\mathbf{A})_{ji} \geq 0$ if $j \neq i$, and $\mathbf{A}$ is irreducible (cf. (6.4.4)) on $\mathbb{S}$. In practice, the selection of $\mathbf{A}$ should be made so that the evaluation of $H(j) - H(i)$ when $(\mathbf{A})_{ij} > 0$ is as "easy as possible." For example, if $\mathbb{S}$ has some sort of natural neighborhood structure with respect to which $\mathbb{S}$ is connected and the computation of $H(j) - H(i)$ when j is a neighbor of i requires very little time, then it is reasonable to take $\mathbf{A}$ so that $(\mathbf{A})_{ij} = 0$ unless $j \neq i$ is a neighbor of i.

Now define $\boldsymbol{\gamma}(\beta)$ as in (6.4.2) and $\mathbf{Q}(\beta)$ as in (6.4.7). Clearly, $\boldsymbol{\gamma}(0)$ is just the normalized, uniform distribution on $\mathbb{S}$: $(\boldsymbol{\gamma}(0))_i = L^{-1}$, where $L \equiv \#\mathbb{S} \geq 2$ is the number of elements in $\mathbb{S}$. On the one hand, as β gets larger, $\boldsymbol{\gamma}(\beta)$ becomes more concentrated on $\mathbb{S}_0$. More precisely, since $Z(\beta) \geq \#\mathbb{S}_0 \geq 1$

$$\langle \mathbf{1}_{\mathbb{S}_0 \complement} \rangle_{\boldsymbol{\gamma}(\beta)} \leq L e^{-\beta\delta}, \quad \text{where } \delta \equiv \min\{H(j) : j \notin \mathbb{S}_0\}. \tag{6.5.1}$$

On the other hand, as β gets larger, Theorem 6.4.11 says that, at least when $\mathfrak{e} > 0$, $\lambda(\beta)$ will be getting smaller. Thus, we are confronted by a conflict.

In view of the introductory discussion, this conflict between the virtues of taking β large, which is tantamount to adopting an approximately steepest decent strategy, versus those of taking β small, which is tantamount to keeping things fluid and thereby diminishing the danger of getting trapped, should be expected. Moreover, a resolution is suggested at the end of that discussion. Namely, in order to maximize the advantages of each, one should start with $\beta = 0$ and allow β to increase with time.[5] That is, we will make β an increasing, continuous function $t \rightsquigarrow \beta(t)$ with $\beta(0) = 0$. In the interest of unencumbering our formulae, we will adopt the notation

$$\begin{aligned} &Z(t) = Z(\beta(t)), \quad &&\boldsymbol{\gamma}_t = \boldsymbol{\gamma}(\beta(t)), \quad &&\langle \cdot \rangle_t = \langle \cdot \rangle_{\boldsymbol{\gamma}_t}, \quad &&\|\cdot\|_{2,t} = \|\cdot\|_{2,\boldsymbol{\gamma}_t}, \\ &\mathrm{Var}_t = \mathrm{Var}_{\boldsymbol{\gamma}_t}, \quad &&\mathbf{Q}(t) = \mathbf{Q}(\beta(t)), \quad &&\mathcal{E}_t = \mathcal{E}_{\beta(t)}, \quad \text{and} \quad &&\lambda_t = \lambda_{\beta(t)}. \end{aligned}$$

Because, in the physical model, β is proportional to the reciprocal of temperature and β increases with time, $t \rightsquigarrow \beta(t)$ is called the *cooling schedule*.

---

[5]Actually, there is good reason to doubt that monotonically increasing β may not be the best strategy. Indeed, the name "simulated annealing" derives from the idea that what one wants to do is simulate the annealing process familiar to chemists, material scientists, skilled carpenters, and followers of Metropolis. Namely, what these people do is alternately heat and cool to achieve their goal, and there is reason to believe we should be following their example. However, I have chosen not to follow them on the unforgivable, but understandable, grounds that my analysis is capable of handling only the monotone case.

6.5.2 Construction of the Transition Probabilities

The Q-matrix here being time dependent means that the associated transition probabilities will be *time-inhomogeneous*. Thus, instead of $t\rightsquigarrow\mathbf{Q}(t)$ determining a one parameter family of transition probability matrices, for each $s\in[0,\infty)$ it will determine a map $t\rightsquigarrow\mathbf{P}(s,t)$ from $[s,\infty)$ into transition probability matrices by the *time-inhomogeneous Kolmogorov forward equation*

$$\frac{d}{dt}\mathbf{P}(s,t)=\mathbf{P}(s,t)\mathbf{Q}(t)\quad\text{on }(s,\infty)\text{ with }\mathbf{P}(s,s)=\mathbf{I}. \tag{6.5.2}$$

Although (6.5.2) is not exactly covered by our earlier analysis of Kolmogorov equations, it nearly is. To see this, we solve (6.5.2) via an approximation procedure in which $\mathbf{Q}(t)$ is replaced on the right hand side by

$$\mathbf{Q}^{(N)}(t)\equiv\mathbf{Q}\big(\lfloor t\rfloor_N\big)\quad\text{where }\lfloor t\rfloor_N=\frac{m}{N}\text{ for }t\in\left[\frac{m}{N},\frac{(m+1)}{N}\right).$$

The solution $t\rightsquigarrow\mathbf{P}^{(N)}(s,t)$ to the resulting equation is then given by the prescription $\mathbf{P}^{(N)}(s,s)=\mathbf{I}$ and

$$\mathbf{P}^{(N)}(s,t)=\mathbf{P}^{(N)}\big(s,s\vee\lfloor t\rfloor_N\big)\sum_{m=0}^{\infty}\frac{(t-s\vee\lfloor t\rfloor_N)^m}{m!}\mathbf{Q}\big(s\vee\lfloor t\rfloor_N\big)^m\quad\text{for }t>s.$$

As this construction makes obvious, $\mathbf{P}^{(N)}(s,t)$ is a transition probability matrix for each $N\ge 1$ and $t\ge s$, and $(s,t)\rightsquigarrow\mathbf{P}^{(N)}(s,t)$ is continuous. Moreover,

$$\begin{aligned}&\big\|\mathbf{P}^{(N)}(s,t)-\mathbf{P}^{(M)}(s,t)\big\|_{\mathrm{u,v}}\\&\quad\le\int_s^t\big\|\mathbf{Q}\big(\lfloor\tau\rfloor_N\big)-\mathbf{Q}\big(\lfloor\tau\rfloor_M\big)\big\|_{\mathrm{u,v}}\big\|\mathbf{P}^{(N)}(s,\tau)\big\|_{\mathrm{u,v}}\,d\tau\\&\qquad+\int_s^t\big\|\mathbf{Q}\big(\lfloor\tau\rfloor_M\big)\big\|_{\mathrm{u,v}}\big\|\mathbf{P}^{(N)}(s,\tau)-\mathbf{P}^{(M)}(s,\tau)\big\|_{\mathrm{u,v}}\,d\tau.\end{aligned}$$

But

$$\big\|\mathbf{Q}(\tau)\big\|_{\mathrm{u,v}}\le\|\mathbf{A}\|_{\mathrm{u,v}}\quad\text{and}\quad\big\|\mathbf{Q}(\tau')-\mathbf{Q}(\tau)\big\|_{\mathrm{u,v}}\le\|\mathbf{A}\|_{\mathrm{u,v}}\|H\|_{\mathrm{u}}\big|\beta(\tau')-\beta(\tau)\big|,$$

and so

$$\begin{aligned}\big\|\mathbf{P}^{(N)}(s,t)-\mathbf{P}^{(M)}(s,t)\big\|_{\mathrm{u,v}}&\le\|\mathbf{A}\|_{\mathrm{u,v}}\|H\|_{\mathrm{u}}\int_s^t\big|\beta\big(\lfloor\tau\rfloor_N\big)-\beta\big(\lfloor\tau\rfloor_M\big)\big|\,d\tau\\&\quad+\|\mathbf{A}\|_{\mathrm{u,v}}\int_s^t\big\|\mathbf{P}^{(N)}(\tau)-\mathbf{P}^{(M)}(\tau)\big\|_{\mathrm{u,v}}\,d\tau.\end{aligned}$$

Hence, after an application of Gronwall's inequality, we find that

$$\sup_{0\le s\le t\le T}\left\|\mathbf{P}^{(N)}(s,t)-\mathbf{P}^{(M)}(s,t)\right\|_{\mathrm{u,v}}$$

$$\le \|\mathbf{A}\|_{\mathrm{u,v}}\|H\|_{\mathrm{u}}e^{\|\mathbf{A}\|_{\mathrm{u,v}}T}\int_0^T\left|\beta\big(\lfloor\tau\rfloor_N\big)-\beta\big(\lfloor\tau\rfloor_M\big)\right|d\tau.$$

Because $\tau\rightsquigarrow\beta(\tau)$ is continuous, this proves that the sequence $\{\mathbf{P}^{(N)}(s,t):N\ge 1\}$ is Cauchy convergent in the sense that, for each $T>0$,

$$\lim_{M\to\infty}\sup_{N\ge M}\sup_{0\le s\le t\le T}\left\|\mathbf{P}^{(N)}(s,t)-\mathbf{P}^{(M)}(s,t)\right\|_{\mathrm{u,v}}=0.$$

As a consequence, we know that there exists a continuous $(s,t)\rightsquigarrow\mathbf{P}(s,t)$ to which the $\mathbf{P}^{(N)}(s,t)$'s converge with respect to $\|\cdot\|_{\mathrm{u,v}}$ uniformly on finite intervals. In particular, for each $t\ge s$, $\mathbf{P}(s,t)$ is a transition probability matrix and $t\in[s,\infty)\longmapsto\mathbf{P}(s,t)$ is a continuous solution to

$$\mathbf{P}(s,t)=\mathbf{I}+\int_s^t\mathbf{P}(s,\tau)\mathbf{Q}(\tau)\,d\tau,\quad t\in[s,\infty),$$

which is the equivalent integrated form of (6.5.2). Furthermore, if $t\in[s,\infty)\longmapsto\mu_t\in M_1(\mathbb{S})$ is continuously differentiable, then

$$\dot{\mu}_t\equiv\frac{d}{dt}\mu_t=\mu_t\mathbf{Q}(t)\quad\text{for } t\in[s,\infty)\quad\Longleftrightarrow\quad\mu_t=\mu_s\mathbf{P}(s,t)\quad\text{for } t\in[s,\infty).\tag{6.5.3}$$

Since the "if" assertion is trivial, we turn to the "only if" statement. Thus, suppose that $t\in[s,\infty)\longmapsto\mu_t\in M_1(\mathbb{S})$ satisfying $\dot{\mu}_t=\mu_t\mathbf{Q}(t)$ is given, and set $\omega_t=\mu_t-\mu_s\mathbf{P}(s,t)$. Then

$$\omega_t=\int_s^t\omega_\tau\mathbf{Q}(\tau)\,d\tau,$$

and so,

$$\|\omega_t\|_{\mathrm{v}}\le\|\mathbf{A}\|_{\mathrm{u,v}}\int_s^t\|\omega_\tau\|_{\mathrm{v}}\,d\tau.$$

Hence, after another application of Gronwall's inequality, we see that $\omega_t=\mathbf{0}$ for all $t\ge s$.

Of course, by applying this uniqueness result when $\mu_t=\delta_i\mathbf{P}(s,t)$ for each $i\in\mathbb{S}$, we learn that $(s,t)\rightsquigarrow\mathbf{P}(s,t)$ is the one and only solution to (6.5.2). In addition, it leads to the following time-inhomogeneous version of the Chapman–Kolmogorov equation:

$$\mathbf{P}(r,t)=\mathbf{P}(r,s)\mathbf{P}(s,t)\quad\text{for } 0\le r\le s\le t.\tag{6.5.4}$$

Indeed, set $\mu_t=\delta_i\mathbf{P}(r,t)$ for $t\ge s$, note that $t\rightsquigarrow\mu_t$ satisfies (6.5.3) with $\mu_s=\delta_i\mathbf{P}(r,s)$, and conclude that $\mu_t=\delta_i\mathbf{P}(r,s)\mathbf{P}(s,t)$.

6.5.3 Description of the Markov Process

Given a probability vector $\boldsymbol{\mu}$, we now want to construct a Markov process $\{X(t): t \geq 0\}$ that has $\boldsymbol{\mu}$ as its initial distribution and $(s,t) \rightsquigarrow \mathbf{P}(s,t)$ as its transition mechanism, in the sense that

$$\mathbb{P}\big(X(0)=i\big)=(\boldsymbol{\mu})_i \ \& \ \mathbb{P}\big(X(t)=j \mid X(\sigma),\ \sigma\in[0,s]\big)=\mathbf{P}(s,t)_{X(s)j}. \quad (6.5.5)$$

The idea which we will use is basically the same as the one which we used in Sects. 5.2.1 & 2.1.1. However, life here is made more complicated by the fact that the time inhomogeneity forces us to have an uncountable number of random variables at hand: a pair for each $(t,i) \in [0,\infty)\times\mathbb{S}$. To handle this situation, take, without loss in generality, $\mathbb{S}=\{1,\dots,L\}$ and, for $(t,i,j)\in[0,\infty)\times\mathbb{S}^2$, set $S(t,i,j)=\sum_{\ell=1}^{j} e^{-\beta(t)(H(\ell)-H(i))^+}(\mathbf{A})_{i\ell}$, take $S(t,i,0)=0$, and define

$$\Psi(t,i,u)=\begin{cases} j & \text{if } \frac{S(t,i,j-1)}{S(t,i,L)} \leq u < \frac{S(t,i,j)}{S(t,i,L)} \\ i & \text{if } u>1. \end{cases}$$

Also, determine $\mathcal{T}:[0,\infty)\times\mathbb{S}\times[0,\infty)\longrightarrow[0,\infty)$ by

$$\int_s^{s+\mathcal{T}(s,i,\xi)} S(\tau,i,L)\,d\tau=\xi.$$

Next, take X_0 to be an $\mathbb{S}$-valued random variable with distribution $\boldsymbol{\mu}$, let $\{E_n: n\geq 1\}$ be a sequence of unit exponential random variables that are independent of each other and of X_0, and let $\{U_n: n\geq 1\}$ be a sequence of mutually independent random variables that are uniformly distributed on $[0,1)$ and are independent of $\sigma(\{X_0\}\cup\{E_n: n\geq 1\})$. Finally, set $J_0=0$ and $X(0)=X_0$, and, when $n\geq 1$, use induction to define

$$J_n-J_{n-1}=\mathcal{T}\big(J_{n-1},X(J_{n-1}),E_n\big),\qquad X(J_n)=\Psi\big(J_n,X(J_{n-1}),U_n\big),\quad \text{and}$$
$$X(t)=X(J_{n-1})\quad \text{for } J_{n-1}\leq t<J_n.$$

Without substantial change, the reasoning given in Sect. 2.1.1 combined with that in Sect. 5.2.2 and Exercise 5.5.12 allow one to show that (6.5.5) holds.

6.5.4 Choosing a Cooling Schedule

In this section, we will give a rational basis on which to choose the cooling schedule $t\rightsquigarrow\beta(t)$. For this purpose, it is essential to keep in mind what it is that we are attempting to do. Namely, we are trying to have the Markov process $\{X(t): t\geq 0\}$ seek out the set $\mathbb{S}_0=\{j: H(j)=0\}$ in the sense that, as $t\to\infty$, $\mathbb{P}(X(t)\notin\mathbb{S}_0)$ should tend to 0 as fast as possible, and the way we hope to accomplish this is by making the distribution of $X(t)$ look as much like γ_t as possible. Thus, on the one hand, we need to give $\{X(t) : t \geq 0\}$ enough time to equilibrate, so that the distribution of $X(t)$ will look a lot like γ_t. On the other hand, in spite of the fact that it may inhibit equilibration, unless we make $\beta(t)$ increase to infinity, there is no reason for our wanting to make the distribution of $X(t)$ look like γ_t.

In order to understand how to deal with the concerns raised above, let μ be a fixed initial distribution, and let μ_t be the distribution at time $t \geq 0$ of the Markov process $\{X(t) : t \geq 0\}$ described in the Sect. 6.5.3 with initial distribution μ. Equivalently, $\mu_t = \mu\mathbf{P}(0,t)$, where $\{\mathbf{P}(s,t) : 0 \leq s \leq t < \infty\}$ is the family of transition probability matrices constructed in Sect. 6.5.2. Next, define $f_t : \mathbb{S} \longrightarrow [0,\infty)$ so that

$$f_t(i) = \frac{(\mu_t)_i}{(\gamma_t)_i} \quad \text{for } t \geq 0 \text{ and } i \in \mathbb{S}.$$

It should be obvious that the size of f_t provides a good measure of the extent to which μ_t resembles γ_t. For example, by Schwarz's inequality and (6.5.1),

$$\begin{aligned} \mathbb{P}\big(X(t) \notin \mathbb{S}_0\big) &= \langle \mathbf{1}_{\mathbb{S}_0\complement} \rangle_{\mu_t} = \langle f_t \mathbf{1}_{\mathbb{S}_0\complement} \rangle_t \\ &\leq \|f_t\|_{2,t}\sqrt{\langle \mathbf{1}_{\mathbb{S}_0\complement}\rangle_t} \leq L^{\frac{1}{2}}\|f_t\|_{2,t}e^{-\frac{\beta(t)\delta}{2}}, \end{aligned} \tag{6.5.6}$$

and so we will have made progress if we can keep $\|f_t\|_{2,t}$ under control.

With the preceding in mind, assume that $t \rightsquigarrow \beta(t)$ is continuously differentiable, note that this assumption makes $t \rightsquigarrow \|f_t\|_{2,t}^2$ also continuously differentiable, and, in fact, that

$$\begin{aligned} \frac{d}{dt}\|f_t\|_{2,t}^2 &= \frac{d}{dt}\left(\frac{1}{Z(t)}\sum_{i \in \mathbb{S}} \big(f_t(i)\big)^2 e^{-\beta(t)H(i)}\right) \\ &= 2\langle f_t, \dot{f}_t\rangle_t - \dot{\beta}(t)\big\langle H - \langle H\rangle_t, f_t^2\big\rangle_t, \end{aligned}$$

since, by (6.4.3), $\dot{Z}(t) = -\dot{\beta}(t)Z(t)\langle H\rangle_t$. On the other hand, we can compute this same derivative another way. Namely, because $\langle \mathbf{P}(0,t)g\rangle_\mu = \langle g\rangle_{\mu_t} = \langle f_t, g\rangle_t$ for any function g,

$$\|f_t\|_{2,t}^2 = \langle f_t\rangle_{\mu_t} = \big\langle \mathbf{P}(0,t) f_t\big\rangle_\mu,$$

and so we can use (6.5.2) to see that

$$\frac{d}{dt}\|f_t\|_{2,t}^2 = \big\langle \mathbf{P}(0,t)\mathbf{Q}(t) f_t\big\rangle_\mu + \big\langle \mathbf{P}(0,t)\dot{f}_t\big\rangle_\mu = -\mathcal{E}_t(f_t, f_t) + \langle f_t, \dot{f}_t\rangle_t.$$

Thus, after combining these to eliminate the term containing $\dot{f}_t$, we arrive at

$$\begin{aligned} \frac{d}{dt}\|f_t\|_{2,t}^2 &= -2\mathcal{E}_t(f_t, f_t) + \dot{\beta}(t)\big\langle H - \langle H\rangle_t, f_t^2\big\rangle_t \\ &\leq -\big(2\lambda_t - \|H\|_{\mathrm{u}}\dot{\beta}(t)\big)\|f_t\|_{2,t}^2 + 2\lambda_t, \end{aligned}$$

where, in passing to the second line, we have used the fact that $\langle f_t\rangle_t = 1$ and therefore that $\mathrm{Var}_t(f) = \|f_t\|_{2,t}^2 - 1$. Putting all this together, we now know that

$$\|H\|_{\mathrm{u}}\dot{\beta}(t) \le \lambda_t \quad \Longrightarrow \quad \frac{d}{dt}\|f_t\|_{2,t}^2 \le -\lambda_t\|f_t\|_{2,t}^2 + 2\lambda_t.$$

The preceding differential inequality for $\|f\|_{2,t}^2$ is easy to integrate. Namely, it says that

$$\frac{d}{dt}\big(e^{\Lambda(t)}\|f_t\|_{2,t}^2\big) \le 2\lambda_t e^{\Lambda(t)} \quad \text{where } \Lambda(t) = \int_0^t \lambda_\tau\, d\tau.$$

Hence,

$$\|f_t\|_{2,t}^2 \le e^{-\Lambda(t)}\|f_0\|_{2,0}^2 + 2\big(1 - e^{-\Lambda(t)}\big) \le \|f_0\|_{2,0}^2 \vee 2.$$

Moreover, since $(\gamma_0)_i = L^{-1}$, where $L = \#\mathbb{S} \ge 2$, $\|f_0\|_{2,0}^2 \le \|f_0\|_{\mathrm{u}}\langle f\rangle_0 \le L$, and so

$$\|H\|_{\mathrm{u}}\dot{\beta}(t) \le \lambda_t \quad \Longrightarrow \quad \|f_t\|_{2,t} \le L^{\frac{1}{2}}. \tag{6.5.7}$$

The final step is to figure out how to choose $t\rightsquigarrow\beta(t)$ so that it satisfies the condition in (6.5.7); and, of course, we are only interested in the case when $\mathbb{S}_0 \ne \mathbb{S}$ or, equivalently, $\|H\|_{\mathrm{u}} > 0$. From Theorem 6.4.11 we know that $\lambda_t \ge c_- e^{-\beta(t)\mathfrak{e}}$. Hence, we can take

$$\beta(t) = \begin{cases} \frac{1}{\mathfrak{e}}\log\big(1 + \frac{c_-\mathfrak{e}t}{\|H\|_{\mathrm{u}}}\big) & \text{when } \mathfrak{e} > 0 \\ \frac{c_- t}{\|H\|_{\mathrm{u}}} & \text{when } \mathfrak{e} = 0. \end{cases} \tag{6.5.8}$$

After putting this together with (6.5.7) and (6.5.6), we have now proved that *when* $\beta(t)$ *is given by* (6.5.8), *then*

$$\mathbb{P}\big(X(t) \notin \mathbb{S}_0\big) \le L \begin{cases} \big(1 + \frac{c_-\mathfrak{e}t}{\|H\|_{\mathrm{u}}}\big)^{-\alpha} & \text{with } \alpha = \frac{\delta}{2\mathfrak{e}} \text{ when } \mathfrak{e} > 0 \\ e^{-\frac{\delta c_- t}{2\|H\|_{\mathrm{u}}}} & \text{when } \mathfrak{e} = 0. \end{cases} \tag{6.5.9}$$

Remark The result in (6.5.9) when $\mathfrak{e} = 0$ deserves some further comment. In particular, it should be observed that $\mathfrak{e} = 0$ does not guarantee success for a steepest decent strategy. Indeed, $\mathfrak{e} = 0$ only means that each i can be connected to $\mathbb{S}_0$ by an allowable path along which H is non-increasing (cf. Exercise 6.6.13), it does not rule out the possibility that, when using steepest decent, one will choose a bad path and get stuck. Thus, even in this situation, one needs enough randomness to hunt around until one finds a good path.

6.5.5 Small Improvements

An observation, which is really only of interest when $\mathfrak{e} > 0$, is that one carry out the same sort of analysis to control the $\|f_t\|_{q,t} \equiv (\langle |f_t|^q\rangle_t)^{\frac{1}{q}}$ for each $q \in [2, \infty)$. The reason for wanting to do so is that it allows one to improve the rate of convergence in (6.5.6) and therefore in (6.5.9). Namely, by (6.6.3),

$$\langle \mathbf{1}_{\mathbb{S}_0\complement}\rangle_{\mu_t} = \langle \mathbf{1}_{\mathbb{S}_0\complement}\rangle_t \frac{\langle \mathbf{1}_{\mathbb{S}_0\complement} f_t\rangle_t}{\langle \mathbf{1}_{\mathbb{S}_0\complement}\rangle_t} \le \langle \mathbf{1}_{\mathbb{S}_0\complement}\rangle_t \left(\frac{\langle \mathbf{1}_{\mathbb{S}_0\complement} f_t^q\rangle_t}{\langle \mathbf{1}_{\mathbb{S}_0\complement}\rangle_t}\right)^{\frac{1}{q}} \le L^{\frac{1}{q}} \|f_t\|_{q,t} e^{-\frac{\beta(t)\delta}{q'}},$$

where $q' \equiv \frac{q}{q-1}$ is the Hölder conjugate of q. Hence, if $\theta \in [\frac{1}{2}, 1)$ and $q = \frac{1}{1-\theta}$, then one can choose a cooling schedule which makes $\mathbb{P}(X(t) \notin \mathbb{S}_0)$ go to 0 at least as fast as $t^{-\frac{\theta\delta}{\mathfrak{e}}}$.

To carry this out, one begins by computing $\frac{d}{dt}\|f_t\|_{q,t}^q$ twice, once for each of the following expressions:

$$\|f_t\|_{q,t}^q = \langle f_t^q\rangle_t \quad \text{and} \quad \|f_t\|_{q,t}^q = \langle \mathbf{P}(0,t) f_t^{q-1}\rangle_\mu.$$

One then eliminates $\dot{f}_t$ from the resulting equations and thereby arrives at

$$\frac{d}{dt}\|f_t\|_{q,t}^q = -q\mathcal{E}_t(f_t, f_t^{q-1}) - \frac{\dot\beta(t)}{q'-1}\langle f_t^q, H - \langle H\rangle_t\rangle_t,$$

where $q' = \frac{q}{q-1}$ and

$$\mathcal{E}_t(\varphi,\psi) \equiv \frac{1}{2}\sum_{i\ne j}(\gamma_t)_i(\mathbf{Q}(t))_{ij}(\varphi(j)-\varphi(i))(\psi(j)-\psi(i)) = -\langle \varphi, \mathbf{Q}(t)\psi\rangle_t.$$

At this point one has to show that

$$\mathcal{E}_t(f_t, f_t^{q-1}) \ge \frac{4(q-1)}{q^2}\mathcal{E}_t(f_t^{\frac{q}{2}}, f_t^{\frac{q}{2}}),$$

and a little thought makes it clear that this inequality comes down to checking that, for any pair $(a,b) \in [0,\infty)^2$,

$$(b^{\frac{q}{2}} - a^{\frac{q}{2}})^2 \le \frac{q^2}{4(q-1)}(b-a)(b^{q-1} - a^{q-1}),$$

which, when looked at correctly, follows from the fundamental theorem of calculus plus Schwarz's inequality. Hence, in conjunction with the preceding and (6.3.6), we find that

$$\frac{d}{dt}\|f_t\|_{q,t}^q \le -\frac{1}{q'}\left(4\lambda_t - q\dot\beta(t)\|H\|_{\mathrm{u}}\right)\|f_t\|_{q,t}^q + \frac{4\lambda_t}{q'}\langle f_t^{\frac{q}{2}}\rangle_t^2.$$

In order to proceed further, we must learn how to control $\langle f_t^{\frac{q}{2}}\rangle_t^2$ in terms of $\|f_t\|_{q,t}^q$. In the case when $q=2$, $\langle f_t^{\frac{q}{2}}\rangle_t^2$ caused no problem because we knew it was equal to 1. When $q>2$, we no longer have so much control over it. Nonetheless, by first writing $\langle f_t^{\frac{q}{2}}\rangle_t^2 = \langle f_t^{\frac{q}{2}-1}\rangle_{\mu_t}^2$ and then using part (c) of Exercise 6.6.2 to see that

$$\left\langle f_t^{\frac{q}{2}-1}\right\rangle_{\mu_t}^2 \le \left\langle f_t^{q-1}\right\rangle_{\mu_t}^{\frac{q-2}{q-1}} = \left\langle f_t^q\right\rangle_t^{\frac{q-2}{q-1}},$$

we arrive at $\langle f_t^{\frac{q}{2}}\rangle_t^2 \le \langle f_t^q\rangle_t^{\frac{q-2}{q-1}}$. Armed with this estimate, we obtain the differential inequality

$$\frac{d}{dt}\|f_t\|_{q,t}^q \le -\frac{1}{q'}\big(4\lambda_t - q\|H\|_{\mathrm{u}}\dot\beta(t)\big)\|f_t\|_{q,t}^q + \frac{4\lambda_t}{q'}\left(\|f_t\|_{q,t}^q\right)^{1-\frac{q'}{q}}.$$

Finally, by taking $\beta(t)=\frac{1}{\epsilon}\log(1+\frac{3c_-\epsilon t}{q\|H\|_{\mathrm{u}}})$, the preceding inequality can be replaced by

$$\frac{d}{dt}\|f_t\|_{q,t}^q \le -\frac{\lambda_t}{q'}\|f_t\|_{q,t}^q + \frac{4\lambda_t}{q'}\left(\|f_t\|_{q,t}^q\right)^{1-\frac{q'}{q}},$$

which, after integration, can be made to yield

$$\|f_t\|_{q,t} \le 2^{\frac{1}{q'}} \vee \|f_0\|_{q,0},$$

at which point the rest of the argument is the same as the one when $q=2$.

Remark Actually, it is possible to do even better if one is prepared to combine the preceding line of reasoning, which is basically a consequence of Poincaré's inequality, with analytic ideas which come under the general heading of Sobolev inequalities. The interested reader might want to consult [3], which is the source from which the contents of this whole section are derived.

6.6 Exercises

Exercise 6.6.1 A function $\psi:[0,\infty)\longrightarrow\mathbb{R}$ is said to be *convex* if the graph of ψ lies below the secant connecting any pair of points on its graph. That is, if it satisfies

$$\psi\big((1-\theta)s+\theta t\big) \le (1-\theta)\psi(s)+\theta\psi(t) \quad \text{for all } 0\le s<t \text{ and } \theta\in[0,1]. \quad (*)$$

This exercise deals with various properties of convex functions, all of which turn on the property that the slope of a convex function is non-decreasing.

(a) If $\{\psi_n\}_1^\infty\cup\{\psi\}$ are functions on $[0,\infty)$ and $\psi_n(t)\longrightarrow\psi(t)$ for each $t\in[0,\infty)$, show that ψ is non-increasing if each of the ψ_n's is and that ψ is convex if each of the ψ_n's is.

(b) If $\psi : [0,\infty) \longrightarrow \mathbb{R}$ is continuous and twice continuously differentiable on $(0,\infty)$, show that ψ is convex on $[0,\infty)$ if and only if $\ddot{\psi} \ge 0$ on $(0,\infty)$.

Hint: The "only if" part is an easy consequence of

$$\ddot{\psi}(t) = \lim_{h\searrow 0} \frac{\psi(t+h)+\psi(t-h)-2\psi(t)}{h^2} \quad \text{for } t \in (0,\infty).$$

To prove the "if" statement, let $0 < s < t$ be given, and for $\epsilon > 0$ set

$$\varphi_\epsilon(\theta) = \psi\big((1-\theta)s + \theta t\big) - (1-\theta)\psi(s) - \theta\psi(s) - \epsilon\theta(1-\theta), \quad \theta \in [0,1].$$

Note that $\varphi_\epsilon(0) = 0 = \varphi_\epsilon(1)$ and that $\ddot{\varphi}_\epsilon > 0$ on $(0,1)$. Hence, by the second derivative test, φ_ϵ cannot achieve a maximum value in $(0,1)$. Now let $\epsilon \searrow 0$.

(c) If ψ is convex on $[0,\infty)$, show that, for each $s \in [0,\infty)$,

$$t \in (s,\infty) \longmapsto \frac{\psi(s)-\psi(t)}{t-s} \quad \text{is non-increasing.}$$

(d) If ψ is convex on $[0,\infty)$ and $0 \le s < t \le u < w$, show that

$$\frac{\psi(s)-\psi(t)}{t-s} \ge \frac{\psi(u)-\psi(w)}{w-u}.$$

Hint: Reduce to the case when $u = t$.

Exercise 6.6.2 Given a probability vector $\boldsymbol{\mu} \in [0,1]^{\mathbb{S}}$, there are many ways to prove that $\langle f\rangle_{\boldsymbol{\mu}}^2 \le \langle f^2\rangle_{\boldsymbol{\mu}}$ for any $f \in L^2(\boldsymbol{\mu})$. For example, one can get this inequality as an application of Schwarz's inequality $|\langle f,g\rangle_{\boldsymbol{\mu}}| \le \|f\|_{2,\boldsymbol{\mu}}\|g\|_{2,\boldsymbol{\mu}}$ by taking $g = \mathbf{1}$. Alternatively, one can use $0 \le \mathrm{Var}_{\boldsymbol{\mu}}(f) = \langle f^2\rangle_{\boldsymbol{\mu}} - \langle f\rangle_{\boldsymbol{\mu}}^2$. However, neither of these approaches reveals the essential role that convexity plays here. Namely, the purpose of this exercise is to show that for any non-decreasing, continuous, convex function $\psi : [0,\infty) \longrightarrow [0,\infty)$ and any $f : \mathbb{S} \longrightarrow [0,\infty)$,

$$\psi\big(\langle f\rangle_{\boldsymbol{\mu}}\big) \le \langle \psi \circ f\rangle_{\boldsymbol{\mu}}, \tag{6.6.3}$$

where the meaning of the left hand side when $\langle f\rangle_{\boldsymbol{\mu}} = \infty$ is given by taking $\psi(\infty) \equiv \lim_{t\nearrow\infty}\psi(t)$. The inequality (6.6.3) is an example of more general statement known as *Jensen's inequality* (cf. Theorem 2.4.15 in [8]).

(a) Use induction on $n \ge 2$ to show that

$$\psi\left(\sum_{k=1}^n \theta_k x_k\right) \le \sum_{k=1}^n \theta_k \psi(x_k)$$

$$\text{for all } (\theta_1,\dots,\theta_n) \in [0,1]^n \text{ with } \sum_{m=1}^n \theta_k = 1 \text{ and } (x_1,\dots,x_n) \in [0,\infty)^n.$$

(b) Let $\{F_N\}_1^\infty$ be a non-decreasing exhaustion of $\mathbb{S}$ by finite sets satisfying $\mu(F_N) \equiv \sum_{i\in F_N}(\boldsymbol{\mu})_i > 0$, apply part (a) to see that

$$\psi\left(\frac{\sum_{i\in F_N} f(i)(\boldsymbol{\mu})_i}{\mu(F_N)}\right) \le \frac{\sum_{i\in F_N}\psi(f(i))(\boldsymbol{\mu})_i}{\mu(F_N)} \le \frac{\langle\psi\circ f\rangle_{\boldsymbol{\mu}}}{\mu(F_N)}$$

for each N, and get the asserted result after letting $N\to\infty$.

(c) As an application of (6.6.3), show that, for any $0 < p \le q < \infty$ and $f:\mathbb{S}\longrightarrow [0,\infty)$, $\langle f^p\rangle_{\boldsymbol{\mu}}^{\frac{1}{p}} \le \langle f^q\rangle_{\boldsymbol{\mu}}^{\frac{1}{q}}$.

(d) Let $p\in(0,\infty)$ and $\{a_\ell : 1\le \ell\le m\}\subseteq[0,\infty)$. Show that

$$\left(\sum_{\ell=1}^m a_\ell\right)^p \le m^{(p-1)^+}\sum_{\ell=1}^m a_\ell^p.$$

Hint: When $p\ge 1$, use part (c) with $p=1$ and $q=p$ for the measure that assigns mass $\frac{1}{m}$ to the points $1,\dots,m$ and the function $f(\ell)=a_\ell$. When $p\in(0,1)$, first reduce to the case when $n=2$, and in that case show that it suffices to show that $(1+x)^p\le 1+x^p$ for $x\ge 0$.

Exercise 6.6.4 This exercise deals with the material in Sect. 6.1.2 and demonstrates that, with the exception of (6.1.10), more or less everything in that section extends to general irreducible, positive recurrent $\mathbf{P}$'s, whether or not they are reversible. Again let $\boldsymbol{\pi}=\boldsymbol{\pi}^{\mathbb{S}}$ be the unique $\mathbf{P}$-stationary probability vector. In addition, for the exercise which follows, it will be important to consider the space $L^2(\boldsymbol{\pi};\mathbb{C})$ consisting of those $f:\mathbb{S}\longrightarrow\mathbb{C}$ for which $|f|\in L^2(\boldsymbol{\pi})$.

(a) Define $\mathbf{P}^\top$ as in (6.1.2), show that $1\ge(\mathbf{P}\mathbf{P}^\top)_{ii}=(\boldsymbol{\pi})_i\sum_{j\in\mathbb{S}}\frac{(\mathbf{P})_{ij}^2}{(\boldsymbol{\pi})_j}$, and conclude that the series in the definition $\mathbf{P}f(i)\equiv\sum_{j\in\mathbb{S}}f(j)(\mathbf{P})_{ij}$ is absolutely convergent for each $i\in\mathbb{S}$ and $f\in L^2(\boldsymbol{\pi};\mathbb{C})$.

(b) Show that $\|\mathbf{P}f\|_{2,\boldsymbol{\pi}}\le\|f\|_{2,\boldsymbol{\pi}}$ for all $f\in L^2(\boldsymbol{\pi};\mathbb{C})$, and conclude that (6.1.8) and (6.1.9) extend to the present setting for all $f\in L^2(\boldsymbol{\pi};\mathbb{C})$.

Exercise 6.6.5 Continuing with the program initiated in Exercise 6.6.4, we will now see that reversibility plays only a minor role in Sect. 6.1.4. Thus, let $\mathbf{P}$ be any irreducible transition probability on $\mathbb{S}$ which is positive recurrent, let d be its period, and set $\theta_d=e^{\sqrt{-1}2\pi d^{-1}}$.

(a) Show that for each $0\le m<d$ there is a function $f_m:\mathbb{S}\longrightarrow\mathbb{C}$ with the properties that $|f_m|\equiv 1$ and $\mathbf{P}f_m=\theta_d^m f_m$.

Hint: Choose a cyclic decomposition $(\mathbb{S}_0,\dots,\mathbb{S}_{d-1})$ as in Sect. 4.1.8, and use (4.1.19).

(b) Given $\alpha\in\mathbb{R}\setminus\{0\}$, set $\theta_\alpha=e^{\sqrt{-1}2\pi\alpha^{-1}}$, and show that there exists an $f\in L^2(\boldsymbol{\pi};\mathbb{C})\setminus\{0\}$ satisfying $\mathbf{P}f=\theta_\alpha f$ if and only if $d=m\alpha$ for some $m\in\mathbb{Z}\setminus\{0\}$.

Hint: By part (a), it suffices to show that no f exists unless $d = m\alpha$ for some non-zero $m \in \mathbb{Z}$. Thus, suppose that f exists for some α which is not a rational number of the form $\frac{d}{m}$, choose $i \in \mathbb{S}$ so that $f(i) \neq 0$, and get a contradiction with the fact that $\lim_{n\to\infty} \mathbf{P}^{nd} f(i)$ exists.

(c) Suppose that $f \in L^2(\mathbb{S}; \mathbb{C})$ is a non-trivial, bounded solution to $\mathbf{P}f = \theta_d^m f$ for some $m \in \mathbb{Z}$, and let $(\mathbb{S}_0, \dots, \mathbb{S}_{d-1})$ be a cyclic decomposition of $\mathbb{S}$. Show that, for each $0 \le r < d$, $f \upharpoonright \mathbb{S}_r \equiv \theta_d^{-rm} c_0$, where $c_0 \in \mathbb{C} \setminus \{0\}$. In particular, up to a multiplicative constant, for each integer $0 \le m < d$ there is exactly one non-trivial, bounded $f \in L^2(\mathbb{S}; \mathbb{C})$ satisfying $\mathbf{P}f = \theta_d^m f$.

(d) Let H denote the linear subspace of $f \in L^2(\boldsymbol{\pi}; \mathbb{C})$ satisfying $\mathbf{P}f = \theta f$ for some $\theta \in \mathbb{C}$ with $|\theta| = 1$. By combining parts (b) and (c), show that d is the dimension of H as a vector space over $\mathbb{C}$.

(e) Assume that $(\mathbf{P})_{ij} > 0 \implies (\mathbf{P})_{ji} > 0$ for all $(i, j) \in \mathbb{S}^2$. Show that $d \le 2$ and that $d = 2$ if and only if there is a non-trivial, bounded $f : \mathbb{S} \longrightarrow \mathbb{R}$ satisfying $\mathbf{P}f = -f$. Thus, this is the only property of reversible transition probability matrices of which we made essential use in Theorem 6.1.14.

Exercise 6.6.6 This exercise provides another way to think about the relationship between non-negative definiteness and aperiodicity. Namely, let $\mathbf{P}$ be a not necessarily irreducible transition probability on $\mathbb{S}$, and assume that $\boldsymbol{\mu}$ is a probability vector for which the detailed balance condition $(\boldsymbol{\mu})_i(\mathbf{P})_{ij} = (\boldsymbol{\mu})_j(\mathbf{P})_{ji}$, $(i, j) \in \mathbb{S}^2$ holds. Further, assume that $\mathbf{P}$ is non-negative definite in $L^2(\boldsymbol{\mu})$: $\langle f, \mathbf{P}f\rangle_{\boldsymbol{\mu}} \ge 0$ for all bounded $f : \mathbb{S} \longrightarrow \mathbb{R}$. Show that $(\boldsymbol{\mu})_i > 0 \implies (\mathbf{P})_{ii} > 0$ and therefore that i is aperiodic if $(\boldsymbol{\mu})_i > 0$. What follows are steps which lead to this conclusion.

(a) Define the matrix $\mathbf{A}$ so that $(\mathbf{A})_{ij} = \langle \mathbf{1}_{\{i\}}, \mathbf{P}\mathbf{1}_{\{j\}}\rangle_{\boldsymbol{\mu}}$, and show that $\mathbf{A}$ is symmetric (i.e., $(\mathbf{A})_{ij} = (\mathbf{A})_{ji}$) and non-negative definite in the sense that $\sum_{ij}(\mathbf{A})_{ij}(\mathbf{x})_i(\mathbf{x})_j \ge 0$ for any $\mathbf{x} \in \mathbb{R}^{\mathbb{S}}$ with only a finite number of non-vanishing entries.

(b) Given $i \neq j$, consider the plane $\{\alpha\mathbf{1}_{\{i\}} + \beta\mathbf{1}_{\{j\}} : \alpha, \beta \in \mathbb{R}\}$ in $L^2(\boldsymbol{\pi})$, and, using the argument with which we derived (6.1.5), show that $(\mathbf{A})_{ij}^2 \le (\mathbf{A})_{ii}(\mathbf{A})_{jj}$. In particular, if $(\mathbf{A})_{ii} = 0$, then $(\mathbf{A})_{ij} = 0$ for all $j \in \mathbb{S}$.

(c) Complete the proof by noting that $\sum_{j\in\mathbb{S}}(\mathbf{A})_{ij} = (\boldsymbol{\mu})_i$.

(d) After examining the argument, show that we did not need $\mathbf{P}$ to be non-negative definite but only that, for a given $i \in \mathbb{S}$, each of the 2×2 submatrices

$$\begin{pmatrix} \langle \mathbf{1}_{\{i\}}, \mathbf{P}\mathbf{1}_{\{i\}}\rangle_{\boldsymbol{\mu}} & \langle \mathbf{1}_{\{i\}}, \mathbf{P}\mathbf{1}_{\{j\}}\rangle_{\boldsymbol{\mu}} \\ \langle \mathbf{1}_{\{j\}}, \mathbf{P}\mathbf{1}_{\{i\}}\rangle_{\boldsymbol{\mu}} & \langle \mathbf{1}_{\{j\}}, \mathbf{P}\mathbf{1}_{\{j\}}\rangle_{\boldsymbol{\mu}} \end{pmatrix}$$

be.

Exercise 6.6.7 Let $\mathbf{P}$ be an irreducible, positive recurrent transition probability with stationary probability $\boldsymbol{\pi}$. Refer to (6.1.2), and show that the first construction in (6.1.3) is again irreducible whereas the second one need not be. Also, show that the period of the first construction is never greater than that of $\mathbf{P}$ and that the second construction is always non-negative definite. Thus, if $\mathbf{P}^\top\mathbf{P}$ is irreducible, it is necessarily aperiodic.

Exercise 6.6.8 Assume that $\mathbf{P}$ is irreducible and reversible with respect to $\boldsymbol{\pi}$. Because the estimate in (6.2.5) is so weak, it is better to avoid it whenever possible. One way to avoid it is to note that $\mathbf{P}^2$ is necessarily non-negative definite and therefore that the associated β equals β_+.

(a) Show $\mathbf{P}^2$ is irreducible if and only if $\mathbf{P}$ is aperiodic.
(b) Assume that $\mathbf{P}$ is aperiodic, define

$$\mathcal{E}^{(2)}(f,f) = \frac{1}{2}\sum_{i,j\in\mathbb{S}} (\boldsymbol{\pi})_i \left(\mathbf{P}^2\right)_{ij} \left(f(j) - f(i)\right)^2,$$

and set

$$\beta^{(2)} = \inf\left\{\mathcal{E}(f,f) : f \in L_0^2(\boldsymbol{\pi}) \text{ and } \|f\|_{2,\boldsymbol{\pi}} = 1\right\}.$$

Show that

$$\left\|\mathbf{P}^n f - \langle f\rangle_{\boldsymbol{\pi}}\right\|_{2,\boldsymbol{\pi}} \le \left(1 - \beta^{(2)}\right)^{\lfloor \frac{n}{2}\rfloor}.$$

Exercise 6.6.9 As in Sect. 3.3, let $\Gamma = (V, \mathcal{E})$ be a connected, finite graph in which no vertex has an edge to itself and there are no double edges, and take $\mathbf{P}_{vw} = \frac{1}{d_v}$, where d_v is the degree of v if $\{v,w\} \in \mathcal{E}$ and $\mathbf{P}_{v,w} = 0$ if $\{v,w\} \notin \mathcal{E}$. As was observed earlier, the associated Markov chain is irreducible.

(a) Show that the chain is aperiodic if and only if the graph is not bipartite. That is, if and only if there is no non-empty $V' \subsetneq V$ with the property that every edge connects a point in V' to one in $V \setminus V'$.
(b) Determine the probability vector $\boldsymbol{\pi}$ by $(\boldsymbol{\pi})_v = \frac{d_v}{2\#\mathcal{E}}$, and show that $\boldsymbol{\pi}$ is a reversible probability vector for $\mathbf{P}$.
(c) Choose a set $\mathcal{P} = \{p(v,w) : v \ne w\}$ of allowable paths, as in Sect. 6.2.2, and show that

$$\beta_+ \ge \frac{2\#\mathcal{E}}{D^2 L(\mathcal{P}) B(\mathcal{P})}, \tag{6.6.10}$$

where $D \equiv \max_{v\in V} d_v$, $L(\mathcal{P})$ is the maximal length (i.e., the maximal number of edges in) of the paths in $\mathcal{P}$, and $B(\mathcal{P})$, the *bottleneck coefficient*, is the maximal number of paths $p \in \mathcal{P}$ which cross over an edge $e \in E$. Obviously, if one chooses $\mathcal{P}$ to consist of geodesics (i.e., paths of minimal length connecting their end points), then $L(\mathcal{P})$ is just the diameter of Γ, and, as such, is as small as possible. On the other hand, because it may force there to be bad bottlenecks (i.e., many paths traversing a given edge), choosing geodesics may not be optimal.

(d) Assuming that the graph is not bipartite, choose $\mathcal{P} = \{p(v) : v \in V\}$ to be a set of allowable closed paths of odd length, and show that

$$\beta_- \geq \frac{2}{DL(\mathcal{P})B(\mathcal{P})}. \tag{6.6.11}$$

Exercise 6.6.12 Here is an example to which the considerations in Exercises 6.6.9 and 6.6.8 apply and give pretty good results. Let $N \geq 2$, take V to be the subset of the complex plane $\mathbb{C}$ consisting of the Nth roots of unity, and let $\mathcal{E}$ be the collection of pairs of adjacent roots of unity. That is, pairs of the form $\{e^{\frac{\sqrt{-1}2\pi(m-1)}{N}}, e^{\frac{\sqrt{-1}2\pi m}{N}}\}$. Finally, take $\mathbf{P}$ and π accordingly, as in Exercise 6.6.9.

(a) Show that $\mathbf{P}$ is always irreducible but that it is aperiodic if and only if N is odd.
(b) Assuming that N is odd, use (c) in the Exercise 6.6.9 to show that $\beta_+ \geq \frac{2}{N}$ and $\beta_- \geq \frac{2}{N^2}$.
(c) Again assume that N is odd, and use Exercise 6.6.8 to show that

$$\left\|\mathbf{P}^n f - \langle f\rangle_\pi\right\|_{2,\pi} \leq \left(1 - \frac{4}{N}\right)^{\lfloor \frac{n}{2} \rfloor} \left\|f - \langle f\rangle_\pi\right\|_{2,\pi}.$$

When N is large, this is a far better estimate than the one that comes from (b).

Exercise 6.6.13 Define $\mathfrak{e}$ and $\mathfrak{m}$ as in Theorem 6.4.11, and show that $\mathfrak{e} = -\mathfrak{m}$ if and only if for each $(i, j) \in \mathbb{S} \times \mathbb{S}_0$ there is an allowable path $(i_0, \dots, i_n)$ starting at i and ending at j along which H is non-increasing. That is, $i_0 = i$, $i_n = j$, and, for each $1 \leq m \leq n$, $\mathbf{A}_{i_{m-1}i_m} > 0$ and $H(i_m) \leq H(i_{m-1})$.

Exercise 6.6.14 In this exercise I will give a very cursory introduction to a class of reversible Markov processes which provide somewhat naïve mathematical models of certain physical systems. In the literature, these are often called, for reasons which will be clear shortly, *spin-flip systems*, and they are among the earliest examples of Glauber dynamics. Here the state space $\mathbb{S} = \{-1, 1\}^N$ is to be thought of as the configuration space for a system of N particles, each of which has "spin" $+1$ or -1. Because it is more conventional, I will use $\boldsymbol{\omega} = (\omega_1, \dots, \omega_N)$ or $\boldsymbol{\eta} = (\eta_1, \dots, \eta_N)$ to denote generic elements of $\mathbb{S}$. Given $\boldsymbol{\omega} \in \mathbb{S}$ and $1 \leq k \leq N$, $\hat{\boldsymbol{\omega}}^k$ will be the configuration obtained from $\boldsymbol{\omega}$ by "flipping" its kth spin. That is, the $\hat{\boldsymbol{\omega}}^k = (\omega_1, \dots, \omega_{k-1}, -\omega_k, \omega_{k+1}, \dots, \omega_N)$. Next, given $\mathbb{R}$-valued functions f and g on $\mathbb{S}$, define

$$\Gamma(f, g)(\boldsymbol{\omega}) = \sum_{k=1}^N \big(f(\hat{\boldsymbol{\omega}}^k) - f(\boldsymbol{\omega})\big)\big(g(\hat{\boldsymbol{\omega}}^k) - g(\boldsymbol{\omega})\big),$$

which is a discrete analog of the dot product of the gradient of f with the gradient of g. Finally, given a probability vector $\boldsymbol{\mu}$ with $(\boldsymbol{\mu})_{\boldsymbol{\omega}} > 0$ for all $\boldsymbol{\omega} \in \mathbb{S}$, define

$$\mathcal{E}^{\mu}(f,g)=\frac{1}{2}\big\langle\Gamma(f,g)\big\rangle_{\mu},\qquad \rho_k^{\mu}(\omega)=\frac{(\mu)_{\omega}+(\mu)_{\hat{\omega}^k}}{2(\mu)_{\omega}}\quad\text{and}$$

$$\left(\mathbf{Q}^{\mu}\right)_{\omega\eta}=\begin{cases}\rho_k^{\mu}(\omega) & \text{if } \eta=\hat{\omega}^k\\ 0 & \text{if } \eta\notin\{\omega,\hat{\omega}^1,\dots,\hat{\omega}^N\}\\ -\sum_{k=1}^N\rho_k^{\mu}(\omega) & \text{if } \eta=\omega.\end{cases}$$

(a) Check that $\mathbf{Q}^{\mu}$ is an irreducible Q-matrix on $\mathbb{S}$ and that the detailed balance condition $(\mu)_{\omega}\mathbf{Q}^{\mu}_{\omega\eta}=(\mu)_{\eta}\mathbf{Q}^{\mu}_{\eta\omega}$ holds. In addition, show that

$$-\big\langle g,\mathbf{Q}^{\mu}f\big\rangle_{\mu}=\mathcal{E}^{\mu}(f,g).$$

(b) Let $\boldsymbol{\lambda}$ be the uniform probability vector on $\mathbb{S}$. That is, $(\boldsymbol{\lambda})_{\omega}=2^{-N}$ for each $\omega\in\mathbb{S}$. Show that

$$\operatorname{Var}_{\mu}(f)\le M_{\mu}\operatorname{Var}_{\boldsymbol{\lambda}}(f),\quad\text{where } M_{\mu}\equiv 2^N\max_{\omega\in\mathbb{S}}\mu_{\omega}$$

and

$$\mathcal{E}^{\boldsymbol{\lambda}}(f,f)\le\frac{1}{m_{\mu}}\mathcal{E}^{\mu}(f,f)\quad\text{where } m_{\mu}=2^N\min_{\omega\in\mathbb{S}}\mu_{\omega}.$$

(c) For each $S\subseteq\{1,\dots,N\}$, define $\chi_S:\mathbb{S}\longrightarrow\{-1,1\}$ so that $\chi_S(\omega)=\prod_{k\in S}\omega_k$. In particular, $\chi_{\emptyset}=\mathbf{1}$. Show that $\{\chi_S:S\subseteq\{1,\dots,N\}\}$ is an orthonormal basis in $L^2(\boldsymbol{\lambda})$ and that $\mathbf{Q}^{\boldsymbol{\lambda}}\chi_S=-2(\#S)\chi_S$, and conclude from these that $\operatorname{Var}_{\boldsymbol{\lambda}}(f)\le\frac{1}{2}\mathcal{E}^{\boldsymbol{\lambda}}(f,f)$.

(d) By combining (b) with (c), show that $\beta_{\mu}\operatorname{Var}_{\mu}\le\mathcal{E}^{\mu}(f,f)$ where $\beta_{\mu}\equiv\frac{2m_{\mu}}{M_{\mu}}$. In particular, if $t\rightsquigarrow\mathbf{P}_t^{\mu}$ is the transition probability function determined by $\mathbf{Q}^{\mu}$, conclude that

$$\big\|\mathbf{P}_t^{\mu}f-\langle f\rangle_{\mu}\big\|_{2,\mu}\le e^{-t\beta_{\mu}}\big\|f-\langle f\rangle_{\mu}\big\|_{2,\mu}.$$

Exercise 6.6.15 Refer to parts (b) and (c) in Exercise 6.6.14. It is somewhat surprising that the spectral gap for the uniform probability $\boldsymbol{\lambda}$ is 2, independent of N. In particular, this means that if $\{\mathbf{P}^{(N)}(t):t\ge 0\}$ is the semigroup determined by the Q-matrix

$$\left(\mathbf{Q}^{(N)}\right)_{\omega\eta}=\begin{cases}1 & \text{if } \eta=\hat{\omega}^k\\ -N & \text{if } \eta=\omega\\ 0 & \text{otherwise}\end{cases}$$

on $\{-1,1\}^N$, then $\|\mathbf{P}^{(N)}(t)f-\langle f\rangle_{\boldsymbol{\lambda}^{(N)}}\|_{2,\boldsymbol{\lambda}^{(N)}}\le e^{-2t}\|f\|_{2,\boldsymbol{\lambda}^{(N)}}$ for $t\ge 0$ and $f\in L^2(\boldsymbol{\lambda}^{(N)})$, where $\boldsymbol{\lambda}^{(N)}$ is the uniform probability measure on $\{-1,1\}^N$. In fact, the situation here provides convincing evidence of that the theory developed in this chapter works in situations where Doeblin's theory is doomed to failure. Indeed, the purpose of this exercise is to prove that, for any $t>0$ and $\omega\in\{-1,1\}^N$,

$\lim_{N\to\infty} \|\boldsymbol{\mu}^{(N)}(t,\boldsymbol{\omega}) - \boldsymbol{\lambda}^{(N)}\|_{\mathrm{v}} = 2$ when $(\boldsymbol{\mu}^{(N)}(t,\boldsymbol{\omega}))_{\boldsymbol{\eta}} = (\mathbf{P}^{(N)}(t))_{\boldsymbol{\omega}\boldsymbol{\eta}}$. That is, although the L^2-estimate is independent of N, the variation estimate deteriorates as N gets larger.

(a) Begin by showing that $\|\boldsymbol{\mu}^{(N)}(t,\boldsymbol{\omega}) - \boldsymbol{\lambda}^{(N)}\|_{\mathrm{v}}$ is independent of $\boldsymbol{\omega}$.

(b) Show that for any two probability vectors $\boldsymbol{\nu}$ and $\boldsymbol{\nu}'$ on a countable space $\mathbb{S}$, $2 \ge \|\boldsymbol{\nu} - \boldsymbol{\nu}'\|_{\mathrm{v}} \ge 2|\boldsymbol{\nu}(A) - \boldsymbol{\nu}'(A)|$ for all $A \subseteq \mathbb{S}$.

(c) For $1 \le k \le N$, define the random variable X_k on $\{-1,1\}^N$ so that $X_k(\boldsymbol{\eta}) = \eta_k$ if $\boldsymbol{\eta} = (\eta_1,\dots,\eta_N)$. Show that, under $\boldsymbol{\lambda}^{(N)}$, the X_k's are mutually independent, $\{-1,1\}$-valued Bernoulli random variables with expectation 0. Next, let $t > 0$ be given, and set $\boldsymbol{\mu}^{(N)} = \boldsymbol{\mu}^{(N)}(t,\boldsymbol{\omega})$, where $\boldsymbol{\omega}$ is the element of $\{-1,1\}$ whose coordinates are all -1. Show that, under $\boldsymbol{\mu}^{(N)}$, the X_k's are mutually independent, $\{-1,1\}$-valued Bernoulli random variables with expectation value $-e^{-2t}$.

(d) Continuing in the setting of (c), let $A^{(N)}$ be the set of $\boldsymbol{\eta}$ for which $\frac{1}{N}\sum_1^N X_k(\boldsymbol{\eta}) \le -\frac{1}{2}e^{-2t}$, and show that

$$\boldsymbol{\mu}^{(N)}\big(A^{(N)}\big) - \boldsymbol{\lambda}^{(N)}\big(A^{(N)}\big) \ge 1 - \frac{8e^{4t}}{N}.$$

In fact, by using the sort of estimate developed at the end of Sect. 1.2.4, especially (1.2.16), one can sharpen this and get

$$\boldsymbol{\mu}^{(N)}\big(A^{(N)}\big) - \boldsymbol{\lambda}^{(N)}\big(A^{(N)}\big) \ge 1 - 2\exp\left(-\frac{Ne^{-4t}}{8}\right).$$

Hint: Use the usual Chebyshev estimate with which the weak law is proved.

(e) By combining the preceding, conclude that $\|\boldsymbol{\mu}^{(N)}(t,\boldsymbol{\omega}) - \boldsymbol{\lambda}^{(N)}\|_{\mathrm{v}} \ge 2(1 - \frac{8e^{4t}}{N})$ for all $t > 0$, $N \in \mathbb{Z}^+$, and $\boldsymbol{\omega} \in \{-1,1\}^N$.

Chapter 7
A Minimal Introduction to Measure Theory

On Easter 1933, A.N. Kolmogorov published *Foundations of Probability*, a book which laid the foundations on which most of probability theory has rested ever since. Because Kolmogorov's model is given in terms of Lebesgue's theory of measures and integration, its full appreciation requires a thorough understanding of that theory. Thus, although it is far too sketchy to provide anything approaching a thorough understanding, this chapter is an attempt to provide an introduction to Lebesgue's ideas and Kolmogorov's application of them in his model of probability theory.

7.1 A Description of Lebesgue's Measure Theory

In this section I will introduce the terminology used in Lebesgue's theory. However, I will systematically avoid giving rigorous proofs of any hard results. There are many places in which these proofs can be found, one of them being [8].

7.1.1 Measure Spaces

The essential components in measure theory are a set Ω, the *space*, a collection $\mathcal{F}$ of subsets of Ω, the collection of *measurable, subsets*, and a function μ from $\mathcal{F}$ into $[0,\infty]$, called the *measure*. Being a space on which a measure might exist, the pair $(\Omega,\mathcal{F})$ is called a *measurable space*, and when a measurable space $(\Omega,\mathcal{F})$ comes equipped with a measure μ, the triple $(\Omega,\mathcal{F},\mu)$ is called a *measure space*.

In order to avoid stupid trivialities, we will always assume that the space Ω is non-empty. Also, we will assume that the collection $\mathcal{F}$ of measurable sets forms a σ*-algebra* over Ω:

$$\Omega\in\mathcal{F},\qquad A\in\mathcal{F}\quad\Longrightarrow\quad A\complement\equiv\Omega\setminus A\in\mathcal{F},\quad\text{and}$$

$$\{A_n\}_1^\infty\subseteq\mathcal{F}\quad\Longrightarrow\quad\bigcup_1^\infty A_n\in\mathcal{F}.$$

D.W. Stroock, *An Introduction to Markov Processes*, Graduate Texts in Mathematics 230,
DOI 10.1007/978-3-642-40523-5_7,

It is important to emphasize that, as distinguished from point-set topology, only finite or countable set theoretic operations are permitted in measure theory.

Using elementary set theoretic manipulations, it is easy to check that

$$A, B \in \mathcal{F} \implies A \cap B \in \mathcal{F} \quad \text{and} \quad B \setminus A \in \mathcal{F}$$
$$\{A_n\}_1^\infty \subseteq \mathcal{F} \implies \bigcap_1^\infty A_n \in \mathcal{F}.$$

Finally, the measure μ is a function which assigns[1] 0 to $\emptyset$ and is *countably additive* in the sense that

$$\{A_n\}_1^\infty \subseteq \mathcal{F} \quad \text{and} \quad A_m \cap A_n = \emptyset \quad \text{when } m \neq n$$
$$\implies \mu\left(\bigcup_1^\infty A_n\right) = \sum_1^\infty \mu(A_n). \tag{7.1.1}$$

In particular, for $A, B \in \mathcal{F}$,

$$\begin{aligned} A \subseteq B &\implies \mu(B) = \mu(A) + \mu(B \setminus A) \geq \mu(A) \\ \mu(A \cap B) < \infty &\implies \mu(A \cup B) = \mu(A) + \mu(B) - \mu(A \cap B). \end{aligned} \tag{7.1.2}$$

The first line comes from writing B as the union of the disjoint sets A and $B \setminus A$, and the second line comes from writing $A \cup B$ as the union of the disjoint sets A and $B \setminus (A \cap B)$ and then applying the first line to B and $A \cap B$. The finiteness condition is needed when one moves the term $\mu(A \cap B)$ to the left hand side in $\mu(B) = \mu(A \cap B) + \mu(B \setminus (A \cap B))$. That is, one wants to avoid having to subtract ∞ from ∞.

When the set Ω is finite or countable, one has no problem constructing measure spaces. In this case one can take $\mathcal{F} = \{A : A \subseteq \Omega\}$, the set of all subsets of Ω, make any assignment of $\omega \in \Omega \longmapsto \mu(\{\omega\}) \in [0, \infty]$, at which point countable additivity demands that we take

$$\mu(A) = \sum_{\omega \in A} \mu(\{\omega\}) \quad \text{for } A \subseteq \Omega,$$

where summation over the empty set is taken to be 0. However, when Ω is uncountable, it is far from obvious that interesting measures can be constructed on a non-trivial collection of measurable sets. Indeed, it is reasonable to think that Lebesgue's most significant achievement was his construction of a measure space in which $\Omega = \mathbb{R}$, $\mathcal{F}$ is a σ-algebra of which every interval (open, closed, or semi-closed) is an element, and μ assigns each interval its length (i.e., $\mu(I) = b - a$ if I is an interval whose right and left end points are b and a).

---

[1]In view of additivity, it is clear that either $\mu(\emptyset) = 0$ or $\mu(A) = \infty$ for all $A \in \mathcal{F}$. Indeed, by additivity, $\mu(\emptyset) = \mu(\emptyset \cup \emptyset) = 2\mu(\emptyset)$, and therefore $\mu(\emptyset)$ is either 0 or ∞. Moreover, if $\mu(\emptyset) = \infty$, then $\mu(A) = \mu(A \cup \emptyset) = \mu(A) + \mu(\emptyset) = \infty$ for all $A \in \mathcal{F}$.

Although the terminology is misleading, a measure space $(\Omega, \mathcal{F}, \mu)$ is said to be *finite* if $\mu(\Omega) < \infty$. That is, the "finiteness" here is not determined by the size of Ω but instead by how large Ω looks to μ. Even if a measure space is not finite, it may be decomposable into a countable number of finite pieces, in which case it is said to be *σ-finite*. Equivalently, $(\Omega, \mathcal{F}, \mu)$ is σ-finite if there exists $\{\Omega_n\}_1^\infty \subseteq \mathcal{F}$ such that[2] $\Omega_n \nearrow \Omega$ and $\mu(\Omega_n) < \infty$ for each $n \ge 1$. Thus, for example, both Lebesgue's measure on $\mathbb{R}$ and counting measure on $\mathbb{Z}$ are σ-finite but not finite.

7.1.2 Some Consequences of Countable Additivity

Countable additivity is the *sine qua non* in this subject. In particular, it leads to the following continuity properties of measures:

$$\begin{aligned} &\{A_n\}_1^\infty \subseteq \mathcal{F} \quad \text{and} \quad A_n \nearrow A &&\implies \mu(A_n) \nearrow \mu(A) \\ \{A_n\}_1^\infty \subseteq \mathcal{F}, \quad &\mu(A_1) < \infty, \quad \text{and} \quad A_n \searrow A &&\implies \mu(A_n) \searrow \mu(A). \end{aligned} \tag{7.1.3}$$

Although I do not intend to prove many of the results discussed in this chapter, the proofs of those in (7.1.3) are too basic and easy to omit. To prove the first line, simply take $B_1 = A_1$ and $B_{n+1} = A_{n+1} \setminus A_n$. Then $B_m \cap B_n = \emptyset$ when $m \ne n$, $\bigcup_1^n B_m = A_n$ for all $n \ge 1$, and $\bigcup_1^\infty B_m = A$. Hence, by (7.1.1),

$$\mu(A_n) = \sum_1^n \mu(B_m) \nearrow \sum_1^\infty \mu(B_m) = \mu\left(\bigcup_1^\infty B_m\right) = \mu(A).$$

To prove the second line, begin by noting that $\mu(A_1) = \mu(A_n) + \mu(A_1 \setminus A_n)$ and $\mu(A_1) = \mu(A) + \mu(A_1 \setminus A)$. Hence, since $A_1 \setminus A_n \nearrow A_1 \setminus A$ and $\mu(A_1) < \infty$, we have $\mu(A_1) - \mu(A_n) \nearrow \mu(A_1) - \mu(A)$ and therefore that $\mu(A_n) \searrow \mu(A)$. Just as in the proof of second line in (7.1.2), we need the finiteness condition here to avoid being forced to subtract ∞ from ∞.

Another important consequence of countable additivity is *countable subadditivity*:

$$\{A_n\}_1^\infty \subseteq \mathcal{F} \implies \mu\left(\bigcup_1^\infty A_n\right) \le \sum_1^\infty \mu(A_n). \tag{7.1.4}$$

Like the preceding, this is easy. Set $B_1 = A_1$ and $B_{n+1} = A_{n+1} \setminus \bigcup_1^n A_m$. Then $\mu(B_n) \le \mu(A_n)$ and

$$\mu\left(\bigcup_1^\infty A_n\right) = \mu\left(\bigcup_1^\infty B_n\right) = \sum_1^\infty \mu(B_n) \le \sum_1^\infty \mu(A_n).$$

---

[2] I write $A_n \nearrow A$ when $A_n \subseteq A_{n+1}$ for all $n \ge 1$ and $A = \bigcup_1^\infty A_n$. Similarly, $A_n \searrow A$ means that $A_n \supseteq A_{n+1}$ for all $n \ge 1$ and $A = \bigcap_1^\infty A_n$. Obviously, $A_n \nearrow A$ if and only if $A_n\complement \searrow A\complement$.

A particularly important consequence of (7.1.4) is the fact that

$$\mu\left(\bigcup_1^\infty A_n\right) = 0 \quad \text{if } \mu(A_n) = 0 \text{ for each } n \geq 1. \tag{7.1.5}$$

That is, *the countable union of sets each of which has measure* 0 *is again a set having measure* 0. Here one begins to see the reason for restricting oneself to countable operations in measure theory. Namely, it is certainly not true that the uncountable union of sets having measure 0 will necessarily have measure 0. For example, in the case of Lebesgue's measure on $\mathbb{R}$, the second line of (7.1.3) implies

$$\mu(\{x\}) = \lim_{\delta \searrow 0} \mu\big((x-\delta, x+\delta)\big) = \lim_{\delta \searrow 0} 2\delta = 0$$

for each point $x \in \mathbb{R}$, and yet $(0,1) = \bigcup_{x \in (0,1)} \{x\}$ has measure 1.

7.1.3 Generating σ-Algebras

Very often one wants to make sure that a certain collection of subsets will be among the measurable subsets, and for this reason it is important to know the following constructions. First, suppose that $\mathcal{C}$ is a collection of subsets of Ω. Then there is a smallest σ-algebra $\sigma(\mathcal{C})$, called the σ-algebra *generated by* $\mathcal{C}$, over Ω which contains $\mathcal{C}$. To see that $\sigma(\mathcal{C})$ always exists, consider the collection of all the σ-algebras over Ω which contain $\mathcal{C}$. This collection is non-empty because $\{A : A \subseteq \Omega\}$ is an element. In addition, as is easily verified, the intersection of any collection of σ-algebras is again a σ-algebra. Hence, $\sigma(\mathcal{C})$ is the intersection of all the σ-algebras which contain $\mathcal{C}$. When Ω is a topological space and $\mathcal{C}$ is the collection of all open subsets of Ω, then $\sigma(\mathcal{C})$ is called the *Borel σ-algebra* and is denoted by $\mathcal{B}_\Omega$.

One of the most important reasons for knowing how a σ-algebra is generated is that one can often check properties of measures on $\sigma(\mathcal{C})$ by making sure that the property holds on $\mathcal{C}$. A particularly useful example of such a result is the one in the following uniqueness theorem.

Theorem 7.1.6 *Suppose that $(\Omega, \mathcal{F})$ is a measurable space and that $\mathcal{C} \subseteq \mathcal{F}$ includes Ω and is closed under intersection (i.e., $A \cap B \in \mathcal{C}$ whenever $A, B \in \mathcal{C}$). If μ and ν are a pair of finite measures on $(\Omega, \mathcal{F})$ and if $\mu(A) = \nu(A)$ for each $A \in \mathcal{C}$, then $\mu(A) = \nu(A)$ for all $A \in \sigma(\mathcal{C})$.*

Proof We will say that $\mathcal{S} \subseteq \mathcal{F}$ is good if

(i) $A, B \in \mathcal{S}$ and $A \subseteq B \implies B \setminus A \in \mathcal{S}$.
(ii) $A, B \in \mathcal{S}$ and $A \cap B = \emptyset \implies A \cup B \in \mathcal{S}$.
(iii) $\{A_n\}_1^\infty \subseteq \mathcal{S}$ and $A_n \nearrow A \implies A \in \mathcal{S}$.

Notice that if $\mathcal{S}$ is good, $\Omega \in \mathcal{S}$, and $A, B \in \mathcal{S} \implies A \cap B \in \mathcal{S}$, then $\mathcal{S}$ is a σ-algebra. Indeed, because of (i) and (iii), all that one has to do is check that $A \cup B \in \mathcal{S}$ whenever $A, B \in \mathcal{S}$. But, because $A \cup B = (A \setminus (A \cap B)) \cup B$, this is clear from (i), (ii), and the fact that $\mathcal{S}$ is closed under intersections. In addition, observe that if $\mathcal{S}$ is good, then, for any $\mathcal{D} \subseteq \mathcal{F}$,

$$\mathcal{S}' \equiv \{A \in \mathcal{S} : A \cap B \in \mathcal{S} \text{ for all } B \in \mathcal{D}\}$$

is again good.

Now set $\mathcal{B} = \{A \in \mathcal{F} : \mu(A) = \nu(A)\}$. From the properties of finite measures, in particular, (7.1.2) and (7.1.3), it is easy to check that $\mathcal{B}$ is good. Moreover, by assumption, $\mathcal{C} \subseteq \mathcal{B}$. Thus, if

$$\mathcal{B}' = \{A \in \mathcal{B} : A \cap C \in \mathcal{B} \text{ for all } C \in \mathcal{C}\},$$

then, by the preceding observation, $\mathcal{B}'$ is again good. In addition, because $\mathcal{C}$ is closed under intersection, $\mathcal{C} \subseteq \mathcal{B}'$. Similarly,

$$\mathcal{B}'' \equiv \left\{A \in \mathcal{B}' : A \cap B \in \mathcal{B}' \text{ for all } B \in \mathcal{B}'\right\}$$

is also good, and, by the definition of $\mathcal{B}'$, $\mathcal{C} \subseteq \mathcal{B}''$. Finally, if $A, A' \in \mathcal{B}''$ and $B \in \mathcal{B}'$, then $(A \cap A') \cap B = A \cap (A' \cap B) \in \mathcal{B}'$, and so $A \cap A' \in \mathcal{B}''$. Hence, $\mathcal{B}''$ is a σ-algebra which contains $\mathcal{C}$, $\mathcal{B}'' \subseteq \mathcal{B}$, and therefore μ equals ν on $\sigma(\mathcal{C})$. □

7.1.4 Measurable Functions

Given a pair $(\Omega_1, \mathcal{F}_1)$ and $(\Omega_2, \mathcal{F}_2)$ of measurable spaces, we will say that the map $F : \Omega_1 \longrightarrow \Omega_2$ is *measurable* if the inverse image of sets in $\mathcal{F}_2$ are elements of $\mathcal{F}_1$: $F^{-1}(\Gamma) \in \mathcal{F}_1$ for every $\Gamma \in \mathcal{F}_2$.[3] Notice that if $\mathcal{F}_2 = \sigma(\mathcal{C})$, F is measurable if $F^{-1}(C) \in \mathcal{F}_1$ for each $C \in \mathcal{C}$. In particular, if Ω_1 and Ω_2 are topological spaces and $\mathcal{F}_i = \mathcal{B}_{\Omega_i}$, then every continuous map from Ω_1 to Ω_2 is measurable.

It is important to know that when $\Omega_2 = \mathbb{R}$ and $\mathcal{F}_2 = \mathcal{B}_{\mathbb{R}}$, *measurability is preserved under sequential limit operations.* To be precise, if $\{f_n\}_1^\infty$ is a sequence of $\mathbb{R}$-valued measurable functions from $(\Omega, \mathcal{F})$ to $(\mathbb{R}, \mathcal{B}_{\mathbb{R}})$, then

$$\begin{aligned} &\omega \rightsquigarrow \sup_n f_n(\omega), \qquad \omega \rightsquigarrow \inf_n f_n(\omega), \\ &\omega \rightsquigarrow \varlimsup_{n\to\infty} f_n(\omega), \quad \text{and} \quad \omega \rightsquigarrow \varliminf_{n\to\infty} f_n(\omega) \end{aligned} \quad \text{are measurable.} \tag{7.1.7}$$

For example, the first of these can be proved by the following line of reasoning. Begin with the observation that $\mathcal{B}_{\mathbb{R}} = \sigma(\mathcal{C})$ when $\mathcal{C} = \{(a, \infty) : a \in \mathbb{R}\}$. Hence

[3]The reader should notice the striking similarity between this definition and the one for continuity in terms of inverse images of open sets.

$f : \Omega \longrightarrow \mathbb{R}$ will be measurable if and only if[4] $\{f > a\}$ for each $a \in \mathbb{R}$, and so the first function in (7.1.7) is measurable because

$$\left\{\sup_n f_n > a\right\} = \bigcup_n \{f_n > a\} \quad \text{for every } a \in \mathbb{R}.$$

As a consequence of the second line in (7.1.7), we know that the set Δ of points ω at which the limit $\lim_{n\to\infty} f_n(\omega)$ exists is measurable, and $\omega \rightsquigarrow \lim_{n\to\infty} f_n(\omega)$ is measurable if $\Delta = \Omega$.

Measurable functions give rise to important instances of the construction in the preceding subsection. Suppose that the space Ω_1 and the measurable space $(\Omega_2, \mathcal{F}_2)$ are given, and let $\mathfrak{F}$ be some collection of maps from Ω_1 into Ω_2. Then the *σ-algebra $\sigma(\mathfrak{F})$ over Ω_1 generated by* $\mathfrak{F}$ is the smallest σ-algebra over Ω_1 with respect to which every element of $\mathfrak{F}$ is measurable. Equivalently, $\sigma(\mathfrak{F})$ is equal to the set $\{F^{-1}(\Gamma) : F \in \mathfrak{F} \;\&\; \Gamma \in \mathcal{F}_2\}$. Finally, suppose that $\mathcal{F}_2 = \sigma(\mathcal{C}_2)$, where $\mathcal{C}_2 \subseteq \mathcal{F}_2$ contains Ω_2 and is closed under intersection (i.e., $A \cap B \in \mathcal{C}_2$ if $A, B \in \mathcal{C}_2$), and let $\mathcal{C}_1$ be the collection of sets of the form $F_1^{-1}(A_1) \cap \cdots \cap F_n^{-1}(A_n)$ for $n \in \mathbb{Z}^+$, $\{F_1, \dots, F_n\} \subseteq \mathfrak{F}$, and $\{A_1, \dots, A_n\} \subseteq \mathcal{C}_2$. Then $\sigma(\mathfrak{F}) = \sigma(\mathcal{C}_1)$, $\Omega_1 \in \mathcal{C}_1$, and $\mathcal{C}_1$ is closed under intersection. In particular, by Theorem 7.1.6, *if μ and ν are a pair of finite measures on $(\Omega_1, \mathcal{F}_1)$ and*

$$\begin{aligned}&\mu\big(\{\omega_1 : F_1(\omega_1) \in A_1, \dots, F_n(\omega_1) \in A_n\}\big)\\ &\quad = \nu\big(\{\omega_1 : F_1(\omega_1) \in A_1, \dots, F_n(\omega_1) \in A_n\}\big)\end{aligned}$$

for all $n \in \mathbb{Z}^+$, $\{F_1, \dots, F_n\} \subseteq \mathfrak{F}$, and $\{A_1, \dots, A_n\} \subseteq \mathcal{C}_2$, then μ equals ν on $\sigma(\mathfrak{F})$.

7.1.5 Lebesgue Integration

Given a measure space $(\Omega, \mathcal{F}, \mu)$, Lebesgue's theory of integration begins by giving a prescription for defining the integral with respect to μ of all non-negative, measurable functions (i.e., all measurable maps f from $(\Omega, \mathcal{F})$ into[5] $([0, \infty]; \mathcal{B}_{[0,\infty]})$. His theory says that when $\mathbf{1}_A$ is the indicator function of a set $A \in \mathcal{F}$ and $a \in [0, \infty)$, then the integral of the function[6] $a\mathbf{1}_A$ should be equal to $a\mu(A)$. He then insists that the integral should be additive in the sense that the integral of $f_1 + f_2$ should be sum of the integrals of f_1 and f_2. In particular, this means that if f is a non-negative, measurable function which is *simple*, in the sense that it takes on only a

[4] When it causes no ambiguity, I use $\{F \in \Gamma\}$ to stand for $\{\omega : F(\omega) \in \Gamma\}$.

[5] In this context, we are thinking of $[0, \infty]$ as the compact metric space obtained by mapping $[0, 1]$ onto $[0, \infty]$ via the map $t \in [0, 1] \longmapsto \tan(\frac{\pi}{2}t)$.

[6] In measure theory, the convention which works best is to take $0\infty = 0$.

finite number of values, then the integral $\int f\,d\mu$ of f must be

$$\sum_{x\in[0,\infty)} x\mu\big(f^{-1}(\{x\})\big),$$

where, because $f^{-1}(\{x\})=\emptyset$ for all but a finite number of x, the sum involves only a finite number of non-zero terms. Of course, before one can insist on this additivity of the integral, one is obliged to check that additivity is consistent. Specifically, it is necessary to show that $\sum_1^n a_m\mu(A_m)=\sum_1^{n'} a'_{m'}\mu(A'_{m'})$ when $\sum_1^n a_m\mathbf{1}_{A_m}=f=\sum_1^{n'} a'_m\mathbf{1}_{A'_{m'}}$. However, once this consistency has been established, one knows how to define the integral of any non-negative, measurable, simple function in such a way that the integral is additive and gives the obvious answer for indicator functions. In particular, additivity implies *monotonicity*: $\int f\,d\nu\le\int g\,d\mu$ when $f\le g$.

To complete Lebesgue's program for non-negative functions, one has to first observe that if $f:E\longrightarrow[0,\infty]$ is measurable, then there exists a sequence $\{\varphi_n\}_1^\infty$ of non-negative, measurable, simple functions with the property that $\varphi_n(\omega)\nearrow f(\omega)$ for each $\omega\in\Omega$. For example, one can take

$$\varphi_n=\sum_{m=1}^{4^n}\frac{m}{2^n}\mathbf{1}_{A_{m,n}}\quad\text{where } A_{m,n}=\big\{\omega: m2^{-n}\le f(\omega)<(m+1)2^{-n}\big\}.$$

Given $\{\varphi_n\}_1^\infty$, what Lebesgue says is that $\int f\,d\mu=\lim_{n\to\infty}\int\varphi_n\,d\mu$. Indeed, by monotonicity, $\int\varphi_n\,d\mu$ is non-decreasing in n, and therefore the indicated limit necessarily exists. On the other hand, just as there was earlier, there is a consistency problem which one must resolve before adopting Lebesgue's definition. This time the problem comes from the fact that there are myriad ways in which to construct the approximating simple functions φ_n, and one must make sure that the limit does not depend on which approximation scheme one chooses. That is, it is necessary to check that if $\{\varphi_n\}_1^\infty$ and $\{\psi_n\}_1^\infty$ are two non-decreasing sequences of non-negative, simple, measurable functions such that $\lim_{n\to\infty}\varphi_n(\omega)=\lim_{n\to\infty}\psi_n(\omega)$ for each $\omega\in\Omega$, then $\lim_{n\to\infty}\int\varphi_n\,d\mu=\lim_{n\to\infty}\int\psi_n\,d\mu$; and it is at this step that the full power of countable additivity must be brought to bear.

Having defined $\int f\,d\mu$ for all non-negative, measurable f's, one must check that the resulting integral is homogeneous and additive: $\int af\,d\mu=a\int f\,d\mu$ for $a\in[0,\infty]$ and $\int(f_1+f_2)\,d\mu=\int f_1\,d\mu+\int f_2\,d\mu$. However, both these properties are easily seen to be inherited from the case of simple f's. Thus, the only remaining challenge in Lebesgue's construction is to get away from the restriction to non-negative functions and extend the integral to measurable functions which can take both signs. On the other hand, if one wants the resulting theory to be linear, then there is no doubt about how this extension must be made. Namely, given a measurable function $f:E\longrightarrow[-\infty,\infty]$, it is not hard to show that $f^+\equiv f\vee 0$ and $f^-\equiv-(f\wedge 0)=(-f)^+$ are non-negative measurable functions. Hence, because $f=f^+-f^-$, linearity demands that $\int f\,d\mu=\int f^+\,d\mu-\int f^-\,d\mu$; and this time there are two problems which have to be confronted. In the first place, at the very least, it is necessary to restrict ones attention to functions f for which at most one of the numbers $\int f^+ d\mu$ or $\int f^- d\mu$ is infinite, otherwise one ends up having to deal with $\infty - \infty$. Secondly, one must, once again, check consistency. That is, if $f = f_1 - f_2$, where f_1 and f_2 are non-negative and measurable, then one has to show that $\int f_1 \, d\mu - \int f_2 \, d\mu = \int f^+ d\mu - \int f^- d\mu$.

In most applications, the measurable functions which one integrates are either non-negative or have the property that $\int |f| \, d\mu < \infty$, in which case f is said to be an *integrable function*. Because $|a_1 f_1 + a_2 f_2| \le |a_1||f_1| + |a_2||f_2|$, the set of integrable functions forms a vector space over $\mathbb{R}$ on which integration with respect to μ acts as a linear function. Finally, if f is a measurable function which is either non-negative or integrable and $A \in \mathcal{F}$, then the product $\mathbf{1}_A f$ is again a measurable function which is either non-negative or integrable and so

$$\int_A f \, d\mu \equiv \int \mathbf{1}_A f \, d\mu \tag{7.1.8}$$

is well defined.

7.1.6 Stability Properties of Lebesgue Integration

The power of Lebesgue's theory derives from the stability of the integral it defines, and its stability is made manifest in the following three famous theorems. Throughout, $(\Omega, \mathcal{F}, \mu)$ is a measure space to which all references about measurability and integration refer. Also, the functions here are assumed to take their values in $(-\infty, \infty]$.

Theorem 7.1.9 (Monotone Convergence) *Suppose that $\{f_n\}_1^\infty$ is a non-decreasing sequence of measurable functions, and, for each $\omega \in \Omega$, set $f(\omega) = \lim_{n\to\infty} f_n(\omega)$. If there exists a fixed integrable function g which is dominated by each f_n, then $\int f_n \, d\mu \nearrow \int f \, d\mu$. If, instead, $\{f_n\}_1^\infty$ is non-increasing and if there exists a fixed integrable function g which dominates each f_n, then $\int f_n \, d\mu \searrow \int f \, d\mu$.*

Theorem 7.1.10 (Fatou's Lemmas) *Given any sequence $\{f_n\}_1^\infty$ of measurable functions, all of which dominate some fixed integrable function g,*

$$\varliminf_{n\to\infty} \int f_n \, d\mu \ge \int \varliminf_{n\to\infty} f_n \, d\mu.$$

If, instead, there is some fixed integrable function g which dominates all of the f_n's, then

$$\varlimsup_{n\to\infty} \int f_n \, d\mu \le \int \varlimsup_{n\to\infty} f_n \, d\mu.$$

Theorem 7.1.11 (Lebesgue's Dominated Convergence) *Suppose that* $\{f_n\}_1^\infty$ *is a sequence of measurable functions, and assume that there is a fixed integrable function* g *with the property that* $\mu(\{\omega : |f_n(\omega)| > g(\omega)\}) = 0$ *for each* $n \geq 1$. *Further, assume that there exists a measurable function* f *to which* $\{f_n\}_1^\infty$ *converges in either one of the following two senses*:

$$\text{(a)} \quad \mu\Big(\Big\{\omega : f(\omega) \neq \lim_{n\to\infty} f_n(\omega)\Big\}\Big) = 0$$

or

$$\text{(b)} \quad \lim_{n\to\infty} \mu\big(\big\{\omega : \big|f_n(\omega) - f(\omega)\big| \geq \epsilon\big\}\big) = 0 \quad \textit{for all } \epsilon > 0.$$

Then

$$\left|\int f_n\, d\mu - \int f\, d\mu\right| \leq \int |f_n - f|\, d\mu \longrightarrow 0 \quad \textit{as } n \to \infty.$$

Before proceeding, a word should be said about the conditions in Lebesgue's Dominated Convergence Theorem. Because integrals cannot "see" what is happening on a *set of measure* 0 (i.e., an $A \in \mathcal{F}$ for which $\mu(A) = 0$) it is natural that conditions which guarantee certain behavior of integrals should be conditions which need only hold on the complement of a set of measure 0. Thus, condition (a) says that $f_n(\omega) \longrightarrow f(\omega)$ for all ω outside of a set of measure 0. In the standard jargon, conditions that hold off of a set of measure 0 are said to hold *almost everywhere*. In this terminology, the first hypothesis is that $|f_n| \leq g$ almost everywhere and (a) is saying that $\{f_n\}_1^\infty$ converges to f almost everywhere, and these statements would be abbreviated in the literature by something like $|f_n| \leq g$ a.e. and $f_n \longrightarrow f$ a.e.. The condition in (b) is related to, but significantly different from, the one in (a). In particular, it does not guarantee that $\{f_n(\omega)\}_1^\infty$ converges for any $\omega \in \Omega$. For example, take μ to be the measure of Lebesgue on $\mathbb{R}$ described above, and take $f_{m+2^n}(\omega) = \mathbf{1}_{[0,2^{-n})}(\omega - m2^{-n})$ for $n \geq 0$ and $0 \leq m < 2^n$. Then $\mu(\{\omega : f_{m+2^n}(\omega) \neq 0\}) = 2^{-n}$, and so $\lim_{n\to\infty} \mu(\{\omega : |f_n(\omega)| \geq \epsilon\}) = 0$ for all $\epsilon > 0$. On the other hand, for each $\omega \in [0, 1)$, $\overline{\lim}_{n\to\infty} f_n(\omega) = 1$ but $\underline{\lim}_{n\to\infty} f_n(\omega) = 0$. Thus, (b) most definitely does not imply (a). Conversely, although (a) implies (b) when $\mu(\Omega) < \infty$, (a) does not imply (b) when $\mu(\Omega) = \infty$. To wit, again take μ to be Lebesgue's measure, and consider $f_n = \mathbf{1}_{\mathbb{R}\setminus[-n,n]}$.

In connection with the preceding discussion, there is a basic estimate, known as *Markov's inequality*, that plays a central role in all measure theoretic analysis. Namely, because, for any $\lambda > 0$, $\lambda \mathbf{1}_{[\lambda,\infty]} \circ f \leq f \mathbf{1}_{[\lambda,\infty]} \circ f \leq |f|$,

$$\mu\big(\{\omega : f(\omega) \geq \lambda\}\big) \leq \frac{1}{\lambda} \int_{\{\omega : f(\omega) \geq \lambda\}} f\, d\mu \leq \frac{1}{\lambda} \int |f|\, d\mu. \tag{7.1.12}$$

In particular, this leads to the conclusion that

$$\lim_{n\to\infty} \int |f_n - f|\, d\mu = 0 \quad \Longrightarrow \quad \mu\big(\big\{\omega : \big|f_n(\omega) - f(\omega)\big| \geq \epsilon\big\}\big) = 0$$

for all $\epsilon > 0$. In particular, the condition (b) is necessary for the conclusion in Lebesgue's theorem. In addition, (7.1.12) proves that

$$\int |f|\,d\mu = 0 \quad \Longrightarrow \quad \mu\big(\{\omega : |f(\omega)| \geq \epsilon\}\big) = 0 \quad \text{for all } \epsilon > 0,$$

and therefore, by (7.1.3),

$$\int |f|\,d\mu = 0 \quad \Longrightarrow \quad \mu\big(\{\omega : f(\omega) \neq 0\}\big) = 0. \tag{7.1.13}$$

Finally, the role of the Lebesgue dominant g is made clear by considering $\{f_n\}_1^\infty$ when either $f_n = n\mathbf{1}_{[0,\frac{1}{n})}$ or $f_n = \mathbf{1}_{[n-1,n)}$ and μ is Lebesgue's measure.

7.1.7 Lebesgue Integration on Countable Spaces

In this subsection we will see what Lebesgue's theory looks like in the relatively trivial case when Ω is countable, $\mathcal{F}$ is the collection of all subsets of Ω, and the measure μ is σ-finite. As we pointed out in Sect. 7.1.1, specifying a measure on $(\Omega, \mathcal{F})$ is tantamount to assigning a non-negative number to each element of Ω, and, because we want our measures to be σ-finite, no element is to be assigned ∞.

Because the elements of Ω can be counted, there is no reason to not count them. Thus, in the case when Ω is finite, there is no loss in generality if we write $\Omega = \{1, \dots, N\}$, where $N = \#\Omega$ is the number of elements in Ω, and, similarly, when Ω is countably infinite, we might as well, at least for abstract purposes, think of its being the set $\mathbb{Z}^+$ of positive integers. In fact, in order to avoid having to worry about the finite and countably infinite cases separately, we will embed the finite case into the infinite one by simply noting that the theory for $\{1, \dots, N\}$ is exactly the same as the theory for $\mathbb{Z}^+$ restricted to measures μ for which $\mu(\{\omega\}) = 0$ when $\omega > N$. Finally, in order to make the notation here conform with the notation in the rest of the book, I will use $\mathbb{S}$ in place of Ω, i, j, or k to denote generic elements of $\mathbb{S}$, and will identify a measure μ with the row vector $\boldsymbol{\mu} \in [0,\infty)^{\mathbb{S}}$ given by $(\boldsymbol{\mu})_i = \mu(\{i\})$.

The first thing to observe is that

$$\int f\,d\mu = \sum_{i \in \mathbb{S}} f(i)(\boldsymbol{\mu})_i \tag{7.1.14}$$

whenever either $\int f^+\,d\mu$ or $\int f^-\,d\mu < \infty$ is finite. Indeed, if $\varphi \geq 0$ is simple, $a_1, \dots, a_L$ are the distinct values taken by f, and $A_\ell = \{i : \varphi(i) = a_\ell\}$, then

$$\begin{aligned}\int \varphi\,d\mu &= \sum_{\ell=1}^{L} a_\ell \mu(A_\ell) = \sum_{\ell=1}^{L} a_\ell \sum_{i \in A_\ell} (\boldsymbol{\mu})_i \\ &= \sum_{\ell=1}^{L} \sum_{i \in A_\ell} \varphi(i)(\boldsymbol{\mu})_i = \sum_{i \in \mathbb{S}} \varphi(i)(\boldsymbol{\mu})_i.\end{aligned}$$

Second, given any $f \geq 0$ on $\mathbb{S}$, set $\varphi_n(i) = f(i)$ if $1 \leq i \leq n$ and $\varphi_n(i) = 0$ if $i > n$. Then,

$$\int f\,d\mu = \lim_{n\to\infty} \int \varphi_n\,d\mu = \lim_{n\to\infty} \sum_{1\leq i\leq n} f(i)(\boldsymbol{\mu})_i = \sum_{i\in\mathbb{S}} f(i)(\boldsymbol{\mu})_i.$$

Finally, if either $\int f^+\,d\mu$ or $\int f^-\,d\mu$ is finite, then it is clear that

$$\begin{aligned}\int f\,d\mu &= \int f^+\,d\mu - \int f^-\,d\mu \\ &= \sum_{\{i:f(i)\geq 0\}} f(i)(\boldsymbol{\mu})_i - \sum_{\{i:f(i)\leq 0\}} f(i)(\boldsymbol{\mu})_i = \sum_{i\in\mathbb{S}} f(i)(\boldsymbol{\mu})_i.\end{aligned}$$

We next want to see what the "big three" look like in this context.

The Monotone Convergence Theorem First observe that it suffices to treat the case when $0 \leq f_n \nearrow f$. Indeed, one can reduce each of these statements to that case by replacing f_n with $f_n - g$ or $g - f_n$. When $0 \leq f_n \nearrow f$, it is obvious that

$$0 \leq \sum_{i\in\mathbb{S}} f_n(i)(\boldsymbol{\mu})_i \leq \sum_{i\in\mathbb{S}} f_{n+1}(i)(\boldsymbol{\mu})_i \leq \sum_{i\in\mathbb{S}} f(i)(\boldsymbol{\mu})_i.$$

Thus, all that remains is to note that

$$\lim_{n\to\infty} \sum_{i\in\mathbb{S}} f_n(i)(\boldsymbol{\mu})_i \geq \lim_{n\to\infty} \sum_{i=1}^{L} f_n(i)(\boldsymbol{\mu})_i = \sum_{i=1}^{L} f(i)(\boldsymbol{\mu})_i$$

for each $L \in \mathbb{Z}^+$, and therefore that the desired result follows after one lets $L \nearrow \infty$.

Fatou's Lemma Again one can reduce to the case when $f_n \geq 0$ and the limit being taken is the limit inferior. But in this case,

$$\begin{aligned}\varliminf_{n\to\infty} \sum_{i\in\mathbb{S}} f_n(i)(\boldsymbol{\mu})_i &= \lim_{m\to\infty} \inf_{n\geq m} \sum_{i\in\mathbb{S}} f_n(i)(\boldsymbol{\mu})_i \\ &\geq \lim_{m\to\infty} \sum_{i\in\mathbb{S}} \inf_{n\geq m} f_n(i)(\boldsymbol{\mu})_i = \sum_{i\in\mathbb{S}} \varliminf_{n\to\infty} f_n(i)(\boldsymbol{\mu})_i,\end{aligned}$$

where the last equality follows from the monotone convergence theorem applied to $0 \leq \inf_{n\geq m} f_n \nearrow \varliminf_{n\to\infty} f_n$.

Lebesgue's Dominated Convergence Theorem First note that, if we eliminate those $i \in \mathbb{S}$ for which $(\boldsymbol{\mu})_i = 0$, none of the conclusions change. Thus, we will assume that $(\boldsymbol{\mu})_i > 0$ for all $i \in \mathbb{S}$ from now on. Next observe that, under this assumption, the hypotheses become $\sup_n |f_n(i)| \le g(i)$ and $f_n(i) \longrightarrow f(i)$ for each $i \in \mathbb{S}$. In particular, $|f| \le g$. Further, by considering $f_n - f$ instead of $\{f_n\}_1^\infty$ and replacing g by $2g$, we may and will assume that $f \equiv 0$. Now let $\epsilon > 0$ be given, and choose L so that $\sum_{i>L} g(i)(\boldsymbol{\mu})_i < \epsilon$. Then

$$\left|\sum_{i\in\mathbb{S}} f_n(i)(\boldsymbol{\mu})_i\right| \le \sum_{i\in\mathbb{S}} \big|f_n(i)\big|(\boldsymbol{\mu})_i \le \sum_{i=1}^{L} \big|f_n(i)\big|(\boldsymbol{\mu})_i + \epsilon \longrightarrow \epsilon \quad \text{as } n \to \infty.$$

7.1.8 Fubini's Theorem

Fubini's theorem deals with products of measure spaces. Given measurable spaces $(\Omega_1, \mathcal{F}_1)$ and $(\Omega_2, \mathcal{F}_2)$, the product of $\mathcal{F}_1$ and $\mathcal{F}_2$ is the σ-algebra $\mathcal{F}_1 \times \mathcal{F}_2$ over $\Omega_1 \times \Omega_2$ which is generated by the set $\{A_1 \times A_2 : A_1 \in \mathcal{F}_1 \ \&\ A_2 \in \mathcal{F}_2\}$ of *measurable rectangles*. An important technical fact about this construction is that, if f is a measurable function on $(\Omega_1 \times \Omega_2, \mathcal{F}_1 \times \mathcal{F}_2)$, then, for each $\omega_1 \in \Omega_1$ and $\omega_2 \in \Omega_2$, $\omega_2 \rightsquigarrow f(\omega_1, \omega_2)$ and $\omega_1 \rightsquigarrow f(\omega_1, \omega_2)$ are measurable functions on, respectively, $(\Omega_2, \mathcal{F}_2)$ and $(\Omega_1, \mathcal{F}_1)$.

In the following statement, it is important to emphasize exactly which variable is being integrated. For this reason, I will use the more detailed notation $\int_\Omega f(\omega)\,\mu(d\omega)$ instead of the more abbreviated $\int f\,d\mu$.

Theorem 7.1.15 (Fubini)[7] *Let $(\Omega_1, \mathcal{F}_1, \mu_1)$ and $(\Omega_2, \mathcal{F}_2, \mu_2)$ be a pair of σ-finite measure spaces, and set $\Omega = \Omega_1 \times \Omega_2$ and $\mathcal{F} = \mathcal{F}_1 \times \mathcal{F}_2$. Then there is a unique measure $\mu = \mu_1 \times \mu_2$ on $(\Omega, \mathcal{F})$ with the property that $\mu(A_1 \times A_2) = \mu_1(A_1)\mu_2(A_2)$ for all $A_1 \in \mathcal{F}_1$ and $A_2 \in \mathcal{F}_2$. Moreover, if f is a non-negative, measurable function on $(\Omega, \mathcal{F})$, then both*

$$\omega_1 \rightsquigarrow \int_{\Omega_2} f(\omega_1, \omega_2)\,\mu_2(d\omega_2) \quad \textit{and} \quad \omega_2 \rightsquigarrow \int_{\Omega_1} f(\omega_1, \omega_2)\,\mu_1(d\omega_1)$$

are measurable functions, and

$$\begin{aligned}
&\int_{\Omega_1}\left(\int_{\Omega_2} f(\omega_1, \omega_2)\,\mu_2(d\omega_2)\right)\mu_1(d\omega_1) \\
&\quad = \int_\Omega f\,d\mu = \int_{\Omega_2}\left(\int_{\Omega_1} f(\omega_1, \omega_2)\,\mu_1(d\omega_1)\right)\mu_2(d\omega_2).
\end{aligned}$$

---

[7]Although this theorem is usually attributed Fubini, it seems that Tonelli deserves, but seldom receives, a good deal of credit for it.

Finally, for any integrable function f on $(\Omega, \mathcal{F}, \mu)$,

$$A_1 = \left\{\omega_1 : \int_{\Omega_2} \big|f(\omega_1, \omega_2)\big| \, \mu_2(d\omega_2) < \infty\right\} \in \mathcal{F}_1,$$

$$A_2 = \left\{\omega_2 : \int_{\Omega_1} \big|f(\omega_1, \omega_2)\big| \, \mu_1(d\omega_1) < \infty\right\} \in \mathcal{F}_2,$$

$$\omega_1 \rightsquigarrow f_1(\omega_1) \equiv \mathbf{1}_{A_1}(\omega_1) \int_{\Omega_2} f(\omega_1, \omega_2) \, \mu_2(d\omega_2)$$

and

$$\omega_2 \rightsquigarrow f_2(\omega_2) \equiv \mathbf{1}_{A_2}(\omega_2) \int_{\Omega_1} f(\omega_1, \omega_2) \, \mu_1(d\omega_1)$$

are integrable, and

$$\int_{\Omega_1} f_1(\omega_1) \, \mu_1(d\omega_1) = \int_{\Omega} f(\omega) \, \mu(d\omega) = \int_{\Omega_2} f_2(\omega_2) \, \mu_2(d\omega_2).$$

In the case when Ω_1 and Ω_2 are countable, all this becomes very easy. Namely, in the notation that we used in Sect. 7.1.6, the measure $\mu_1 \times \mu_2$ corresponds to row vector $\boldsymbol{\mu}_1 \times \boldsymbol{\mu}_2 \in [0, \infty)^{\mathbb{S}_1 \times \mathbb{S}_2}$ given by $(\boldsymbol{\mu}_1 \times \boldsymbol{\mu}_2)_{(i_1, i_2)} = (\boldsymbol{\mu})_{i_1} (\boldsymbol{\mu})_{i_2}$, and so, by (7.1.14), Fubini's Theorem reduces to the statement that

$$\sum_{i_1 \in \mathbb{S}_1} \left(\sum_{i_2 \in \mathbb{S}_2} a_{i_1 i_2} \right) = \sum_{(i_1, i_2) \in \mathbb{S}_1 \times \mathbb{S}_2} a_{i_1 i_2} = \sum_{i_2 \in \mathbb{S}_2} \left(\sum_{i_i \in \mathbb{S}_1} a_{i_1 i_2} \right)$$

when $\{a_{i_1 i_2} : (i_1, i_2) \in \mathbb{S}_1 \times \mathbb{S}_2\} \subseteq (-\infty, \infty]$ satisfies either $a_{i_1 i_2} \geq 0$ for all (i_1, i_2) or $\sum_{(i_1, i_2)} |a_{i_1 i_2}| < \infty$. In proving this, we may and will assume that $\mathbb{S}_1 = \mathbb{Z}^+ = \mathbb{S}_2$ throughout and will start with the case when $a_{i_1 i_2} \geq 0$. Given any pair $(n_1, n_2) \in (\mathbb{Z}^+)^2$,

$$\sum_{(i_1, i_2) \in \mathbb{S}_1 \times \mathbb{S}_2} a_{i_1 i_2} \geq \sum_{\substack{(i_1, i_2) \in \mathbb{S}_1 \times \mathbb{S}_2 \\ i_1 \leq n_1 \,\&\, i_2 \leq n_2}} a_{i_1 i_2} = \sum_{i_2 = 1}^{n_2} \left(\sum_{i_1 = 1}^{n_1} a_{i_1 i_2} \right).$$

Hence, by first letting $n_1 \to \infty$ and then letting $n_2 \to \infty$, we arrive at

$$\sum_{(i_1, i_2) \in \mathbb{S}_1 \times \mathbb{S}_2} a_{i_1 i_2} \geq \sum_{i_2 \in \mathbb{S}_2} \left(\sum_{i_1 \in \mathbb{S}_1} a_{i_1 i_2} \right).$$

Similarly, for any $n \in \mathbb{Z}^+$,

$$\sum_{\substack{(i_1, i_2) \mathbb{S}_1 \times \mathbb{S}_2 \\ i_1 \vee i_2 \leq n}} a_{i_1 i_2} = \sum_{i_2 = 1}^{n} \left(\sum_{i_1 = 1}^{n} a_{i_1 i_2} \right) \leq \sum_{i_2 \in \mathbb{S}_1} \left(\sum_{i_2 \in \mathbb{S}_2} a_{i_1 i_2} \right),$$

and so, after letting $n \to \infty$, we get the opposite inequality. Next, when $\sum_{(i_1,i_2)} |a_{i_1 i_2}| < \infty$,

$$\sum_{i_1 \in \mathbb{S}_1} |a_{i_1 i_2}| < \infty \quad \text{for all } i_2 \in \mathbb{S}_2,$$

and so

$$\begin{aligned}\sum_{(i_1,i_2)\in\mathbb{S}_1\times\mathbb{S}_2} a_{i_1 i_2} &= \sum_{(i_1,i_2)\in\mathbb{S}_1\times\mathbb{S}_2} a^+_{i_1 i_2} - \sum_{(i_1,i_2)\in\mathbb{S}_1\times\mathbb{S}_2} a^-_{i_1 i_2} \\ &= \sum_{i_2\in\mathbb{S}_2}\Bigl(\sum_{i_1\in\mathbb{S}_1} a^+_{i_1 i_2}\Bigr) - \sum_{i_2\in\mathbb{S}_2}\Bigl(\sum_{i_1\in\mathbb{S}_1} a^-_{i_1 i_2}\Bigr) \\ &= \lim_{n\to\infty} \sum_{\substack{i_2\in\mathbb{S}_2 \\ i_2\le n}}\Bigl(\sum_{i_1\in\mathbb{S}_1} a_{i_1 i_2}\Bigr) = \sum_{i_2\in\mathbb{S}_2}\Bigl(\sum_{i_1\in\mathbb{S}_1} a_{i_1 i_2}\Bigr).\end{aligned}$$

Finally, after reversing the roles 1 and 2, we get the relation with the order of summation reversed.

7.2 Modeling Probability

To understand how these considerations relate to probability theory, think about the problem of modeling the tosses of a fair coin. When the game ends after the nth toss, a Kolmogorov model is provided by taking $\Omega = \{0, 1\}^n$, $\mathcal{F}$ the set of all subsets of Ω, and setting $\mu(\{\omega\}) = 2^{-n}$ for each $\omega \in \Omega$. More generally, any measure space in which Ω has total measure 1 can be thought of as a model of probability, for which reason such a measure space is called a *probability space*, the measure μ is called a *probability measure* and it is often denoted by $\mathbb{P}$. In this connection, when dealing with probability spaces, ones intuition is aided by extending the metaphor to other objects. For example, one calls Ω the *sample space*, its elements are called *sample points*, the elements of $\mathcal{F}$ are called *events*, and the number that $\mathbb{P}$ assigns an event is called the *probability* of that event. In addition, a measurable map is called a *random variable*, it tends to be denoted by X instead of F, and, when it is $\mathbb{R}$-valued, its integral is called its *expected value*. Moreover, the latter convention is reflected in the use of $\mathbb{E}[X]$, or, when more precision is required, $\mathbb{E}^{\mathbb{P}}[X]$ to denote $\int X\,d\mathbb{P}$. Also, $\mathbb{E}[X, A]$ or $\mathbb{E}^{\mathbb{P}}[X, A]$ is used to denote $\int_A X\,d\mathbb{P}$, the expected value of X on the event A. Finally, the *distribution* of a random variable whose values lie in the measurable space $(E, \mathcal{B})$ is the probability measure $\mu = X_*\mathbb{P}$ on $(E, \mathcal{B})$ given by $\mu(B) = \mathbb{P}(X \in B)$. In particular, when X is $\mathbb{R}$-valued, its *distribution function* F_X is defined so that

$$F_X(x) = (X_*\mathbb{P})\bigl((-\infty, x]\bigr) = \mathbb{P}(X \le x).$$

Obviously, F_X is non-increasing. Moreover, as an application of (7.1.3), one can show that F_X is continuous from the right, in the sense that $F_X(x) = \lim_{y \searrow x} F_X(y)$ for each $x \in \mathbb{R}$, $\lim_{x \searrow -\infty} F_X(x) = 0$, and $\lim_{x \nearrow \infty} F_X(x) = 1$. At the same time, one sees that $\mathbb{P}(X < x) = F_X(x-) \equiv \lim_{y \nearrow x} F_X(y)$, and, as a consequence, that $\mathbb{P}(X = x) = F_X(x) - F_X(x-)$ is the jump in F_X at x.

7.2.1 Modeling Infinitely Many Tosses of a Fair Coin

Just as in the general theory of measure spaces, the construction of probability spaces presents no analytic (as opposed to combinatorial) problems as long as the sample space is finite or countable. The only change from the general theory is that the assignment of probabilities to the sample points must satisfy the condition that $\sum_{\omega \in \Omega} \mathbb{P}(\{\omega\}) = 1$. The technical analytic problems arise when Ω is uncountable. For example, suppose that, instead of stopping the game after n tosses, one thinks about a coin tossing game of indefinite duration. Clearly, the sample space will now be $\Omega = \{0, 1\}^{\mathbb{Z}^+}$. In addition, when $A \subseteq \Omega$ depends only on tosses 1 through n, then A should be a measurable event and the probability assigned to A should be the same as the probability that would have been assigned had the game stopped after the nth toss. That is, if $\Gamma \subseteq \{0, 1\}^n$ and[8] $A = \{\omega \in \Omega : (\omega(1), \dots, \omega(n)) \in \Gamma\}$, then $\mathbb{P}(A)$ should equal $2^{-n} \# \Gamma$.

Continuing with the example of an infinite coin tossing game, one sees (cf. (7.1.3)) that, for any fixed $\eta \in \Omega$, $\mathbb{P}(\{\eta\})$ is equal to

$$\lim_{n \to \infty} \mathbb{P}\big(\{\omega : \big(\omega(1), \dots, \omega(n)\big) = \big(\eta(1), \dots, \eta(n)\big)\}\big) = \lim_{n \to \infty} 2^{-n} = 0.$$

Hence, in this case, nothing is learned from the way in which probability is assigned to points: every sample points has probability 0. In fact, it is far from obvious that there exists a probability measure on Ω with the all these properties. Nonetheless, as is proved in Sect. 2.2 of [8], one does. To be precise, what is proved there is the following theorem.

Theorem 7.2.1 *Let $\Omega = \{0, 1\}^{\mathbb{Z}^+}$ and let $\mathcal{A}$ be the set of all subsets A of Ω of the form*

$$A = \big\{\omega \in \Omega : \omega(m) = \epsilon_m \text{ for } 1 \le m \le n\big\}, \tag{7.2.2}$$

where $n \ge 1$ and $(\epsilon_1, \dots, \epsilon_n) \in \{0, 1\}^n$. If $\mathcal{B} = \sigma(\mathcal{A})$, then there is a unique probability measure $\mathbb{P}$ on $(\Omega, \mathcal{B})$ such that $\mathbb{P}(A) = 2^{-n}$ if $A \in \mathcal{A}$ is the one in (7.2.2).

---

[8] It is convenient here to identify Ω with the set a mappings ω from $\mathbb{Z}^+$ into $\{0, 1\}$. Thus, we will use $\omega(n)$ to denote the "nth coordinate" of ω.

7.3 Independent Random Variables

In Kolmogorov's model, independence is best described in terms of σ-algebras. If $(\Omega, \mathcal{F}, \mathbb{P})$ is a probability space and $\mathcal{F}_1$ and $\mathcal{F}_2$ are σ-subalgebras (i.e., are σ-algebras which are subsets) of $\mathcal{F}$, then we say that $\mathcal{F}_1$ and $\mathcal{F}_2$ are *independent* if $\mathbb{P}(\Gamma_1 \cap \Gamma_2) = \mathbb{P}(\Gamma_1)\mathbb{P}(\Gamma_2)$ for all $\Gamma_1 \in \mathcal{F}_1$ and $\Gamma_2 \in \mathcal{F}_2$. It should be comforting to recognize that, when $A_1, A_2 \in \mathcal{F}$ and, for $i \in \{1, 2\}$, $\mathcal{F}_i = \sigma(\{A_i\}) = \{\emptyset, A_i, A_i\complement, \Omega\}$, then, as is easily checked, $\mathcal{F}_1$ is independent of $\mathcal{F}_2$ precisely when, in the terminology of elementary probability theory, "A_1 is independent of A_2": $\mathbb{P}(A_1 \cap A_2) = \mathbb{P}(A_1)\mathbb{P}(A_2)$.

The notion of independence gets inherited by random variables. Namely, the members of a collection $\{X_\alpha : \alpha \in \mathcal{I}\}$ of random variables on $(\Omega, \mathcal{F}, \mathbb{P})$ are said to be *mutually independent* if, for each pair of disjoint subsets $\mathcal{J}_1$ and $\mathcal{J}_2$ of $\mathcal{I}$, the σ-algebras $\sigma(\{X_\alpha : \alpha \in \mathcal{J}_1\})$ and $\sigma(\{X_\alpha : \alpha \in \mathcal{J}_2\})$ are independent. One can use Theorem 7.1.6 to show that this definition is equivalent to saying that if X_α takes its values in the measurable space $(E_\alpha, \mathcal{B}_\alpha)$, then, for every finite subset $\{\alpha_m\}_1^n$ of distinct elements of $\mathcal{I}$ and choice of $B_{\alpha_m} \in \mathcal{B}_{\alpha_m}$, $1 \le m \le n$,

$$\mathbb{P}(X_{\alpha_m} \in B_{\alpha_m} \text{ for } 1 \le m \le n) = \prod_1^n \mathbb{P}(X_{\alpha_m} \in B_{\alpha_m}).$$

As a dividend of this definition, it is essentially obvious that if $\{X_\alpha : \alpha \in \mathcal{I}\}$ are mutually independent and if, for each $\alpha \in \mathcal{I}$, F_α is a measurable map on the range of X_α, then $\{F_\alpha(X_\alpha) : \alpha \in \mathcal{I}\}$ are again mutually independent. Finally, by starting with simple functions, one can show that if $\{X_m\}_1^n$ are mutually independent and, for each $1 \le m \le n$, f_m is a measurable $\mathbb{R}$-valued function on the range of X_m, then

$$\mathbb{E}\big[f_1(X_1) \cdots f_n(X_n)\big] = \prod_1^n \mathbb{E}\big[f_m(X_m)\big]$$

whenever the f_m's are all bounded or are all non-negative.

7.3.1 Existence of Lots of Independent Random Variables

As a consequence of Theorem 7.2.1, we know that there exist an infinite family of mutually independent random variables. Namely, if $(\Omega, \mathcal{B}, \mathbb{P})$ is the probability space discussed in that theorem and $X_m(\omega) = \omega(m)$ is the mth coordinate of ω, then, for any choice of $n \ge 1$ and $(\epsilon_1, \dots, \epsilon_n) \in \{0, 1\}^n$,

$$\mathbb{P}(X_m = \epsilon_m,\ 1 \le m \le n) = 2^{-n} = \prod_1^n \mathbb{P}(X_m = \epsilon_m).$$

Thus, the random variables $\{X_m\}_1^\infty$ are mutually independent. Mutually independent random variables, like these, which take on only two values are called *Bernoulli random variables*.

As we are about to see, Bernoulli random variables can be used as building blocks to construct many other families of mutually independent random variables. The key to such constructions is contained in the following lemma.

Lemma 7.3.1 *Given any family $\{B_m\}_1^\infty$ of mutually independent, $\{0,1\}$-valued Bernoulli random variables satisfying $\mathbb{P}(B_m = 0) = \frac{1}{2} = \mathbb{P}(B_m = 1)$ for all $m \in \mathbb{Z}^+$, set $U = \sum_1^\infty 2^{-m} B_m$. Then U is uniformly distributed on $[0,1)$. That is, $\mathbb{P}(U \le u)$ is 0 if $u < 0$, u if $u \in [0,1)$, and 1 if $u \ge 1$.*

Proof Given $N \ge 1$ and $0 \le n < 2^N$, we want to show that

$$\mathbb{P}\big(n2^{-N} < U \le (n+1)2^{-N}\big) = 2^{-N}. \tag{$*$}$$

To this end, note that $n2^{-N} < U \le (n+1)2^{-N}$ if and only if $\sum_1^N 2^{-m} B_m = (n+1)2^{-N}$ and $B_m = 0$ for all $m > N$, or $\sum_1^N 2^{-m} B_m = n2^{-N}$ and $B_m = 1$ for some $m > N$. Hence, since $\mathbb{P}(B_m = 0 \text{ for all } m > N) = 0$, the left hand side of $(*)$ is equal to the probability that $\sum_1^N 2^{-m} B_m = n2^{-N}$. However, elementary considerations show that, for any $0 \le n < 2^N$ there is exactly one choice of $(\epsilon_1, \dots, \epsilon_N) \in \{0,1\}^N$ for which $\sum_1^N 2^{-m}\epsilon_m = n2^{-N}$. Hence,

$$\mathbb{P}\left(\sum_1^N 2^{-m} B_m = n2^{-N}\right) = \mathbb{P}(B_m = \epsilon_m \text{ for } 1 \le m \le N) = 2^{-N}.$$

Having proved $(*)$, the rest is easy. Since

$$\mathbb{P}(U = 0) = \mathbb{P}(B_m = 0 \text{ for all } m \ge 1) = 0,$$

$(*)$ tells us that, for any $1 \le k \le 2^N$

$$\mathbb{P}\big(U \le k2^{-N}\big) = \sum_{m=0}^{k-1} \mathbb{P}\big(m2^{-N} < U \le (m+1)2^{-N}\big) = k2^{-N}.$$

Hence, because, by (7.1.3), $u \rightsquigarrow \mathbb{P}(U \le u)$ is continuous from the right, it is now clear that $F_U(u) = u$ for all $u \in [0,1)$. Finally, since $\mathbb{P}(U \in [0,1]) = 1$, this completes the proof. □

Now let $\mathcal{I}$ a non-empty, finite or countably infinite set. Then $\mathcal{I} \times \mathbb{Z}^+$ is countable, and so we can construct a 1-to-1 map $(\alpha, n) \rightsquigarrow N(\alpha, n)$ from $\mathcal{I} \times \mathbb{Z}^+$ onto $\mathbb{Z}^+$. Next, for each $\alpha \in \mathcal{I}$, define $\omega \in \Omega = \{0,1\}^{\mathbb{Z}^+} \longmapsto X_\alpha(\omega) \in \Omega$ so that the nth coordinate of $X_\alpha(\omega)$ is $\omega(N(\alpha, n))$. Then, as random variables on the probability space

$(\Omega, \mathcal{B}, \mathbb{P})$ in Theorem 7.2.1, $\{X_\alpha : \alpha \in \mathcal{I}\}$ are mutually independent and each has distribution $\mathbb{P}$. Hence, if $\Phi : \Omega \longrightarrow [0,1]$ is the continuous map given by

$$\Phi(\eta) \equiv \sum_{m=1}^{\infty} 2^{-m}\eta(m) \quad \text{for } \eta \in \Omega$$

and if $U_\alpha \equiv \Phi(X_\alpha)$, then the random variables $\{U_\alpha : \alpha \in \mathcal{I}\}$ are mutually independent and, by Lemma 7.3.1, each is uniformly distributed on $[0,1]$.

The final step in this program is to combine the preceding construction with the well-known fact that any $\mathbb{R}$-valued random variable can be represented in terms of a uniform random variable. More precisely, a map $F : \mathbb{R} \longrightarrow [0,1]$ is called a *distribution function* if F is non-decreasing, continuous from the right, and tends to 0 at $-\infty$ and 1 at $+\infty$. Given such an F, define

$$F^{-1}(u) = \inf\{x \in \mathbb{R} : F(x) \geq u\} \quad \text{for } u \in [0,1].$$

Notice that, by right continuity, $F(x) \geq u \iff F^{-1}(u) \leq x$. Hence, if U is uniformly distributed on $[0,1]$, then $F^{-1}(U)$ is a random variable whose distribution function is F.

Theorem 7.3.2 *Let $(\Omega, \mathcal{B}, \mathbb{P})$ be the probability space in Theorem 7.2.1. Given any finite or countably infinite index set $\mathcal{I}$ and a collection $\{F_\alpha : \alpha \in \mathcal{I}\}$ of distribution functions, there exist mutually independent random variables $\{X_\alpha : \alpha \in \mathcal{I}\}$ on $(\Omega, \mathcal{B}, \mathbb{P})$ with the property that, for each $\alpha \in \mathcal{I}$, F_α is the distribution of X_α.*

7.4 Conditional Probabilities and Expectations

Just as they are to independence, σ-algebras are central to Kolmogorov's definition of conditioning. To be precise, given a probability space $(\Omega, \mathcal{F}, \mathbb{P})$, a sub-$\sigma$-algebra Σ, and a random variable X which is non-negative or integrable, Kolmogorov says that the random variable X_Σ is a *conditional expectation* of X given Σ if X_Σ is a non-negative or integrable random variable which is measurable with respect to Σ (i.e., $\sigma(\{X_\Sigma\}) \subseteq \Sigma$) and satisfies

$$\mathbb{E}[X_\Sigma, \Gamma] = \mathbb{E}[X, \Gamma] \quad \text{for all } \Gamma \in \Sigma. \tag{7.4.1}$$

When X is the indicator function of a set $B \in \mathcal{F}$, then the term conditional expectation is replaced to *conditional probability*.

To understand that this definition is an extension of the one given in elementary probability courses, begin by considering the case when Σ is the trivial σ-algebra $\{\emptyset, \Omega\}$. Because only constant random variables are measurable with respect to $\{\emptyset, \Omega\}$, it is clear that the one and only conditional expectation of X will be $\mathbb{E}[X]$. Next, suppose that $\Sigma = \sigma(\{A\}) = \{\emptyset, A, A\complement, \Omega\}$ for some $A \in \mathcal{F}$ with $\mathbb{P}(A) \in (0, 1)$. In this situation, it is an easy matter to check that, for any $B \in \mathcal{F}$,

$$\omega \rightsquigarrow \frac{\mathbb{P}(B \cap A)}{\mathbb{P}(A)} \mathbf{1}_A(\omega) + \frac{\mathbb{P}(B \cap A\complement)}{\mathbb{P}(A\complement)} \mathbf{1}_{A\complement}(\omega)$$

is a conditional probability of B given Σ. That is, the quantity $\mathbb{P}(B|A) \equiv \frac{\mathbb{P}(B\cap A)}{\mathbb{P}(A)}$, which in elementary probability theory would be called "the conditional probability of B given A," appears here as the value that Kolmogorov's conditional expectation value of $\mathbf{1}_B$ given $\sigma(\{A\})$ takes on A. More generally, if Σ is generated by a finite or countable partition $\mathcal{P} \subseteq \mathcal{F}$ of Ω, then, for any non-negative or integrable random variable X,

$$\omega \rightsquigarrow \sum_{\{A \in \mathcal{P}: \mathbb{P}(A) > 0\}} \frac{\mathbb{E}[X, A]}{\mathbb{P}(A)} \mathbf{1}_A(\omega)$$

will be a conditional expectation of X given Σ.

Of course, Kolmogorov's definition brings up two essential questions: existence and uniqueness. A proof of existence in general can be done in any one of many ways. For instance, when $\mathbb{E}[X^2] < \infty$, one can easily see that, just as $\mathbb{E}[X]\mathbf{1}$ is the X' at which $\mathbb{E}[(X - X')^2]$ achieves its minimum among constant random variables X', so X_Σ is the Σ-measurable X' at which $\mathbb{E}[(X - X')^2]$ achieves its minimum value. In this way, the problem of existence can be related to a problem of orthogonal projection in the space of all square-integrable random variables, and, although they are outside the scope of this book, such projection results are familiar to anyone who has studied the theory of Hilbert spaces.

Uniqueness, on the other hand, is both easier and more subtle than existence. Specifically, there is no naïve uniqueness statement here, because, in general, there will be uncountably many ways to take X_Σ.[9] On the other hand, every choice differs from any other choice on a set of measure at most 0. To see this, suppose that X'_Σ is a second non-negative random variable which satisfies (7.4.1). Then $A = \{X'_\Sigma > X_\Sigma\} \in \Sigma$, and so the only way that (7.4.1) can hold is if $\mathbb{P}(A) = 0$. Similarly, $\mathbb{P}(X_\Sigma > X'_\Sigma) = 0$, and so $\mathbb{P}(X_\Sigma \neq X'_\Sigma) = 0$.

In spite of the ambiguity caused by the sort of uniqueness problems just discussed, it is common to ignore, in so far as possible, this ambiguity and proceed as if a random variable possesses only one conditional expectation with respect to a given σ-algebra. In this connection, the standard notation for a conditional expectation of X given Σ is $\mathbb{E}[X|\Sigma]$ or, when $X = \mathbf{1}_B$, $\mathbb{P}(B|\Sigma)$, which is the notation that I adopted in the earlier chapters of this book.

[9]This non-uniqueness is the reason for my use of the article "a" instead of "the" in front of "conditional expectation."

7.4.1 Conditioning with Respect to Random Variables

In this book, essentially all conditioning is done when $\Sigma = \sigma(\mathfrak{F})$ (cf. Sect. 7.1.4) for some family $\mathfrak{F}$ of measurable functions on $(\Omega, \mathcal{F}, \mathbb{P})$. When Σ has this form, the conditional expectation of a random variable X will be a measurable functions of the functions in $\mathfrak{F}$. For example, if $\mathfrak{F} = \{F_1, \dots, F_n\}$ and the functions F_m all take their values in a countable space $\mathbb{S}$, a conditional expectation value $\mathbb{E}[X \mid \sigma(\mathfrak{F})]$ of X given $\sigma(\mathfrak{F})$ has the form $\Phi(F_1, \dots, F_n)$, where $\Phi(i_1, \dots, i_n)$ is equal to

$$\frac{\mathbb{E}[X, F_1 = i_1, \dots, F_n = i_n]}{\mathbb{P}(F_1 = i_1, \dots, F_n = i_n)} \quad \text{or} \quad 0$$

according to whether $\mathbb{P}(F_1 = i_1, \dots, F_n = i_n)$ is positive or 0. In order to emphasize that conditioning with respect to $\sigma(\mathfrak{F})$ results in a function of $\mathfrak{F}$, I use the notation $\mathbb{E}[X|\mathfrak{F}]$ or $\mathbb{P}(B|\mathfrak{F})$ instead of $\mathbb{E}[X|\sigma(\mathfrak{F})]$ or $\mathbb{P}(B|\sigma(\mathfrak{F}))$.

To give more concrete examples of what we are talking about, first suppose that X and Y are independent random variables with values in some countable space $\mathbb{S}$, and set $Z = F(X, Y)$, where $F : \mathbb{S}^2 \longrightarrow \mathbb{R}$ is bounded. Then

$$\mathbb{E}[Z|X] = v(X) \quad \text{where } v(i) = \mathbb{E}\big[F(i, Y)\big] \text{ for } i \in \mathbb{S}. \tag{7.4.2}$$

A less trivial example is provided by our discussion of Markov chains. In Chap. 2 we encoded the Markov property in equations like

$$\mathbb{P}(X_{n+1} = j \mid X_0, \dots X_n) = (\mathbf{P})_{X_n j},$$

which displays this conditioning as a function, namely $(i_0, \dots, i_n) \rightsquigarrow (\mathbf{P})_{i_n j}$, of the random variables in terms of which the condition is made. (Of course, the distinguishing feature of the Markov property is that the function depends only on i_n and not $(i_0, \dots, i_{n-1})$.) Similarly, when we discussed Markov processes with a continuous time parameter, we wrote

$$\mathbb{P}\big(X(t) = j \mid X(\sigma), \sigma \in [0, s]\big) = \big(\mathbf{P}(t - s)\big)_{X(s) j},$$

which again makes it explicit that the conditioned quantity is a function of the random variables on which the condition is imposed.

References

1. Diaconis, P., Stroock, D.: Geometric bounds for eigenvalues of Markov chains. Ann. Appl. Probab. **1**(1), 36–61 (1991)
2. Dunford, N., Schwartz, J.: Linear Operators, Part I. Wiley Classics Lib. Wiley-Interscience, New York (1988)
3. Holley, R., Stroock, D.: Simulated annealing via Sobolev inequalities. Commun. Math. Phys. **115**(4), 553–569 (1988)
4. Karlin, S., Taylor, H.: A First Course in Stochastic Processes, 2nd edn. Academic Press, San Diego (1975)
5. Norris, J.R.: Markov Chains. Cambridge Series in Statistical & Probabilistic Mathematics. Cambridge University Press, Cambridge (1997)
6. Revuz, D.: Markov Chains. Mathematical Library, vol. 11. North-Holland, Amsterdam (1984)
7. Riesz, F., Sz.-Nagy, B.: Functional Analysis. Dover, New York (1990). Translated from the French edition by F. Boron, reprint of 1955 original
8. Stroock, D.: Mathematics of Probability. GSM, vol. 149. AMS, Providence (2013)

D.W. Stroock, *An Introduction to Markov Processes*, Graduate Texts in Mathematics 230, DOI 10.1007/978-3-642-40523-5, © Springer-Verlag Berlin Heidelberg 2014

Index

D.W. Stroock, *An Introduction to Markov Processes*, Graduate Texts in Mathematics 230,
DOI 10.1007/978-3-642-40523-5, © Springer-Verlag Berlin Heidelberg 2014

Printed in the United States
By Bookmasters